D1391331

BROADCASTING EMPIRE

Broadcasting Empire

The BBC and the British World, 1922–1970

SIMON J. POTTER

OXFORD
UNIVERSITY PRESS

Great Clarendon Street, Oxford OX2 6DP
United Kingdom

Oxford University Press is a department of the University of Oxford.
It furthers the University's objective of excellence in research, scholarship,
and education by publishing worldwide. Oxford is a registered trade mark of
Oxford University Press in the UK and in certain other countries

First Edition published in 2012
Reprinted 2013

Published in the United States of America by Oxford University Press
198 Madison Avenue, New York, NY 10016, United States of America

British Library Cataloguing in Publication Data
Data available

Library of Congress Cataloging in Publication Data
Data available

ISBN 978-0-19-956896-3

For Thomas and Ciara, who thought (or hoped)
that this book would be about Peppa Pig and radiators.

Acknowledgements

I have received generous financial support for my research from a number of sources: an Irish Research Council for the Humanities and Social Sciences Research Fellowship and Small Project grant; a Millennium Fund Minor Project grant and a Library Special Research Fund grant from the National University of Ireland, Galway; a Harold White Fellowship at the National Library of Australia; a Menzies Centre Australian Bicentennial Fellowship; an Ireland–Canada University Foundation Scholarship; an International Council for Canadian Studies Faculty Research Award; and a Royal Irish Academy/British Academy Exchange Fellowship. During research trips I was offered welcome hospitality by the University of Sydney, the Australian National University, and the Stout Centre at Victoria University, Wellington.

Numerous libraries and archives provided invaluable assistance in locating and accessing primary and secondary material. I would like to express my thanks to those at the BBC Written Archives Centre, especially Jeff Walden, who vetted previously unseen material and pointed me in the direction of further archival riches. Rachel Lord, Blair Parkes, and John Kelcher devoted much time to guiding me through the Radio New Zealand Sound Archive. The Bodleian Library of Commonwealth and African Studies at Rhodes House, Oxford, provided a calm place to read and write, and Lucy McCann helped locate hard-to-find material in remote storage. Elizabeth Smith and her colleagues allowed me open access to the archives of the Commonwealth Broadcasting Association, and Guy Tranter opened up previously unreleased files from the ABC archives. Unpublished archival material is quoted with the permission of the BBC Written Archives Centre and the Master and Fellows of Massey College.

Lawrence Constable, Peter Downes, Kent Fedorowich, Mark Hampton, Len Kuffert, Eddy Vickery, and the late Neville Petersen all generously shared research material. Steven Ellis, Ged Martin, and Ann Barry offered vital assistance during difficult times. I would like to thank my parents Joan and David, my brother Matthew, and my wife Maria Scott, for all their support. My children, to whom this book is dedicated, arrived about half-way through the endeavour, and provided many welcome distractions.

SJP

Contents

List of Abbreviations and Acronyms

ABC	Australian Broadcasting Commission
ABCB	Australian Broadcasting Control Board
AIR	All India Radio
ANZ	Archives New Zealand, Wellington, New Zealand
ATL	Alexander Turnbull Library, National Library of New Zealand, Wellington, New Zealand
AWA	Amalgamated Wireless (Australasia) Ltd
BBC	British Broadcasting Corporation (before 1927, British Broadcasting Company)
BBG	Board of Broadcast Governors, Canada
BCINA	British Commonwealth International Newsfilm Agency Ltd (also known as 'Visnews')
BFEBS	British Far Eastern Broadcasting Service, Singapore
CBA	Commonwealth Broadcasting Association, London, UK
CBC	Canadian Broadcasting Corporation
CBS	Columbia Broadcasting System, US
CNR	Canadian National Railways
CPD	*Commonwealth Parliamentary Debates*, Australia
CPR	Canadian Pacific Railway
CRBC	Canadian Radio Broadcasting Commission
CRL	Canadian Radio League
DoI	Department of Information, Australian Commonwealth
ESBS	Egyptian State Broadcasting Service
FAO	Food and Agriculture Organisation, United Nations
GOS	General Overseas Service, BBC
GPO	General Post Office, UK
HJFRT	*Historical Journal of Film, Radio and Television*
HL	Hocken Library, Dunedin, New Zealand
ITA	Independent Television Authority, UK
ITN	Independent Television News, UK
JICH	*Journal of Imperial and Commonwealth History*
LAC	Library and Archives Canada, Ottawa, Canada
LTS	London Transcription Service
MALS	Manchester Archives and Library Services, Manchester Central Library, UK
MBS	Malayan Broadcasting Service
ML	Mitchell Library, State Library of New South Wales, Sydney, Australia
MoI	Ministry of Information, UK
NAA	National Archives of Australia
NARA	National Archives and Records Administration, College Park, Maryland, US
NAS	North American Service, BBC
NBC	National Broadcasting Company, US
NBS	National Broadcasting Service, New Zealand
NCBS	National Commercial Broadcasting Service, New Zealand

NFSA	National Film and Sound Archive, Canberra, Australia
NLA	National Library of Australia, Canberra, Australia
NZBA	New Zealand Broadcasting Authority
NZBB	New Zealand Broadcasting Board
NZBC	New Zealand Broadcasting Corporation
NZBS	New Zealand Broadcasting Service
NZPD	*New Zealand Parliamentary Debates*
OHBE	*Oxford History of the British Empire*
OWI	Office of War Information, US
P&T Department	Post and Telegraph Department, New Zealand
PBS	Palestine Broadcasting Service
PMG's Department	Postmaster-General's Department, Commonwealth of Australia
RBC	Radio Broadcasting Company, New Zealand
RNZSA	Radio New Zealand Sound Archive, Christchurch, New Zealand
SABC	South African Broadcasting Corporation
UBCSCD	Library of the University of British Columbia Special Collections Division, Vancouver, Canada
UKNA	United Kingdom National Archives, Kew, UK
UMA	University of Melbourne Archives, Melbourne, Australia
UNESCO	United Nations Educational, Scientific, and Cultural Organization
UPMT	United Press Movietone Television
UTARM	University of Toronto Archives and Records Management, Toronto, Canada
WAC	Written Archives Centre, BBC, Caversham Park, Reading, UK

Glossary

Actuality—the presentation of non-fictional events or people in radio and television programmes and films. **Actualities** present scenes from 'real life'.

Features—programmes combining different production methods and types of material (actuality, dramatization, music, spoken word, etc.) to present an idea or a theme in a striking way. Features were seen as a way to bring out radio's special characteristics, yet had something in common with the filmed documentary. They were generally shaped by the artistic values of a single producer.

Long wave—low-frequency radio transmissions, capable of travelling very long distances, but requiring very high-powered transmitters.

Medium wave—medium-frequency radio transmissions, suitable for reaching receivers located several hundred miles away (further at night).

FM—transmissions using frequency modulation to provide high-fidelity sound signals, in the Very High Frequency (VHF) band of the radio spectrum. FM signals cannot bend around the curvature of the earth, and FM transmitters thus have a quite limited broadcast range.

Outside broadcast—a programme involving live transmission or recording from locations beyond the fixed studio.

Rediffusion (also known as 'wired wireless')—a system by which radio signals are relayed from a central station to loudspeakers in people's homes and/or public places, using landlines. Later, **rediffusion** was also used to describe similar ways of providing television services.

Relay station—an intermediate installation of receivers and transmitters, used for boosting and redirecting short-wave transmissions and thus improving signal strength.

Short wave—high-frequency transmissions which can be generated by relatively low-powered transmitters, and yet travel very long distances. **Direct listening** is possible for members of the public who have their own short-wave receiving sets. Short-wave signals can also be **rebroadcast** on other frequencies by local stations equipped with short-wave receivers, and thus made available to listeners who do not possess short-wave sets of their own.

Transcriptions—programmes recorded on disc for broadcasting.

Introduction

During the middle decades of the twentieth century the British Broadcasting Corporation (BBC) deployed radio and television as tools of empire. It sought to promote enthusiasm at home for Britain's imperial role, and to link Britons in these islands with a wider British diaspora in the 'white settler dominions' of Canada, Australia, New Zealand, and South Africa. Audiences in Britain and the dominions were encouraged to imagine themselves as part of a global Britannic community. In many ways, this marked an attempt to preserve an imperial order inherited from the Victorians: it was not part of any twentieth-century attempt to create a new British-led global system of power and influence. The BBC sought to reinforce the unity of the British world in a pervasive atmosphere of imperial weakness, and in the face of major challenges, not least the prospect of Americanization.

This was a task that was probably doomed to failure from the outset. So why did the BBC undertake it? The explanation lies, in part, in the fact that while weaknesses in Britain's world-system were apparent throughout this period, contemporaries did not predict the end of empire until it was almost upon them. As recent work on British 'decolonization' has shown, only in the late 1950s did the contraction of British power and influence overseas come to seem irreversible. The BBC also undertook its imperial mission because it reckoned that, by doing so, it could strengthen its position as a domestic British institution. Overseas activities bolstered the corporation's role and prestige at home, and protected it from the possibility that another British broadcasting authority would emerge to serve imperial purposes. The BBC was able to extend its monopoly of broadcasting within Britain into a monopoly of broadcasting from Britain to the dominions and tropical colonies. From an early stage, the BBC was thus engaged in broadcasting to and about the empire, and in building up a broadcasting empire of its own. It was motivated by a mixture of institutional self-interest and ideological commitment.

However, the BBC also discovered that in order to discharge its imperial mission, it required the collaboration of public broadcasting authorities overseas and especially in the dominions. Cooperation from these organizations was often forthcoming but, like the BBC, they possessed institutional, national, and Britannic agendas of their own. In seeking to further those agendas, they often attempted to curb the BBC's centralizing and dominating tendencies. A history of the BBC and empire that only encompassed policies and programmes made in Britain would be misleading. We need to consider the interactions of the BBC with a range of other organizations, across the empire's internal boundaries. These interactions shaped the audiences available for BBC programmes overseas, and how those audiences responded to what they heard and saw.

By examining the relationships that were established between public broadcasting authorities, and by charting how these relationships waxed and waned, we can learn much about wider currents of cultural exchange in the twentieth-century British world. We can also comprehend something of the nature of the transnational history of broadcasting, not as a benign story of the flow of ideas and influences from country to country and their integration into an egalitarian new global order, but as a field of geopolitical contest. In crude terms, through conflict and compromise, during the twentieth century the balance of global media power shifted from Britain to the US, and with it changed the source and nature of the programmes that people around the British world listened to and watched. Did this signify liberation, or subjection to a new form of empire? As Raymond Williams put it:

> Many British working-class people welcomed American culture, or the Americanised character of British commercial television, as an alternative to a British 'public' version which, from a subordinate position, they already knew too well. In many parts of the world this apparently free-floating and accessible culture was a welcome alternative to dominant local cultural patterns and restrictions. Young people all over Europe welcomed the pirate broadcasters, as an alternative to authorities they suspected or distrusted or were simply tired of. The irony was that what came free and easy and accessible was a planned operation by a distant and invisible authority—the American corporations.[1]

In spite of the prevailing contemporary rhetoric of English-speaking unity, and the reality of substantial borrowings and cross-border traffic that over decades had created many similarities between British and American culture, contemporaries did not look on this change with equanimity.

COMMUNICATION AND EMPIRE

In the early 1950s, the Canadian economic historian Harold Innis published his ideas about how empires shape, and are shaped by, the nature of human communication. He mixed contemporary references with examples from the distant past, reaching back into the history of ancient Egypt and Babylon.[2] Previous generations had pondered similar themes, albeit with a more obvious emphasis on recent developments. During the nineteenth century railways and telegraphs had helped unify powerful, continental nation states, such as Germany and the US. Similarly, steamships and undersea telegraph cables had seemed to bind together the widely dispersed British maritime empire. To contemporary eyes, new communications technologies offered the prospect of an easier and more reciprocal flow of

[1] Raymond Williams, *Television: Technology and Cultural Form* (Abingdon, 2005 [1974]), 136.

[2] Harold A. Innis, *Empire and Communications* (Toronto, 2007 [1950]) and *The Bias of Communication* (Toronto, 1995 [1951]).

information and ideas between metropole and periphery, facilitating commerce and political consultation without compromising settler self-government. As one newspaperman remarked,

> these triumphs over space and time have made not only possible, but almost inevitable, the cohesion in a single vast political organisation of Dominions and Dependencies sundered by half the world's girth, and displaying every diversity of race and climate and production.[3]

In the wake of the First World War, some felt that airborne communication could play much the same imperial role. Encouraging the development of civil aviation, the Prince of Wales claimed that,

> as the roads of the Roman Empire failed to keep pace with the requirements of the times, so the modern communications are quite insufficient for a great Commonwealth of Nations which extends to all parts of the globe. The British Empire has more to gain than any other nation from efficient air communications.[4]

Radio might similarly help draw together the British world in closer union.

Yet aviation and wireless emerged in less auspicious imperial circumstances than had the communications technologies of the previous century. For the Victorians, steam propulsion and telegraphy had come as a bonus, at a time of wider, confident imperial expansion. The integrative effects of communication, it could be assumed, would merely reinforce the more profound attractions of British economic, political, and cultural power and influence. However, this underlying strength had ebbed somewhat by the 1920s. British economic and military predominance could no longer be taken for granted. Although it was during this very period that the British Empire reached its greatest geographical extent, new acquisitions (in the Middle East, for example) could be interpreted as a sign of weakness rather than strength. Unable to rely on a dominating 'informal' British economic presence, policymakers felt obliged to establish 'formal' empire, even though the advantages could never outweigh the cost and trouble of direct rule in these areas. Moreover, in other core territories British policymakers were actually obliged to retreat: most notably in Ireland, Egypt, and Iraq. The threat of communal and nationalist political mobilization also reared its head in inter-war India.[5] At home and abroad this was the beginning of a 'morbid age', characterized by generalized fears of disorder, decline, and the collapse of established authority.[6]

Before 1914 new communications technologies had often seemed to develop in such a way as to support British expansion overseas. After 1918 this was no longer

[3] J. Saxon Mills, *The Press and Communications of the Empire* (London, 1924), quote at 3. Maj. A. V. T. Wakely, *Some Aspects of Imperial Communications* (London, 1924). Robert Kubicek, 'British Expansion, Empire, and Technological Change', in Andrew Porter (ed.), *Oxford History of the British Empire* (*OHBE*), iii, *The Nineteenth Century* (Oxford, 1999), 247–8. Duncan Bell, *The Idea of Greater Britain: Empire and the Future of World Order, 1860–1900* (Princeton and Oxford, 2007), 81–91.

[4] *The Times* (London), 1 July 1921. For many similar examples, see Gordon Pirie, *Air Empire: British Imperial Civil Aviation, 1919–39* (Manchester, 2009).

[5] John Darwin, *The Empire Project: The Rise and Fall of the British World-System, 1830–1970* (Cambridge, 2009), 305–417.

[6] Richard Overy, *The Morbid Age: Britain Between the Wars* (London, 2009).

assured. Nevertheless, contemporaries still wondered if communications technologies could be harnessed so as actively to shape, rather than merely reflect, the destiny of the empire. Could imperial communication links bind the British Empire together in the face of countervailing tendencies towards disintegration? Perhaps it would require state intervention in order to achieve this. Even during the nineteenth century the powerful engine of Britain's developing industrial economy had not been capable, without guidance, of precisely shaping overseas communications links to suit perceived imperial interests. In accordance with contemporary worship of the principles of laissez faire, the provision of an imperial communications infrastructure was largely left to private enterprise. However, when private companies failed to satisfy public requirements, government action was contemplated. In order to harness steam-powered shipping to the empire's needs, the British government provided subsidies for the carriage of mail, troops, migrants, and other cargoes. Private telegraph and cable companies were similarly given financial support by British and colonial governments to ensure that key outlying areas of the British world-system were adequately connected with the imperial core. Even Reuters, the empire's news agency, was supported by British government funding at key moments. If required, direct state control of the means of imperial communication was also sometimes contemplated. In 1902 the state-owned and -managed Pacific Cable was opened to connect Canada, Australia, and New Zealand, a collaborative project involving the British and dominion governments.[7] During the twentieth century the state became a major player in many of the empire's railway systems, and particularly in civil aviation. In Britain, Imperial Airways was established in 1924 as a government-subsidized monopoly.[8]

Would unassisted private enterprise be capable of developing wireless as a means of imperial communication, or would state intervention prove necessary? In the decade preceding the First World War, radio was used primarily for 'point-to-point' communication between one transmitter and one receiver (or a small number of receivers), rather than for broadcasting to thousands, or millions, of receivers. The UK Marconi Company and its overseas affiliates created and operated much of the infrastructure necessary for ship-to-shore wireless communication and long-distance wireless telegraphy. Yet state intervention was still required: in 1913 Westminster ratified an agreement with Marconi to construct a publicly financed chain of imperial wireless telegraph stations, a project that foundered amid accusations of corruption and the outbreak of war. The stations were eventually built in the 1920s, operated by Marconi and its affiliates in conjunction with the UK General

[7] Peter J. Hugill, *Global Communications since 1844: Geopolitics and Technology* (Baltimore, MD and London, 1999), 39–46. Dwayne R. Winseck and Robert M. Pike, *Communication and Empire: Media, Markets, and Globalization, 1860–1930* (Durham, NC and London, 2007), 157–67.

[8] Kubicek, 'British Expansion'. Andrew Porter, *Victorian Shipping, Business and Imperial Policy: Donald Currie, the Castle Line and Southern Africa* (Woodbridge, 1986). Freda Harcourt, 'British Oceanic Mail Contracts in the Age of Steam, 1838–1914', *Journal of Transport History*, 3rd ser., 9 (March 1988), 1–18. Peter Fearon, 'The Growth of Aviation in Britain', *Journal of Contemporary History*, 20/1 (January 1985), 21–40. Pirie, *Air Empire*. Donald Read, *The Power of News: The History of Reuters*, 2nd edn (Oxford, 1999). Simon J. Potter, *News and the British World: The Emergence of an Imperial Press System, 1876–1922* (Oxford, 2003).

Post Office (GPO). State intervention in point-to-point wireless communication continued thereafter. In 1929 British public and private cable and wireless telegraph interests were merged into Imperial and International Communications Ltd, 'nominally private but in fact government regulated and monopolistic'.[9]

In 1922 the BBC was established with a monopoly over broadcasting in the UK, under remote state control. It soon became apparent that it would operate on a 'public service', non-commercial basis. Public ownership and control were responses to domestic rather than imperial requirements: the BBC developed as a means to tame broadcasting in Britain, and to help broadcasting tame the new, potentially unstable democracy that Britain had become after the First World War. Nevertheless, a decision made for domestic reasons had significant, if unintended, imperial consequences.

First, the principle of non-commercial operation came to be seen as a key virtue, and was extended from the BBC's domestic operations to its overseas role. Although private enterprise was given considerable scope in broadcasting in the dominions and in some of the tropical colonies, it was not until the 1950s that it was allowed to play a significant role in broadcasting across the internal boundaries of the empire. Broadcasting from Britain to the dominions was thus not driven by commercial motives, funded by advertising, or regulated by the market mechanism. There were obvious reasons for this: if British broadcasting could not hope to compete commercially with American broadcasting in overseas markets, then why not shun the idea of competition and markets entirely, and operate on a different footing? However, the problem remained that, if the BBC did not sell its services overseas on a commercial basis, then how could it fund them? Listeners abroad could not easily be obliged to pay a licence fee to hear BBC programmes, as home listeners were. If the BBC accepted direct government subsidies instead, then its autonomy from the day-to-day demands of politicians and civil servants might be undermined. This dilemma was not definitively resolved before 1970, or indeed thereafter.

Second, the BBC also sought to extend from domestic to overseas contexts a range of more profound ideas about the purpose of broadcasting. The most basic tenet of 'public service broadcasting' (as it evolved during the 1920s under the high-minded, visionary leadership of the BBC's first director-general, John Reith) was that radio could and indeed must seek actively to shape society. Wireless should not just entertain, but also educate and inform its listeners, allowing them to appreciate the high-cultural fruits of civilization and participate in the newly democratic political system in a reasoned and rational manner. The BBC would help restore national unity and social, cultural, and political stability. Applied overseas, these ideas came to mean that the BBC would also seek actively to reinforce the bonds of empire. In this way, it was believed, broadcasting could encourage international peace and order and, to a lesser extent, the spread of democratic values, thus helping Britain retain its influence in the wider world.

[9] Daniel R. Headrick, *The Invisible Weapon: Telecommunications and International Politics, 1851–1945* (New York and Oxford, 1991), 202–8, quote at 208. W. J. Baker, *A History of the Marconi Company* (London, 1970). Aitor Anduaga, *Wireless and Empire: Geopolitics, Radio Industry, and Ionosphere in the British Empire, 1918–1939* (Oxford, 2009).

Historians have tended to follow the BBC's own organizational patterns, and separate its domestic and overseas operations out into distinct categories for research and analysis.[10] This obscures links between the insular and expansive aspects of British broadcasting, and makes it difficult fully to comprehend either one of them. The benefits to be gained from viewing the BBC's domestic and imperial functions together in a single 'analytic field'[11] can be demonstrated by considering, for example, the BBC's role in the projection of Britishness.

THE BBC AND BRITISHNESS, AT HOME AND OVERSEAS

At home, from an early stage the BBC sought to reinforce a sense of overarching British identity, in the hope that this would ameliorate conflict among different social groups, and render less traumatic the perceived disintegration of barriers between hitherto distinct regional communities. Posted to Manchester in the 1920s, the BBC Features Producer Geoffrey Bridson later recalled how

> only occasional [soccer] Cup Finals at Wembley then brought Northerners down to London in force, cloth-capped hordes of rude barbarians, talking incomprehensible dialects, and gawping round the West End like so many Goths in the Roman forum.[12]

The BBC might push the boundaries of British civilization out from London and into the provinces.

Some contemporaries, and later some historians, alleged that in construing its domestic duty in this way (until the 1960s at least) the BBC neglected the tastes and true interests of the working classes and catered predominantly to middle-class requirements. Moreover, it is argued, the BBC suppressed the development of Britain's regional cultures, imposing stultifying, London-centred perspectives. Yet such accusations are not entirely fair. Particularly after broadcasting's first decade, some BBC planners and producers (Bridson among them) really did seek to present programmes that the majority of the population would value, and that would air their voices and concerns. BBC programmes also acknowledged and even celebrated Britain's regional diversity. Indeed, the BBC allowed Welsh, Scottish, and Northern Irish identities to attain new coherence and vigour, by devolving some of its authority to branches in the 'national regions' and mandating them to serve each region as a separate unit.[13] The

[10] With the notable exception of Asa Briggs in his institutional history of the BBC, *The History of Broadcasting in the United Kingdom* (Oxford, 5 vols, 1961–95).

[11] On this point more generally, see Ann Laura Stoler and Frederick Cooper, 'Between Metropole and Colony: Rethinking a Research Agenda', in Cooper and Stoler (eds), *Tensions of Empire: Colonial Cultures in a Bourgeois World* (Berkeley and Los Angeles, 1997), 15.

[12] D. G. Bridson, *Prospero and Ariel: The Rise and Fall of Radio, a Personal Recollection* (London, 1971), 26.

[13] For discussion of and contributions to this debate see: Paddy Scannell and David Cardiff, *A Social History of British Broadcasting*, i, *1922–1939: Serving the Nation* (Oxford, 1991); Thomas Hajkowski, *The BBC and National Identity in Britain, 1922–53* (Manchester, 2010); Gillian McIntosh, *The Force of Culture: Unionist Identities in Twentieth-Century Ireland* (Cork, 1999), 69–95; Jamie Medhurst, 'Television in Wales, *c.*1950–1970' and Daniel Day, '"Nation Shall Speak Peace

BBC was, and remains, a large, complex, and contradictory organization, employing people from and in all parts of the UK, and from many other countries.[14] It was never likely to generate a uniform or unchallenged interpretation of its nation-building duty.

The BBC also projected Britishness beyond the shores of the 'home' islands. This was partly because, from the late 1930s onwards, the BBC was used by the state as a subcontractor for purposes of 'cultural diplomacy'. Direct state funding was provided to supplement the BBC's independent sources of revenue, allowing the corporation to promote Britain's image and interests overseas and present news and information from a British perspective to listeners in foreign countries, along with other subtle forms of propaganda. However, from an even earlier point, and without a direct government subsidy, the BBC voluntarily participated in the overseas projection of Britishness as part of its public-service remit, in order to link the British nation at home with its diasporic branches outside the home islands. Much of this diaspora was located in the dominions of Canada, Australia, New Zealand, and South Africa. These people, and these places, were perceived to be part of a white, English-speaking 'British world'. The dominions were not regarded as 'foreign': they were 'overseas', in the very specific British-imperial sense of the word. They were also seen as being quite different from Britain's tropical, dependent colonies in Africa, Asia, the West Indies, and the Pacific.[15]

The British world was the product of late nineteenth- and early twentieth-century 'explosive colonization'. Rapid demographic and economic growth, fuelled by migration primarily from Britain and Ireland, and capital investment primarily from Britain, created outward-looking, export-oriented economies in Canada, Australia, and New Zealand that thrived on selling food and raw materials in British markets. South Africa experienced a somewhat similar boom at the very end of the nineteenth century, driven by diamonds and especially by gold. Britain's older empire in India and the Caribbean, and its newer colonies in Africa and Asia seized during the frantic European carve-ups of the *fin de siècle*, seemed to languish by comparison.[16]

By the early twentieth century, the constitutional autonomy of the settler dominions had grown considerably to match their ability to shoulder the costs of

Unto Nation": the BBC and the Projection of a New Britain, 1967–1982', both in Michael Bailey (ed.), *Narrating Media History* (London and New York, 2009).

[14] Tom Burns, *The BBC: Public Institution and Private World* (London, 1977). Georgina Born, *Uncertain Vision: Birt, Dyke and the Reinvention of the BBC* (London, 2004), 1–39.

[15] On the concept of the British world (which has been the focus of a great deal of recent historical research and discussion) see Phillip A. Buckner and Carl Bridge, 'Reinventing the British World', *Round Table*, 92/368 (January 2003), 77–88. Carl Bridge and Kent Fedorowich (eds), *The British World: Diaspora, Culture and Identity* (London, 2003), also published as a special issue of *Journal of Imperial and Commonwealth History* (*JICH*), 31/2 (May 2003). Phillip Buckner and R. Douglas Francis (eds), *Rediscovering the British World* (Calgary, 2005). Kate Darian-Smith, Patricia Grimshaw, and Stuart Macintyre (eds), *Britishness Abroad: Transnational Movements and Imperial Cultures* (Carlton, Vic., 2007).

[16] James Belich, *Replenishing the Earth: The Settler Revolution and the Rise of the Anglo-World, 1783–1939* (Oxford, 2009). Darwin, *The Empire Project*. B. R. Tomlinson, 'Economics and Empire: the Periphery and the Imperial Economy' in Porter (ed.), *OHBE*, iii.

self-rule, again in marked contrast to the tropical dependencies. In Canada, Australia, and South Africa, separate colonies came together into large regional groupings, based on various forms of federal organization. In New Zealand, power was centralized in a more unitary polity. The settler states governed their own internal affairs, and assumed increasing authority over trade, foreign relations, and defence. Yet these states were not independent. Not only did Britain reserve significant powers over their external affairs, but close economic, military, and cultural connections with Britain also remained. In pursuing their own agendas, settlers and their political leaders might clash with the imperial government. However, very seldom did this lead them to question the overarching ties of Britannic identity. As the historian Douglas Cole put it, dominion 'patriotism' (loyalty to individual settler states) and Britannic 'nationalism' (a sense of belonging to a global British nation) were complementary.[17]

In Britain, 'constructive imperialists' attempted to harness the future economic and demographic potential of the dominions and thus combat the perceived weaknesses of the mother country in the face of foreign competition. Their campaign to forge the empire into a tariff-protected, semi-autarkic trading bloc stalled in the face of fiscal orthodoxy in the years before the First World War. However, other measures to encourage unity through improved communication, such as the imperial penny-post, subsidized shipping lines, and state-run telegraph cables were adopted, and seemed to reinforce the play of market forces in linking together the component parts of the British world.[18]

The First World War marked a high point in Britannic cooperation, but simultaneously pushed relations between Britain and the dominions into uncertain new territory. Voluntarily, the dominions made a massive contribution to the imperial war effort, in terms of soldiers, war matériel, and food. The dominions became the most important and reliable economic and military heartlands of empire, outside Britain itself. However, not all were satisfied with the terms upon which wartime contributions were made. In South Africa (where, only fifteen years earlier, Britain had gone to war to impose its will on the Boer republics of the interior) the Union government had to suppress an Afrikaner rising. In Canada, attempts to introduce compulsory wartime military service widened the rift between Anglophones and Francophones, and in Australia the threat of conscription similarly divided society along class and sectarian lines. Even fiercely loyal political leaders from the dominions were dissatisfied with the British government's management of the war effort, and some demanded closer consultation regarding the deployment of imperial resources. Initially, it seemed that this might lead to the creation of joint policymaking

[17] Douglas L. Cole, 'The Problem of "Nationalism" and "Imperialism" in British Settlement Colonies', *Journal of British Studies*, 10/2 (May 1971), 160–82. See also: Neville Meaney, 'Britishness and Australian Identity: the Problem of Nationalism in Australian History and Historiography', *Australian Historical Studies*, 32/116 (April 2001), 76–90; and Simon J. Potter, 'Richard Jebb, John S. Ewart, and the Round Table, 1898–1926', *English Historical Review*, 122/495 (February 2007), 105–32.

[18] E. H. H. Green, 'The Political Economy of Empire, 1880–1914' in Porter (ed.), *OHBE*, iii. Gary B. Magee and Andrew S. Thompson, *Empire and Globalisation: Networks of People, Goods and Capital in the British World, c. 1850–1914* (Cambridge, 2010).

bodies. However, it soon became clear that governments would instead demand greater constitutional autonomy, at least in Canada, South Africa, and the Irish Free State (granted dominion status in 1921, but never really fitting the mould).[19]

Frustrated by the difficulties of negotiating common policies, British officials and statesmen were happy to acquiesce. Yet this did not mean that they gave up hope of future cooperation; rather, devolution of full sovereignty to settler governments might actually allow the component parts of what was increasingly called the 'British Commonwealth of Nations' to work together voluntarily, and thus more effectively. Continuing economic, military, cultural, and sentimental connections would encourage such cooperation. Constructive imperialists, in the ascendant in Britain after the war, believed that a natural complementarity of British and dominion interests could be reinforced by planned programmes of migration to the dominions, and by economic policies such as state-assisted marketing of empire produce. Common 'Britannic solutions' might thus be found to the problems facing the British world. Australian and New Zealand policymakers, conscious of their continued military and economic dependency on Britain, generally shared this vision for the future. Canadian and South African responses were often more ambivalent.[20]

With hindsight, as a means of compensating for the collapse of British economic and military might, empire marketing and migration schemes might appear paltry and deluded. Yet during the 1920s, despite the morbid fears of many, the profound and irreversible nature of British decline was not necessarily apparent. The massive damage inflicted on Britain's rivals by the First World War temporarily suppressed imperial competition. Moreover, after the initial economic downturn that followed the war, Britain successfully encouraged the rebuilding of liberal international commercial and financial structures. The globalization that had benefited Britain so much in the past thus resumed, for a short while at least. The US was Britain's most significant economic rival in this period, exporting huge volumes of manufactured goods, and investing heavily abroad. Yet much of its overseas power remained potential rather than actual, particularly as it continued to shun any leadership role in world affairs. In cultural terms, the US quickly established global influence through Hollywood's dominance of the film industry.[21] Yet only in Canada did 'Americanization' seem an overwhelming threat to British culture and its British-world variants.

British constructive imperialists were 'high modernists', who believed that wise state intervention, led by experts working according to scientific principles, could

[19] Robert Holland, 'The British Empire and the Great War, 1914–1918' in Judith M. Brown and Wm. Roger Louis (eds), *OHBE*, iv, *The Twentieth Century* (Oxford, 1999). C. P. Stacey, *Canada and the Age of Conflict: A History of Canadian External Policies*, i, *1867–1921* (Toronto, 1977). Philip G. Wigley, *Canada and the Transition to Commonwealth: British-Canadian Relations, 1917–1926* (Cambridge, 1977). Ged Martin, 'The Irish Free State and the Evolution of the Commonwealth' in Ronald Hyam and Ged Martin, *Reappraisals in British Imperial History* (London, 1975).

[20] John Darwin, 'A Third British Empire? The Dominion Idea in Imperial Politics' in Brown and Louis (eds), *OHBE*, iv.

[21] Robert Boyce, *The Great Interwar Crisis and the Collapse of Globalization* (Basingstoke, 2009), 3–4, 78–9, 179.

reshape the world for the better and allow Britain to face the challenges of the 1920s.[22] The BBC was developed according to a very similar philosophy, in order to meet domestic requirements: public-service broadcasting involved the creation of an autonomous state body, directed by cultural and technical experts, with the aim of improving society. The way that Reith built the BBC at home meant that it was also naturally, if incidentally, suited to discharging a constructive imperialist function overseas. Indeed, constructive imperialist politicians scarcely needed to encourage Reith and his subordinates to use broadcasting to promote the unity of the British world, for this was a task that the BBC took upon itself, largely unprompted.

As with its nation-building broadcasts for domestic audiences, in its overseas services the BBC emphasized the overarching high culture shared by Britain's cultivated elites, while also reflecting the regional diversity of the home islands and their traditional folk cultures. It seemed logical to apply the same approach to projecting Britishness at home and overseas: Britain and the British world had, after all, been 'forged' at much the same time, with the English, Scottish, Irish, and Welsh all playing a significant role in both national and imperial developments. The composition of, and identities contained within, the twentieth-century British diaspora thus reflected the varied regional origins of nineteenth-century migrants. Around the British world Britishness tended to sit alongside, rather than replace, other forms of felt community.[23]

Yet in practice, neither at home nor abroad, did Britishness form the basis for an unchallenged consensus, or indeed for an effective and lasting sense of unity. Here, the failures of British identity-building again illuminate similarities and connections between domestic and overseas developments. In the dominions and colonies, different groups could certainly turn Britannic identities and cultural exports from Britain to their own purposes, but doing so often involved significant adaptation and transformation. Isolated and nostalgic British expatriates (administrators, planters, and traders) in Africa and Asia might eagerly lap up anything that reinforced their connection with the distant centre of empire, but might also reject programmes that failed to harmonize with their memories of 'home'. In the settler dominions appreciation of sophisticated British cultural exports offered would-be cultivated elites a means to assert their own authority, and to devalue and thus tame the rough-and-ready folk cultures of the frontier and the small town.[24] But how would such audiences react to the suggestive, lowbrow humour enjoyed by most of the BBC's listeners at home? In overseas settings, Britishness was thus always liable to become something subtly different from any of its UK variants.

Broadcasting also had its own particular limitations as a means of projecting a widely acceptable form of Britishness overseas. Here, a comparison with the press is instructive. The fact that newspapers around the British world were often dependent upon the UK-based Reuters news agency seldom provoked much

[22] James C. Scott, *Seeing Like a State: How Certain Schemes to Improve the Human Condition have Failed* (New Haven and London, 1998).

[23] Linda Colley, *Britons: Forging the Nation, 1707–1837* (New Haven and London, 1992).

[24] Jonathan F. Vance, *A History of Canadian Culture* (Don Mills, Ont., 2009), 105–34.

hostility. Indeed, when arrangements for transmitting news around the empire were criticized, it was generally on the grounds that not enough news was reaching the dominions from Britain, or that news was only travelling from Britain through a single channel, rather than multiple (British) channels. That Reuters news was not deemed to be an irritating symbol of Britain's dominant position within the British world was at least in part because it was mediated and 'indigenized' by locally produced newspapers, which subedited reports to suit the particular requirements of their own readers. Reuters news thus did not speak with an unambiguously metropolitan voice. Meanwhile, British newspapers like *The Times* might reach the dominions, but they did not do so in sufficient volume, or sufficiently quickly, to compete with locally produced newspapers.[25]

Radio, on the other hand, relied upon the human voice. In some cases, this did not really matter: Shakespearean actors spoke in similar tones in Stratford-Upon-Avon and Stratford, Ontario, and as one Australian remarked at the proposed imposition of blanket quotas on the import of recorded radio programmes from Britain, 'Putting a tariff on Shakespeare doesn't advance our own literary endeavour.'[26] Yet much of the time, radio did remind listeners that people in different parts of the British world had different accents. This could in turn draw attention to the fact that social contexts, geopolitical perspectives, and the very ways in which Britannic and other forms of identity were conceived of and felt, varied considerably from place to place. BBC programmes did, literally, carry a metropolitan voice. In Britain as well as in the dominions, BBC English might seem to some a marker of cultural quality and authority, but to others it evoked privilege, hierarchy, and the narrow interests of Britain's elite. At home and overseas, audiences thus responded to BBC attempts at broadcasting Britishness with an unpredictable mixture of enthusiasm and resistance.

CULTURAL IMPERIALISM IN A DECLINING EMPIRE

Yet until the early 1940s at least, the BBC had little way of gauging audience responses. Prior to the Second World War, systematic audience research was purposefully neglected, even where home listeners were concerned. It was even more difficult to find out about overseas reactions: the BBC lacked offices in the dominions capable of feeding information back to London, and few of its staff possessed detailed knowledge, or indeed any experience at all, of life in the dominions and colonies.

This did change somewhat during the war, but initially, for advice about local tastes and requirements, the BBC had to rely on the dominion public broadcasters that it hoped would carry its programmes as part of their own services. These organizations were hardly impartial sources. They had their own agendas, relating to building national communities, purveying particular kinds of Britishness, and

[25] Potter, *News and the British World*.

[26] Keith Barry to Charles Moses, 3 September 1945, National Archives of Australia, Sydney, SP613/1, 8/1/29 pt. 1, 3188849.

protecting their own positions in local broadcasting systems, often in the face of commercial competition. From an early stage collaborative relationships among the British world's public broadcasting authorities were extremely important and yet intensely complicated. These relationships developed and changed according to complex rhythms over the decades that followed. The BBC remained the dominating authority in broadcasting in the British world. This was thanks to a resource base that reflected the demographic weight of UK audiences relative to their dominion counterparts, and the fact that the BBC did not have to share with any competitors the resources provided by its home territory, because until the mid-1950s it enjoyed a domestic monopoly. Nevertheless, the BBC could not ride rough-shod over its collaborators in the dominions, for it was too dependent on them for assistance.

The uneven spatial distribution of people always made it unlikely that the exchange of cultural influences within the English-speaking world would take place on equal terms. In the 1930s, Britons in the UK outnumbered their Australian or Canadian Anglophone counterparts approximately seven-to-one, and English-speaking settlers in New Zealand or South Africa thirty-to-one. The population of the US was meanwhile more than two and a half times that of the UK. New Zealand had about as many people as the State of Washington. Relative to their overall population sizes, the dominions produced substantial urban centres, but these were nevertheless small compared to cities such as London and New York, which exerted power and influence on a global scale. London was an imperial cultural centre, and from the 1930s it became an imperial broadcasting hub, sucking in the raw material of talent, and exporting the fruits of concentrated cultural activity. From the perspective of the BBC in London, Australia and Canada contained potential audiences similar in size to one of the larger UK broadcasting regions. New Zealand and South African audiences were comparable to those present in the smallest UK regions. Their claims on the BBC's attention were weighed accordingly.

Given these inescapable demographic realities, and the fact that improving transnational communications links seemed to reduce the possibility of separate development, could the dominions realistically expect to produce distinctive 'national' cultures of their own? As one Canadian critic argued in 1932,

> Surely it is a mistake to expect Canadian culture to assume a form and expression wholly different from the sources of its inspiration... distinctive national achievements in the realm of culture belong in the main to an earlier period when countries were compelled by comparative isolation to depend upon their native resources.[27]

Indeed, perhaps the obstacles to cultural exchange would prove so great that the dominions would not even be able to make much of a contribution to the wider whole? Such questions lurked at the back of the minds of public broadcasters around the British world.

[27] Norman Rogers quoted in Maria Tippett, *Making Culture: English-Canadian Institutions and the Arts before the Massey Commission* (Toronto, 1990), 128.

From the 1960s, academics and others wrote of 'cultural imperialism' and 'media imperialism'. The contemporary US, like Victorian Britain, was so powerful economically that it seemed able to foist its cultural influence on the outside world. In so doing, it further reinforced its global economic might. One of the defining features of cultural imperialism, according to critics, was the fact that communication was generally one way rather than reciprocal. Information and influence only flowed outwards, from New York and Hollywood. Private media concerns, grown fat on the enormous, lightly regulated domestic market, and indeed with the support of American policymakers, could flood foreign countries with cheap exports, swamping indigenous producers and audiences, and making money while serving broader geopolitical interests. The cultural diversity and independence of other countries was thus compromised.[28] In a deregulated world information order, in which American companies would always be the biggest players, '[i]t is extremely difficult for a society to practice free flow of media and enjoy a national culture at the same time—unless it happens to be the United States of America.'[29]

These arguments were undoubtedly somewhat simplistic, and have since been heavily critiqued. However, they reflected genuine and widely shared contemporary anxieties. Public broadcasters in Britain and the dominions explicitly discussed the threat of Americanization, particularly after the Second World War. In the dominions during the 1930s some had alleged that the BBC was perpetuating a cultural imperialism of its own, encouraging an unreciprocated flow of material out from London and refusing to broadcast programmes from the dominions to listeners in Britain. Yet the BBC was never able to export programmes in sufficient volume to threaten to swamp dominion broadcasters and listeners in the way that American producers later did. In the British world, certainly from the 1940s onwards, anxieties about the cultural influence of Britain generally stemmed from the fact that it was too weak, not too strong. The BBC failed to offer other public broadcasters sufficient support in their efforts to prevent the dominions being absorbed into the new 'homogenized culture of the American Empire'.[30]

During the 1950s and especially the 1960s contemporaries had to adjust to the reality of Britain's diminished world role, and growing American influence.[31] This transition involved accommodation more than confrontation, and was eased somewhat by the fact that the boundaries between the US and the British world had always been porous, with America a familiar 'other', or even imaginable as part of a

[28] Herbert I. Schiller, *Mass Communication and American Empire* (New York, 1969); Alan Wells, *Picture-Tube Imperialism? The Impact of U.S. Television on Latin America* (Maryknoll, NY, 1972); Jeremy Tunstall, *The Media are American: Anglo-American Media in the World* (London, 1977); Thomas L. McPhail, *Electronic Colonialism: The Future of International Broadcasting and Communication* (Beverly Hills, CA and London, 1981); and John Tomlinson, *Cultural Imperialism: A Critical Introduction* (Baltimore, MD, 1991).

[29] Anthony Smith, *The Geopolitics of Information: How Western Culture Dominates the World* (New York, 1980), 53. For a more historically oriented analysis, see Emily S. Rosenberg, *Spreading the American Dream: American Economic and Cultural Expansion, 1890–1945* (New York, 1982).

[30] George Grant, *Lament for a Nation: The Defeat of Canadian Nationalism* (Ottawa, 1995 [1965]), 26.

[31] Ronald Hyam, *Britain's Declining Empire: The Road to Decolonisation, 1918–1968* (Cambridge, 2006).

single English-speaking world. Contemporaries turned to such ideas for comfort during and after the Second World War. Yet they provided limited reassurance. An underlying sense that commercialized US culture was alien, inferior, and dangerous remained, and informed negative reactions to the fall of the British world and the rise of the American superpower. During the 1960s the BBC staged something of a comeback, exporting substantial amounts of television programming. Yet in the new Anglo-American global media order, Britain was at best a very junior partner.[32]

In thinking about the BBC's imperial role, we also need to factor in another important bias. Seeking to reflect and ensure the continued existence of a transnational community of Britons, the BBC inevitably assigned a secondary status to others. Its approach towards Francophones in Canada and Afrikaans-speakers in South Africa was hesitant, particularly in the 1930s, reflecting broader uncertainty as to whether Britannic identities were based on an exclusively Anglo-Saxon 'British race patriotism', or a more inclusive shared British culture that could potentially attract and integrate others. Meanwhile, non-whites in Africa, Asia, and the Caribbean, and indigenous peoples in the dominions, were largely ignored. This reflected, in part, practical constraints. In the dependent colonies, linguistic fragmentation was the norm and, except in India and the West Indies, English-speakers represented a microscopic minority. Broadcasting in a myriad tongues would have required resources that the inter-war BBC simply did not command. More significantly, in every colony, poverty restricted the ownership of radio receiving sets. Yet the artificial constraints imposed by contemporary thinking about racial difference also played a role. The white British world was perceived to be the most important part of the empire, from which most could be gained by strengthening imperial bonds, and in which the need to do such work seemed most pressing. Africans and Asians might claim to be British, and had periodically been told by the imperial and colonial states that they were British, and should make sacrifices accordingly, particularly in times of war. Yet more frequently, non-whites were excluded from the British imagined community, both at home and overseas.[33] Before the Second World War, with the exception of India, the dependent colonies seemed economically marginal, and politically more or less quiescent. There was little to be gained from making them the focus of the BBC's overseas broadcasting efforts. This only changed with the sudden British 'colonial development' drive of the early post-war years.

BROADCASTING, EMPIRE, AND BRITAIN'S CULTURES

As the chapters that follow seek to demonstrate, despite all its ambitions, the BBC had only limited success in projecting domestically derived forms of Britishness

[32] Tunstall, *The Media are American.*

[33] Sean Hawkins and Philip D. Morgan, 'Blacks and the British Empire: an Introduction' in Morgan and Hawkins (eds), *Black Experience and the Empire* (Oxford, 2004). Specifically on broadcasting, see Darrell Newton, 'Calling the West Indies: the BBC World Service and *Caribbean Voices*', *Historical Journal of Film, Radio and Television* (*HJFRT*), 28/4 (October 2008), 489–97.

overseas. Public broadcasting was, in the end, simply not sufficient to the task of reinforcing Britannic identities in the dominions at a time of imperial decline. Similarly, the BBC largely failed in its domestic imperial mission of making audiences at home more aware of the empire and of the British nation overseas. The gap between intention and achievement was significant.

Historians do not agree about the nature and extent of empire's impact on cultures and identities in Britain. Recent historiographical debates have centred on the effects of overseas expansion on British attitudes towards race and belonging, looking at evidence concerning the intellectual and political elite, and also 'ordinary' people. Some scholars have emphasized the profound significance of imperial influences, while others have claimed that the effects were only superficial. The debate is hard to resolve, as it centres on difficult issues of reception and audience response, and takes us into areas for which we lack representative and reliable, first-hand evidence, particularly regarding working-class attitudes towards empire. However, it seems safe to assume that responses to empire in Britain were varied, complex, and changing, influenced by the interaction of a range of factors including social class, gender, and diverse regional patterns of imperial engagement.[34]

What can a study of broadcasting add to our knowledge of the impact of empire on British culture? First, it should illustrate the extent and nature of imperial influences on one of the key institutions in twentieth-century British cultural life: the BBC itself. As P. J. Marshall has noted, if there was broad support at home for Britain's imperial role, then we might expect to see this demonstrated not in political debates (which registered controversy rather than consensus) but rather in the

> 'imperializing' of British institutions, as they adapted to the needs of empire and perhaps absorbed values associated with it. The record suggests some changes along these lines, but also much resistance to change... certain British institutions were very willing to add an imperial role to their role within Britain. Whether British institutions were prepared to accept fundamental change as the price of an imperial role was another matter.[35]

Gauging whether the BBC's imperial role fundamentally reshaped its institutional nature, or merely expressed and reinforced existing domestically generated features might thus offer an important contribution to our understanding of Britain's imperial experience.

Second, studying the BBC ought to provide us with some indication of perceived audience attitudes and responses to empire-related programmes, shedding light upon popular engagement with empire. Granted, the BBC's own particular view of its relationship with its audiences might make such an exercise less rewarding than anticipated. Most BBC officers did not think they should simply give audiences what they wanted, and the BBC was thus slow to adopt any systematic policy of listener research. Nevertheless, many BBC producers and planners wanted

[34] For a survey of the historiography see Simon J. Potter, 'Empire, Cultures and Identities in Nineteenth- and Twentieth-Century Britain', *History Compass*, 5/1 (January 2007), 51–71.

[35] P. J. Marshall, 'Imperial Britain', *JICH*, 23/3 (September 1995), 379–94, quotes at 380 and 382.

programmes to have a constructive impact on audiences, reshaping preconceptions and knowledge. They were keen to find out whether programmes and policies were having the intended results. Among themselves, when it came to covering the empire, they thus discussed the factors that they thought influenced popular attitudes, devised policies to alter those perceived attitudes in various ways, and over time tried to discover whether their efforts had been successful. The claims that BBC officers made about audience tastes, attitudes, and knowledge cannot be taken as a direct or reliable proxy for popular responses to empire and imperial propaganda. Yet they can show how a key group of opinion formers, seeking to negotiate and shape those responses, worked with certain basic assumptions that in turn influenced their own actions, and the nature of the empire-related material that they produced.

Third, this study offers a more direct insight into what the BBC's own employees, who constituted a cross-section of Britain's cultivated elite, thought about empire, and about the British world in particular. BBC officers were drawn from a range of upper- and middle-class backgrounds, many were observant and perceptive, and most were able to record their opinions in letters, memoranda, and reports that were carefully preserved in the BBC's archives. These witnesses to the changing British world enjoyed opportunities for travel that many of their contemporaries were denied, and were able to mix not only with prominent artists and politicians, but also with the 'ordinary' people who featured in, listened to, and watched their programmes. BBC officers belonged to an organization with a global reach and sense of purpose that was at the same time at least one step removed from the world of party politics and high officialdom that otherwise provides us with so much of our evidence for understanding public discourses about empire and Britishness.

An early essay on the BBC's imperial role by John M. MacKenzie suggested that, while senior BBC administrators saw it as the corporation's duty to promote imperial sentiment, those who actually made the programmes seldom shared such enthusiasms.[36] However, the evidence deployed by MacKenzie (which focused on broadcast celebrations of Empire Day) could also be interpreted as suggesting the producers' dissatisfaction with the poverty of material available for use, rather than their more general antipathy to the idea of empire. This would help explain the arguments advanced more recently by Thomas Hajkowski and Siân Nicholas, who have both surveyed BBC empire-related radio programmes and programme policies, particularly during the Second World War. Hajkowski concentrates on the programmes themselves, and emphasizes the persuasive power of the BBC over its audiences. He concludes that millions of listeners found comfort in the 'rosy vision of Empire' projected by an enthusiastic BBC, a vision 'that implicitly confirmed British superiority'. Nicholas meanwhile shares the more pessimistic outlook of those who made the programmes, the BBC producers who seldom believed that they had succeeded in overcoming massive listener apathy or resistance to

[36] John M. MacKenzie, 'In Touch With the Infinite: the BBC and the Empire, 1923–53' in MacKenzie (ed.), *Imperialism and Popular Culture*.

empire-related material. According to this view, the BBC might have projected splendid imperial imagery, but listeners simply switched off.[37]

Wendy Webster has meanwhile examined BBC programmes in the context of a wide-ranging study of the images of empire that were deployed in Britain during and after the Second World War. She argues that these images reflected the ambivalent and shifting nature of (specifically) English attitudes towards Africans and Asians, and the simultaneous operation of domestic and other international (particularly American) influences. Wartime propaganda sought to promote the image not just of a 'people's war', with a British nation working together in unity, but also of a 'people's empire', drawn together by shared ideals of welfare, development, and egalitarianism. At the same time, Webster argues, more negative ideas about Africans and Asians endured. While images of a racially inclusive empire predominated during the coronation celebrations of 1953, for example, the combined post-war shocks of African and Asian anti-colonial violence and large-scale Commonwealth migration to Britain together acted to intensify perceptions and representations of racial difference.[38]

This book builds on these insights, arguing that the idea of a white British world was central to the BBC's projection of empire to domestic audiences. This bias had formidable institutional roots. Connections with broadcasters in the dominions were much stronger than with their counterparts in the dependent colonies. Certainly until the 1950s BBC officers tended to view their imperial role primarily in terms of linking domestic listeners with a white Britannic community overseas. However, the chapters that follow also argue that it would be wrong to exaggerate the relative significance of this imperial role, for the BBC, or for British popular attitudes more generally. Engagement with the British world was only a minor element of the BBC's work. Other domestic priorities and other transnational connections (particularly with the US and Europe) absorbed a much greater proportion of the attention of listeners, producers, and administrators. Perceived UK audience resistance to empire-related themes remained marked, not just in relation to programmes concerning Africa and Asia, but also to those seeking to stimulate a sense of connection with Britannic communities overseas. BBC officers continually cited this stubborn resistance as an obstacle to the reciprocal flow of material from the dominions back to Britain. Claims about popular apathy were not an excuse to overload BBC schedules with empire-related programmes, for the amount of such material carried on the domestic services remained tiny, relative to the rest of the material that went to make up broadcasting schedules. By 1970, with Britain's battered world-system well into the final stages of collapse, BBC officers had largely given up on their sporadic and generally unsuccessful attempts to make UK audiences empire- or Commonwealth-minded.

[37] Thomas Hajkowski, 'The BBC, the Empire, and the Second World War, 1939–1945', *HJFRT*, 22/2 (June 2002), 135–55. Hajkowski, *BBC and National Identity*. Siân Nicholas, '"Brushing Up Your Empire": Dominion and Colonial Propaganda on the BBC's Home Services, 1939–45', in Bridge and Fedorowich (eds), *British World*.

[38] Wendy Webster, *Englishness and Empire, 1939–1965* (Oxford, 2005).

1

Diversity, 1922–31

In 1932 the first London representative of the US National Broadcasting Company (NBC) paid his first visit to the director-general of the British Broadcasting Corporation (BBC). With unconscious symbolism, he presented as a gift 'a magnificent hand-painted globe of the world'.[1] It would be gross exaggeration to characterize subsequent rivalries between British and American broadcasters as an outright struggle for world domination. The relationship between the BBC and the US networks involved much collaboration and compromise, with ideas, programmes, and people travelling in both directions. Nevertheless, from an early stage, BBC officers perceived the Americanization of broadcasting around the world, and particularly within the British world, as a potential threat, appearing over time in various guises. During broadcasting's first decade BBC officers worried not so much about programmes, which did not yet travel in any quantity, but rather that the basic American approach to organizing radio services would spread overseas.

As the British literary critic F. R. Leavis argued, British fears of 'Americanization' were to some extent misguided. The worrying changes apparent in US culture were not a contagion that threatened to spread across the Atlantic, but rather symptoms of a profound underlying transformation occurring in all 'modern' societies, a 'levelling-down' that was merely more advanced (and therefore more noticeable) in the US than elsewhere.[2] At least as far as radio was concerned, there was an element of truth to this assertion. During the 1920s broadcasters and regulators in Britain and the US faced the same fundamental question: how should the new mass medium operate in an environment in which old regional and communal barriers were breaking down under the pressure of social, economic, and technological change, and in which previously excluded classes and social groups were now being admitted to the political and cultural life of the nation-state? In a potentially unstable 'mass' society, how could the number of voices on air be restricted, and how could limitations be placed on what they would say? In both countries the solution seemed in part to lie in national networking, the linking up of transmitters in order to coordinate and control what was broadcast.[3] Broadcasting was financed and operated in Britain and America in distinct ways, but arguably with the same aim of taming the medium and harnessing it as a force for national integration.

[1] John Reith diary, 26 September 1932, BBC Written Archives Centre (WAC), S60/5/3/2.

[2] F. R. Leavis, *Mass Civilisation and Minority Culture* (Cambridge, 1930), 7.

[3] Christopher Anderson and Michael Curtin, 'Writing Cultural History: the Challenge of Radio and Television' in Niels Brügger and Søren Kolstrup (eds), *Media History: Theories, Methods, Analysis* (Aarhus, 2002).

In the US private enterprise dominated, with minimal state regulation. Radio came to play a predominantly conservative social function, thanks to the logic of market forces. By the end of the 1920s American radio was generally financed by advertising, and networking had become big business. NBC and the Columbia Broadcasting System (CBS) emerged as the two largest networks, purchasing time in which to air their programmes on privately run stations around the country. These network programmes were produced live and fed to affiliate stations using landlines: a costly business. The networks sold on the right to sponsor their programmes to large corporations, mainly producers of branded household goods, which could afford to pay the amounts needed to support network broadcasting, and which wanted to market their goods on a national basis. These corporations employed advertising agencies to manage such activities. Increasingly, the agencies took responsibility for actually producing the network programmes. All the way down the line, the commercial interests involved in broadcasting tended to shy away from controversial or adventurous programming, and instead to favour popular but unexceptionable entertainment that would guarantee audiences and advertising revenues.[4] The American approach did at least offer listeners, as consumers, the appearance of choice, and encouraged a friendly, accessible style of radio that was welcome in an increasingly impersonal world.[5]

In Britain, by contrast, broadcasting was tamed in a more heavy-handed fashion through the principle of public ownership and operation. A single authority provided programmes to listeners on a not-for-profit basis, without commercial sponsorship of programmes or on-air advertising. That authority, the BBC, remained under 'remote' state control: answerable to parliament, but protected from day-to-day government interference. BBC officers thought that the British way of network broadcasting compared favourably with the American approach: not only did it allow the production of culturally superior programming, but it also seemed a more effective means of disciplining the medium. British broadcasting, which rejected the cult of celebrity in favour of a more impersonal ethos, did not produce unpredictable demagogues of the kind that occasionally found a voice on air in the US.[6] The BBC warned dominion governments that, if they failed to restrict commercial broadcasting, then 'in due course they would find themselves up against an Australian Huey Long or Father Coughlin'.[7]

[4] Eric Barnouw, *A History of Broadcasting in the United States*, i, *A Tower in Babel: to 1933* (New York, 1966). See also Barnouw, *History of Broadcasting in the United States*, ii, *The Golden Web: 1933 to 1953* (New York, 1968), 57–8; Alice Goldfarb Marquis, 'Written on the Wind: the Impact of Radio during the 1930s', *Journal of Contemporary History*, 19/3 (July, 1984), 385–415; and Susan Smulyan, *Selling Radio: the Commercialization of American Broadcasting, 1920–1934* (Washington, DC and London, 1994).

[5] Bruce Lenthall, *Radio's America: the Great Depression and the Rise of Modern Mass Culture* (Chicago, 2007).

[6] Michael Bailey, 'Rethinking Public Service Broadcasting: the Limits to Publicness', in Richard Butsch (ed.), *Media and Public Spheres* (Basingstoke, 2007).

[7] J. C. W. Reith to H. D. Vickery, 8 January 1936, WAC, E1/1095. Reith diary, 6 June 1935, WAC, S60/5/4/2. On Long and Coughlin see Lenthall, *Radio's America*, 115–41.

During the inter-war years the BBC sought to export the principles of public broadcasting overseas. This was not just because they believed those principles to be superior to American commercial methods. Spreading the British model overseas would help perpetuate British cultural influence, and contain Americanization.

In the dominions, as in many other places, the American and British approaches certainly became key reference points for those seeking to regulate broadcasting, or to use it to make money, or serve a particular political, social, or cultural purpose. However, the two models were seldom reproduced overseas in exactly their original forms. 'Hybridization' occurred, as aspects of the different approaches were selected, blended together, and mixed with local adaptations and innovations. Indeed, the distinctiveness of the two approaches began to blur even in their original settings, as broadcasting practices and policies in the UK and US gradually influenced one another.[8] Local adaptation was encouraged by the fact that neither the US nor UK approaches emerged fully formed. Both developed gradually, as regulators and broadcasters began better to understand their work. The 'public service' principles that directed the operations of the BBC may have coalesced relatively early, but it was some time before they were recognized and enshrined in the organization's royal charter. Similarly, in the US experimentation was necessary before businessmen saw how to make advertising and networking pay. To understand how broadcasting developed in the dominions and colonies, we thus need to examine decision-making in the exact contemporary context in which it took place. Timing was important, as emergent local, national, and Britannic considerations interacted with constantly shifting overseas models. Sometimes decisions were made that left an unanticipated but near-indelible impact on the broadcasting landscape, and helped determine the prospects for success of the attempts of the BBC, and others, to develop imperial broadcasting structures and relationships over the decades that followed.

BROADCASTING IN BRITAIN

In Britain, broadcasting was pioneered by the Marconi Company, already a major player in the field of wireless telegraphy, with imperial and indeed global interests. Marconi began experimental transmissions in 1920 (including a recital by the Australian *prima donna*, Dame Nellie Melba), and inaugurated Britain's first regular broadcasting service in February 1922. Other private companies also began to enter the field. The task of regulating broadcasting in the UK was assigned to the General Post Office (GPO), as an extension of its established responsibilities in the fields of telegraphy and wireless telegraphy. After initial hesitation, in 1922 the

[8] Michele Hilmes, *Network Nations: A Transnational History of British and American Broadcasting* (New York and London, 2012). Bridget Griffen-Foley, 'The Birth of a Hybrid: The Shaping of the Australian Radio Industry', *Radio Journal*, 2/3 (2004), 153–69. On the broader context of connections between the mass media in Britain and America, see Joel H. Wiener and Mark Hampton (eds), *Anglo-American Media Interactions, 1850–2000* (Basingstoke, 2007).

GPO encouraged a decisive new departure, motivated not by what broadcasting had become in Britain, but rather by what it threatened to become in the US. Private radio stations had proliferated in America: 219 had been registered by 1 May 1922. Many operated on overlapping wavelengths, resulting in the 'jamming' not only of private broadcasts but also of military signals. A senior GPO official visited the US in the winter of 1921–2: on his return, following consultation with the major manufacturers of radio equipment, it was agreed that the number of stations to be operated in the UK, and the wavelengths to be made available, would both be strictly limited. The manufacturers were asked to propose a collaborative structure for providing an appropriate broadcasting service, funded by listener licence fees and royalties on sales of receiving sets. In keeping with the desire to control exactly who was running the business of broadcasting, the selling of airtime to advertisers was prohibited. After some hard bargaining, it was agreed to form a single company that would produce and transmit all programmes: the British Broadcasting Company. Any British radio manufacturer could become a shareholder. Entry to the world of broadcasting was thus not exclusive; but that world consisted entirely of the BBC. Existing broadcasting infrastructure was either decommissioned or absorbed into the new company, which began broadcasting on 14 November 1922.[9]

Two key public enquiries subsequently helped clarify the relationship between the BBC, its shareholders, and the state. The Sykes Committee reported in August 1923 that, while 'control of such a potential power over public opinion and the life of the nation ought to remain with the State', direct management by politicians or civil servants would mean that broadcasts would either be partisan or 'intolerably dull'. Instead, the day-to-day running of the service should be delegated to an autonomous operating authority.[10] This was an early formulation of what became known as the principle of 'remote state control'. The state was to retain ultimate ownership and control of broadcasting. Authority was only devolved to the BBC, and parliament reserved the right to legislate to change the terms under which the company operated. However, at the same time, politicians and civil servants were prevented from intervening in the day-to-day running of the authority or from directing programme policy. This was not direct state control as it developed in continental Europe.

In accordance with these new principles, and in adopting the recommendations of the Crawford Committee (published in March 1926), the GPO subsequently surrendered most of its notional powers of intervention in matters of general and programme policy, although it retained regulatory powers over technical matters. Meanwhile, on the committee's recommendation, rather than renew the BBC's licence, the government opted to transform the company into a public corporation, established by royal charter with effect from 1 January 1927. The shareholdings of the radio manufacturers were liquidated, and a new board was created,

[9] W. J. Baker, *A History of the Marconi Company* (London, 1970). Asa Briggs, *The History of Broadcasting in the United Kingdom*, i, *The Birth of Broadcasting* (Oxford, 1995 [1961]), 33–114.

[10] *The Broadcasting Committee: Report* (London, 1923 [Cmd. 1951]), 6, 14.

made up of five 'independent' part-time governors, including a chairman, appointed by the government. The exact powers of the governors, in terms of formulating policy and aiding or supervising the chief executive officer, the director-general, in the day-to-day running of the BBC, would be clarified over the years that followed and remain open to renegotiation. This mirrored the broader process by which the relationship between the BBC and the state was defined and modified over time.[11]

The emergence of the BBC as a public broadcasting authority controlled remotely by the state and enjoying a monopoly of domestic broadcasting seems strange in the context of the untrammelled private enterprise and competition that had come to characterize the traditional medium of mass communication in Britain and the dominions, the newspaper press. However, for contemporaries, broadcasting did not seem to belong in a category with the press, but rather with cable and wireless telegraphy, and thus required a similar degree of state involvement. At the same time as the GPO and Marconi and the other wireless manufacturers were hammering out the basis upon which the BBC would operate, they were also discussing the construction of the long-delayed imperial wireless telegraphy stations. In broadcasting, the decision to exercise public control also reflected the intersection of a range of domestic interests. Wireless manufacturers welcomed what appeared to be a chance to block foreign imports of equipment. Music and entertainment industry interests wished to limit the damage they feared broadcasting might do to the profitability of live entertainment. The press was eager to see radio established on a non-commercial, and thus non-competing, basis. GPO officials regarded a broadcasting monopoly as a convenient and efficient solution to the problems of waveband scarcity and regulation. Politicians showed relatively little interest in the whole issue: those on the Left who did think about it supported public control, while to the Right it seemed a suitably paternalistic way of dealing with a potentially transformative new technology.[12]

The Sykes Committee reported in 1923 that the fledgling BBC already considered itself a 'public utility service'.[13] This reflected the terms under which the wireless companies had formed their combine, and the influence of John Reith, the young Scottish engineer appointed as the first general manager (later director-general). Reith championed the idea of public service in order to maintain the broad consensus upon which the BBC's continued existence depended. A superb administrator, Reith could inspire enduring loyalty and affection among his subordinates. He found it more difficult to deal with his equals, superiors, and those with whom he disagreed. Reith presented an austere and authoritarian façade to the world, which he developed self-consciously, to the point of theatricality. He

[11] Briggs, *History of Broadcasting in the UK*, i, 299–323. Paddy Scannell, 'Public Service Broadcasting: the History of a Concept' in Andrew Goodwin and Garry Whannel (eds), *Understanding Television* (London, 1990).

[12] D. L. LeMahieu, 'John Reith, 1889–1971: Entrepreneur of Collectivism' in Susan Pederson and Peter Mandler (eds), *After the Victorians: Private Conscience and Public Duty in Modern Britain* (London, 1994), 193–4.

[13] *Broadcasting Committee: Report*, 27.

played on his formidable physical presence: six feet six inches tall, with a striking scar running across his left cheek acquired during active service on the Western Front. A reviewer of one of his efforts at amateur dramatics thought he resembled 'Boris Karloff at his most sinister'. His colleague Lionel Fielden wrote that, whenever he had a meeting with Reith, he had to suppress 'an insane desire... to clamber on to my chair before speaking'.[14]

Although Reith eventually made enough political enemies to blight his career, he steered the BBC successfully through many of its early challenges. Reith defused opposition by emphasizing the BBC's national role and reach. With a linked network of transmitters capable of serving almost all of the nation's listeners with the same material, the BBC would disseminate programmes of the highest standards and thus encourage cultural 'uplift'.

> The Broadcasting Service should bring into the greatest possible number of homes in the fullest degree all that is best in every department of human knowledge, endeavour and achievement. Rightly developed and controlled it will become a world influence with immense potentialities for good—equally for harm, if its function is wrongly or loosely conceived. The preservation of a high moral tone is obviously of paramount importance. Everything definitely vulgar or hurtful must of course be avoided... Popularity must not be sought in ways where it is soonest found... He who prides himself on giving what he thinks the public want is often creating a fictitious demand for lower standards which he himself will then satisfy.[15]

Reith's inspiration was partly Victorian: he echoed Matthew Arnold's claim that 'national' and 'public' cultural institutions should be used to promote 'the best knowledge, the best ideas'.[16] A stern Calvinist, he also shared Thomas Arnold's 'public distrust of the frivolous and the sensual'.[17] Nevertheless, Reith was clearly a product of his own time. Early twentieth-century cultivated elites claimed that a new, commercialized mass culture threatened to engulf Britain in a wave of materialistic, titillating, sentimental, often American, dross. Cinema, popular newspapers, and gramophone recordings were the carriers of the disease. The cure was active promotion of forms of culture that appealed to the intellect and imagination, or so it was claimed. In particular, elite cultural institutions that policed standards had to be respected, and public-spirited experts empowered to determine what was produced and how it should be disseminated to a mass audience.[18]

Reith was a high modernist, advocating the application of technical and scientific methods by the state to encourage human progress. Such progress did not

[14] Ian McIntyre, *The Expense of Glory: A Life of John Reith* (London, 1993), 224. Lionel Fielding, *The Natural Bent* (London, 1960), 128.

[15] J. C. W. Reith, 'Memorandum of Information on the Scope and Conduct of the Broadcasting Service, Submitted as Evidence to the Broadcasting Committee, 1925', 2–3, WAC, R4/27/1. See also Reith, *Broadcast over Britain* (London, 1924).

[16] Matthew Arnold, *Culture and Anarchy and Other Writings*, ed. Stefan Collini (Cambridge, 1993), 79.

[17] LeMahieu, 'John Reith', 195.

[18] D. L. LeMahieu, *A Culture for Democracy: Mass Communication and the Cultivated Mind in Britain between the Wars* (Oxford, 1988), 103–7.

seem guaranteed in the wake of the war: old certainties had been shaken by the widening of the franchise and the inclusion of women for the first time; by the decline of the Liberal Party and the rise of Labour; and by the generalized threat of political violence and social upheaval that had accompanied demobilization.[19] Public broadcasting would help govern democracy, and also help democracy govern itself. It would guide and moderate working-class participation in public life, and provide individuals with the means to evaluate political arguments and affairs rationally for themselves. Reith thought that otherwise the raw 'democratic method' would be 'hopeless'.[20] His approach was elitist, paternalistic, and snobbish, but also progressive: enlightened administrators would help people build a more just and inclusive society. The democratic method might be hopeless, but the 'democratic aim' was not.[21]

Reith believed that 'cultural standards' (i.e. what was deemed good and appropriate for broadcast) should be set by BBC experts, on behalf of an audience that, while hopefully appreciative, would remain passive in terms of the making of those standards. Reith thus saw little point in any systematic form of listener research. If the audience did not at first like what it was given, it would, through continued exposure, learn to appreciate it. Ideas about protecting cultural standards provided a crucial weapon for Reith in his attempts to justify and defend the BBC's 'unified control' of all domestic broadcasting, to explain the virtues of what he later called 'the brute force of monopoly'.[22] The legitimacy of the BBC's monopoly rested upon its claims to supremacy in all fields of programming and administration. Often, the comparison was drawn with the degraded nature of broadcasting in America. After the Second World War, the need to maintain standards was one of the BBC's main arguments against the introduction of commercial competition. In broadcasting, it was claimed, competition could only mean a race to the bottom. The obsession with 'standards' also influenced how the BBC, well into the post-war years, approached collaboration with dominion broadcasters. Inferior contributions from overseas could not be accepted out of a desire to promote empire unity. Such material had to match BBC standards, cultural as well as technical.

Reith's philosophy set the tone at the BBC, arguably until the 1960s at least, but it was never unchallenged. Other officers, particularly those employed to commission and produce programmes, had more radical, or at least less proscriptive, visions of the possibilities of broadcasting. Fielden, who joined the BBC in 1927,

[19] Joseph T. Stuart, 'The Question of Human Progress in Britain after the Great War', *British Scholar*, 1/1 (September 2008), 53–78. Jon Lawrence, 'Forging a Peaceable Kingdom: War, Violence, and Fear of Brutalization in Post-First World War Britain', *Journal of Modern History*, 75/3 (September 2003), 557–89. Ross McKibbin, *Parties and People: England, 1914–1951* (Oxford, 2010), 1–68.

[20] Reith diaries, 18 November 1930, WAC, S60/5/3/1. Reith, 'Memorandum of Information', 3–4, WAC, R4/27/1. Bailey, 'Rethinking Public Service Broadcasting'. Such ideas also drew on much older ways of thinking about the role of the press in society: see Mark Hampton, *Visions of the Press in Britain, 1850–1950* (Urbana and Chicago, 2004), 13.

[21] Andrew Boyle, *Only the Wind Will Listen: Reith of the BBC* (London, 1972), 229.

[22] J. C. W. Reith, *Into the Wind* (London, 1949), 99. Mark Pegg, *Broadcasting and Society, 1918–1939* (London, 1983), 92–146.

later described the approach of the organization's middle echelons, and the limited ability of senior administrators to control them:

> The atmosphere was one-third boarding school, one-third Chelsea party, one-third crusade. Or possibly the crusade bulked a little larger. There was the same feeling of dedication and hope which had characterised the League of Nations in its earliest days...
>
> The elemental fact about broadcasting is its tremendous output. You may have all the authorities and restrictions and committees and regulations: but they are all defeated by the rapidity of successive programmes... in the last resort public opinion will be formed by the men who actually produce programmes. The men who sit at the top, the ageing generals, the chairmen of gas boards, the ineffective professors, the uninspired journalists... know almost nothing about what is going on under their noses.[23]

Maurice Gorham, who joined the BBC a year before Fielden, similarly suspected that 'BBC executives dreamt of a BBC untroubled by broadcasts. Luckily there were always plenty of obscure people on the production side who did their jobs according to their lights and not according to the book.'[24]

The BBC did in fact broadcast a reasonably wide range of programmes during the 1920s, both in terms of music (much of it performed live by the BBC's own bands and orchestras) and the spoken word (including talks, drama, news, and children's and religious programmes). From 1930 listeners even had a choice of two BBC services: 'National' (broadcast to the whole country) and 'Regional' (broadcast to discrete regions of the UK). Later in the 1930s the BBC also began to devote more resources to entertainment, particularly in the field of variety. This willingness to schedule an increasing amount of 'popular' material partly reflected the necessity to compete with stations broadcasting to British listeners from Europe. It was also a response to the widening scope of the BBC's audience, which gradually came to encompass all regions and social classes (even if the southern half of England, and the middle classes more generally, remained over-represented). Over two million households had listener licences by the end of 1926, over five million by the end of 1932, and over nine million by 1939. Perhaps inevitably, the BBC was deemed too 'highbrow' by some and too 'lowbrow' by others. Maintaining a safe 'middlebrow' approach in the face of such criticism could stifle creativity.[25]

The BBC also faced complaints that, as a monopoly broadcaster under remote state control, it could not be critical in its coverage of political or controversial issues. Until 1928 'controversial broadcasting' was effectively banned by the GPO. This meant that non-mainstream opinions were scarcely heard. To some, this seemed particularly obvious, and galling, during the General Strike of 1926: rather than providing an open forum in which diverse political viewpoints might be

[23] Lionel Fielden, *The Natural Bent* (London, 1960), 103–5.

[24] Maurice Gorham, *Sound and Fury: Twenty-one Years in the BBC* (London, 1948), 54.

[25] Briggs, *History of Broadcasting in the UK*, i, 15–18, 228–58 and ii, *The Golden Age of Wireless* (Oxford, 1995 [1965]), 42–3, 235–7, 253, 417. LeMahieu, *Culture for Democracy*, 274–91.

expressed, the BBC was by default on the side of established authority. Early constraints would cast a long shadow: 'For over thirty years, throughout the era of the BBC's monopoly, political broadcasting was structured in deference to the state.'[26]

More subtly, a common social background continued to bind BBC officers to their counterparts in politics and the civil service. Membership of the same 'clubbable' world meant that, if debates occurred over the autonomy and policies of the BBC, they generally did so against a shared background of core values, assumptions, and attitudes. And that clubbable world was almost exclusively male. At the inter-war BBC, women were largely employed only as secretaries: many were 'pin-money girls' from relatively wealthy backgrounds, working as a prelude to marriage.[27] The predominance of men in the upper echelons of the BBC was unsurprising, given contemporary attitudes towards gender and employment. More unusual was the fact that, between 1924 and 1946, the BBC did not regularly produce programmes aimed at women as a discrete audience. In other parts of the English-speaking world, commercial stations were quick to see women as an important daytime audience (whether this was a liberating phenomenon is open to question: the soap operas and other programmes made for women tended to emphasize home, family, and consumerism). In Germany in the 1920s and 1930s programmes were made by as well as for women, although the tone remained generally conservative and domestic.[28]

BROADCASTING IN THE BRITISH EMPIRE

The poverty of many parts of the British Empire militated against any rapid development of broadcasting during the inter-war years. In India and the dependent colonies few people could afford receiving sets, or access an electricity supply. Those who could were generally white expatriates, and they were assumed to be more interested in tuning in to programmes from 'home' than to locally originated material. In India, after some experimentation, in 1927 the Indian Broadcasting Company began operating under government licence. This was an offshoot of a company run by two Parsee entrepreneurs, in which the Marconi Company owned a majority holding. Low-powered stations were opened at Bombay and Calcutta. General Manager Eric Dunstan was ex-BBC, as were four other senior staff members. The company received 80 per cent of the listener licence fees collected by the government and a 10 per cent royalty on imported receivers, and also generated some revenue through the sale of on-air advertising. However, by the end of 1929 fewer than 8,000 listener licences had been purchased. The slow uptake of licences

[26] Paddy Scannell and David Cardiff, *A Social History of British Broadcasting*, i, *1922–1939: Serving the Nation* (London, 1991), 101–2. See also Jean Seaton and Ben Pimlott, 'The Struggle for "Balance"', in Jean Seaton and Ben Pimlott (eds), *The Media in British Politics* (Aldershot, 1987).

[27] Gorham, *Sound and Fury*, 23.

[28] Kate Lacey, *Feminine Frequencies: Gender, German Radio, and the Public Sphere, 1923–1945* (Ann Arbor, 1996).

starved the company of revenue. At the end of 1928 the five staff members who had been brought out from Britain resigned en masse in protest at the lack of resources, and in early 1930 the company went into liquidation. Reluctantly, the government of India took over its assets. Although Reith urged the India Office and the government of India to establish broadcasting on a sounder footing, there was little enthusiasm among administrators for a state-owned system. This was due partly to the expense of such a scheme, but also reflected the fear that, as Indian nationalists gained a greater voice in government, state-controlled radio might be used as a tool of political mobilization against the Raj.[29]

In other dependent colonies, officials and private entrepreneurs were similarly slow to act. By the end of the 1920s modest stations had been established in Burma and Ceylon under government supervision. Services in Singapore and Hong Kong remained embryonic, and Kenya was the only British colony in Africa with a station operating within its borders. A 'rediffusion' or 'wired wireless' system had been installed at Port Stanley in the Falklands Islands, linking loudspeakers in people's homes with a central transmission point. In the British West Indies there was no medium-wave broadcasting at all.[30]

In the dominions, broadcasting meanwhile developed more rapidly, but seemingly along American rather than British lines. To some extent this reflected conscious imitation of US practices, but it was also a response to technological and geographic realities. The dispersed nature of the population in each of the dominions meant that, compared to the UK, a relatively large number of individual stations could operate on the limited wavelengths available without jamming one another. Population dispersal also made it more expensive than in the UK to link multiple transmitters to shared production centres by wired landlines, in order to provide a single national BBC-style networked service. In the dominions, as in the US, there thus seemed less need for state regulation or involvement, a freer hand was given to private enterprise, and networking developed later than in Britain.[31] When the BBC began to present itself as the authority that should coordinate broadcasting in the British Empire, it thus found itself operating in alien territory. Indeed, the divergences that had occurred in these crucial early years would shape and limit opportunities for collaboration among public broadcasters around the British world for decades to follow.

In the dominions, again in marked contrast to the UK experience, there was little consensus about how broadcasting should be organized. Debates concerning the relative merits of private and public involvement in broadcasting often involved discussion of the suitability of British and American practices to local

[29] 'Broadcasting in South Africa—Progress in the East', *The Times* (London), 12 April 1926. P. J. Edmunds, 'Notes on Radio Broadcasting in India', 2 July 1928, Library and Archives Canada (LAC), RG42 1077/227-3-11. 'Broadcasting in India', *The Times* (London), 24 February 1930. H. R. Luthra, *Indian Broadcasting* (New Delhi, 1986), 21–36, 49–61, 367. Partha Sarathi Gupta, *Radio and the Raj, 1921–47* (Calcutta, 1995), 2–17. Government of India, Office of the Controller of Broadcasting, *Report on the Progress of Broadcasting in India* (Delhi, 1940).

[30] 'The British Broadcasting Corporation—Empire Broadcasting', November 1929, United Kingdom National Archives (UKNA), DO35/198/2.

[31] Smulyan, *Selling Radio*.

circumstance. In crude terms, supporters of public broadcasting tended to argue that the Americanization of the dominions should be resisted. Champions of private radio meanwhile stressed that what was good for Britain was not necessarily appropriate in Canada, Australia, New Zealand, or South Africa. Both sides claimed to have the best interests of 'the nation' at heart, and debates about broadcasting became linked with broader discussions about US cultural influence and Britannic and national identities.

Among regulatory authorities in the dominions, at the outset it was the New Zealand Post and Telegraph (P&T) Department that followed the British model most closely. After surveying developments in Britain, Europe, and the US, from 1923 the department restricted private stations to certain wavelengths and operating hours. Most significantly, it imposed a ban on the selling of time to advertisers. Subsequently, inspired by both British and Australian precedents, the government created a new broadcasting franchise. A private company, the Radio Broadcasting Company (RBC), established by an unholy alliance of dairying and engineering interests, was licensed to broadcast from selected 'A-class' stations in the four main urban centres, funded by listener licence fees and supervised by the P&T Department. Other private 'B-class' stations were meanwhile left to struggle on, denied either a share of licence fee revenue or authorization to sell time to advertisers.[32] Defending the new arrangement, the New Zealand postmaster-general argued that,

> we have followed, so far as this agreement is concerned, the principle adopted in Great Britain and... in India, Italy, and other countries. The Government retains very strict control of the broadcasting business, but delegates to a company... a duty which, in my opinion, can be more satisfactorily discharged by private enterprise than by a Government Department.

A strong 'central authority' would 'attain efficiency', educating and entertaining listeners even in remote rural areas.[33] However, while such rhetoric clearly echoed the arguments that had been deployed in Britain by both the Sykes Committee and Reith, in fact the RBC only vaguely resembled the BBC. The principles of public ownership, public-service operation, and remote state control were reproduced only to a limited extent. In the field of music alone did the RBC adopt something approaching a Reithian mission of cultural uplift. Moreover, it was not possible to broadcast a networked service to the entire nation, as the P&T Department could not supply the requisite high-quality landlines. Instead, the four RBC stations broadcast separate schedules, each heavily dependent on live performances by local, mainly amateur, musicians. In some parts of the country, listeners could not pick up a single RBC station.[34]

Broadcasting in Australia, Canada, and South Africa diverged even more markedly from the British model. On Empire Day, 1923, Australian private broadcasters

[32] Patrick Day, *A History of Broadcasting in New Zealand*, i, *The Radio Years* (Auckland, 1994), 48–66.

[33] *New Zealand Parliamentary Debates*, 208, 25 September 1925, 849–51.

[34] Day, *History of Broadcasting in New Zealand*, i, 67–90, 104.

met at a conference in Sydney to work out how to make radio pay. While loudly proclaiming their loyalty to the crown and empire, they disregarded the Australian postmaster-general's advice that they heed the 'satisfactory arrangement' reached in Britain. Britannic nationalism did not dictate that commercial interests should be sacrificed out of a desire to emulate Britain. One of the most persuasive speakers at the conference was Ernest Fisk, a pioneer of broadcasting in Australia and managing director of Amalgamated Wireless (Australasia) Ltd (AWA), a large company that was part owned by Marconi. Fisk had just returned from a trip to the UK where, he claimed, the BBC was failing to cater to the tastes of the British public, particularly in terms of entertainment. According to Fisk, Australia needed to avoid a British-style broadcasting monopoly. However, Fisk also warned of the problems that had been encountered by broadcasters in the lightly regulated and highly competitive American market: many privately owned stations in the US had failed and gone under. Australia needed to establish its own system, Fisk argued, in which private stations would be given freedom to compete, but would also be guaranteed a revenue stream sufficient to allow their continued existence. The conference agreed on a licensing system whereby listeners would pay to subscribe to the stations of their choice, and have their sets physically 'sealed' so that they could only listen to those stations.[35]

This Australian solution quickly proved unworkable. Listeners could easily unseal and modify their sets, tuning in to stations that they had not paid to hear. A 'bifurcated' system was thus introduced, incorporating elements of the British and American approaches and influencing New Zealand. The PMG's Department would henceforth issue private stations with one of two types of transmitting licence. 'A-class' stations would be funded by listeners' licence fees, collected, divided up, and distributed by the PMG's Department. They would also be allowed to sell some airtime to advertisers. 'B-class' stations would not receive a share of listener licence revenue, and would (it was envisaged) operate low-power transmitters, funded mainly by advertising.[36]

Fisk meanwhile proposed to take hybridization a stage further, and win control of technical networking arrangements for AWA. He argued that private competition was only necessary in the field of programme production: technical operations could be managed more efficiently as a monopoly, and Fisk lobbied the Australian postmaster-general to grant AWA this privilege. He claimed that, as the Australian federal government was a substantial AWA shareholder, his was a semi-public company that could be relied upon to serve national interests.[37]

Fisk was British-born and a devoted proponent of imperial unity throughout his life. However, he was clearly not a slavish advocate of the BBC model: he would follow the British approach only as far as was compatible with his company's commercial interests. Fisk might best be described as a broadcasting 'sub-imperialist',

[35] 'The First Australian Conference—Verbatim Report', LAC, RG33-14, 5.

[36] Griffen-Foley, 'Birth of a Hybrid', 156–8.

[37] 'Memorandum re Broadcasting', 15 September 1926; 'Memorandum no. 2—re Broadcasting: General Outline of Proposals, 23 May 1928'; and Fisk, 'Control and Management of National Broadcasting Service', 5 February 1932, all in Mitchell Library (ML), MSS 6275, E. T. Fisk papers, 25.

seeking to further AWA's interests (often presented as those of the Australian nation) by carving out an expansive media empire in the southern hemisphere. Fisk was happy to work with other broadcasters around the British world, and to emphasize his desire to serve both nation and empire, as long as this helped further AWA's own business agenda.[38]

There were men like Fisk, and companies like AWA, in South Africa and Canada too. I. W. Schlesinger, a financier who already controlled much of the South African theatre business, enjoyed some success in creating a privately owned broadcasting empire. In 1927 Schlesinger took over radio stations in Cape Town, Johannesburg, and Durban, and began operating them under the aegis of the African Broadcasting Company, mainly to serve urban, English-speaking whites. Reith met Schlesinger soon after this coup, and expressed amazement that a private broadcasting monopoly could be established 'in these days after all our experience'. In South Africa, as in New Zealand, the poor quality of available landlines ruled out networking.[39]

In Canada more rapid progress was made in linking stations together, with railway companies playing an important role. An early lead was taken by Canadian National Railways (CNR), a nationalized conglomerate of failing companies. To attract passengers, the CNR began to equip its trains with radio receiving sets and to produce programmes. In 1923 it arranged Canada's first sponsored network broadcast, leasing landlines and airtime on private stations to connect Ottawa and Montreal. The company subsequently established its own stations and arranged more network broadcasts (in English and French) using its own improved landlines. Did broadcasting by the publicly owned CNR amount to a public service, as it was understood in Britain? CNR was clearly broadcasting to advertise itself, but it also presented its radio programmes as a gift to the nation. Broadcasts were designed to reflect the high standard of overall service promised to passengers, and included classical music, dramas, talks, farming sessions and market reports, and even some broadcasts for schools. Less apparent were the public-service credentials of other companies entering the field, including the privately owned Canadian Pacific Railway (CPR), which in 1930 installed its own improved landlines, allowing it to offer sponsored network broadcasts.[40]

As these broadcasting structures developed in the dominions, shaped largely by the interests of private enterprise, alternative visions of the possibilities of radio began to emerge. To serve widely dispersed populations, broadcasting entrepreneurs in Australia and Canada, as in the US, had established a multiplicity of individual, privately owned stations. However, these were clustered in densely settled urban centres, where potential audience numbers and advertising revenues were

[38] On AWA, see Jock Given, 'Another Kind of Empire: the Voice of Australia, 1931–1939', *HJFRT*, 29/1 (March 2009), 41–56.

[39] Reith diaries, 9 October 1927, WAC, S60/5/2/2. 'Broadcasting in Africa' [March 1935], WAC, E1/4. Graham Hayman and Ruth Tomaselli, 'Ideology and Technology in the Growth of South African Broadcasting', in Ruth Tomaselli, Keyan Tomaselli, and Johan Muller (eds), *Broadcasting in South Africa* (Bellville, 1989), 24–30.

[40] E. Austin Weir, *The Struggle for National Broadcasting in Canada* (Toronto, 1965), 1–83.

greatest. Rural areas were left with little or no service. Broadcasting seemed to be exacerbating older geographical inequalities and, in Canada, to be following the well-established north–south pattern of free-market interconnection with the US. Despite the countervailing east–west axis introduced by the railway broadcasting networks, many Canadians were listening to American stations that broadcast from across the border, and by 1932 four major Canadian stations had affiliated with American networks.[41] Policymakers began to question whether this was desirable.

In Australia the 'bifurcated' system failed to deliver profitability for private broadcasting. Initially, the federal government proved reluctant to intervene. Despite seeking and securing evidence from the British GPO and the BBC, a 1927 Australian Royal Commission on Wireless dismissed the case for public broadcasting, arguing that none of the experiments tried in Britain, New Zealand, or Europe had provided listeners with a suitably varied service.[42] However, attempts to distribute licence-fee revenue between private stations in city and country areas on a more equitable basis subsequently failed. The decision was then made to transfer control of the technical operations of A-class stations to the PMG's Department. A contract for providing programmes for the A-class stations was awarded to the Australian Broadcasting Company, a private company under government supervision formed by a combination of theatre, cinema, and music companies. B-class stations continued to operate on a commercial basis.[43]

This new system approximated to Fisk's earlier proposals, but with the PMG's Department taking the place reserved for AWA. It reflected the preferences of the Director-General of Postal Services, H. P. Brown, who had visited Britain and Europe in 1928 to investigate broadcasting structures. Like Fisk, although Brown was British-born he approached the BBC model with a critical eye. Brown approved of the public-service principles on which the BBC operated, and later told Reith that an Australian equivalent of the BBC might eventually be established. However, Brown was also impressed by broadcasting in Weimar Germany. Here, technical responsibilities had been retained by the postal authorities, and programme responsibilities licensed to regional broadcasting companies. Brown's plans for Australia represented a significant modification of the UK approach, drawing on elements of the German model, and adapting the whole to suit local circumstance.[44]

Nevertheless, it was to the BBC that the new Australian Broadcasting Company likened itself in public statements. Its directors promised to 'emulate the very high standard set by the British Broadcasting Corporation . . . We will place our service

[41] Ibid., 98–100.

[42] F. W. Phillips to Reith, 11 January 1927 and BBC to GPO, 10 March 1927, both in WAC, E1/390. 'Report of the Royal Commission on Wireless', *Parliament of the Commonwealth of Australia: Papers Presented to Parliament*, iv (Melbourne, 1926–8).

[43] K. S. Inglis, *This is the ABC: The Australian Broadcasting Commission, 1932–1983* (Carlton, Vic., 1983), 8–12.

[44] Reith diaries, 26 April 1929, WAC, S60/5/2/2. H. P. Brown to Reith, 26 April 1929 and Reith to Brown, 17 June 1929, WAC, E1/341/1. Brown had been an engineer with the UK GPO before he joined the Australian PMG's Department.

to the public before all else.' They announced that they would ask the BBC to send one of its leading producers to Australia in an advisory capacity.[45] They also made claims about broadcasting and public taste that were paternalistic even by Reith's standards: '*what the public wants and has always wanted is to be taught what to want.* The Public has a wavering mind which responds readily to those who have stronger minds than its own...'.[46] However, the invitation to the BBC to send a producer to help shape Australian tastes never materialized, and the onset of the world economic depression, which restricted the growth of licence-fee revenues, placed serious limitations on the company's activities.[47]

While the new system was being introduced in Australia, the Canadian government was also examining the case for public broadcasting. A series of disputes over controversial religious broadcasts by, and allocation of wavelengths to, private stations was the most obvious reason for the establishment of a Royal Commission on Radio Broadcasting. The Liberal Prime Minister W. L. Mackenzie King already had some knowledge of how the CNR and BBC radio networks operated, and he appointed to the commission Charles A. Bowman, the British-born editor of the Liberal *Ottawa Citizen* and one of the most vocal advocates of public broadcasting in Canada. Bowman and the other commissioners (the chairman, Sir John Aird, president of the Canadian Bank of Commerce, and Dr Augustin Frigon, director of *l'École Polytechnique* in Montreal) were told by NBC executives in New York that they expected most Canadian stations to join their network. Shocked, Aird immediately arranged for the commissioners to visit Britain and Europe. On arrival, the commission's secretary announced to the press that Canadian opinion favoured the British model, and the commissioners were given every opportunity to examine BBC operations. On their return to Canada, the commissioners began to evangelize in support of public broadcasting, presenting the BBC as a useful model, and dismissing claims that it was unpopular in Britain or gave listeners insufficient choice. Aird argued that a carefully adapted Canadian counterpart could import British programmes and produce similar ones of its own, thus assisting 'the people in Canada in doing away with a great deal of American material'. US radio was presented as a threat to, and public broadcasting a bulwark of, both Britannic and Canadian identities.[48] As Bowman's *Ottawa Citizen* put it, public broadcasting would help preserve in Canada 'a distinct culture, with preponderant British traditions'.[49]

[45] Statement by Stuart Doyle, *Sydney Morning Herald*, 7 June 1929, quoted in Michael S. Counihan, 'The Construction of Australian Broadcasting: Aspects of Australian Radio in the 1920s' (MA thesis, Monash University, 1981), 220.

[46] *The Australian Broadcasting Company Year Book 1930*, quoted in Counihan, 'Construction of Australian Broadcasting', 225–6, original emphasis.

[47] Brown to Reith, 10 September 1929, WAC, E1/341/1.

[48] For a fuller discussion of the Aird Commission and the BBC model, see Simon J. Potter, 'Britishness, the BBC, and the Birth of Canadian Public Broadcasting, 1928–1936' in Gene Allen and Daniel Robinson (eds), *Communicating in Canada's Past: Essays in Media History* (Toronto, 2009). Aird quoted in summary of public hearings held at Quebec City, 5 June 1929, LAC, RG42, 1077/227-10-8.

[49] 'Canada's Radio Opportunity', *Ottawa Citizen*, 13 September 1928.

The commission's final report praised the BBC, endorsed Reith's emphasis on broadcasting 'as an instrument of education', and recommended that broadcasting be publicly owned and run as a domestic monopoly on a 'public service' basis. But the Aird Report did not explicitly appeal to British identities, or even recommend the creation of a unified national network. Although the report claimed that broadcasting would be 'a great force in fostering a national spirit and interpreting national citizenship', it was influenced by the German practice of devolving responsibility for programming to the regions. This was appealing given strong provincial identities, particularly in Quebec. The report thus recommended the establishment of a 'Canadian Radio Broadcasting Company' to manage the technical side of broadcasting, on a public-service basis. Some programmes might be made by commercial sponsors, to give Canadian advertisers a chance to compete with their US rivals, but overall responsibility for programming would rest with autonomous provincial broadcasting directors and advisory councils. Reith recognized that, as in Australia, the German and British models were being selectively plundered, and that the Aird Report's recommendations diverged significantly from UK principles of centralized control in a single company with a powerful chief executive officer. He was, perhaps, not sorry when the Aird Report was shelved. Strong protests against any sort of interference with private interests led the Liberal government to hesitate until, as in Australia, the onset of the depression derailed plans for the future development of broadcasting. King's Liberals were replaced by R. B. Bennett's Conservatives, who formed a government that was not expected to be friendly to the idea of any form of state enterprise.[50]

EMPIRE BROADCASTING THE BBC WAY

Thus, by 1929, broadcasting in the British Empire did not correspond to a clear or finished pattern. Services in India and the tropical colonies were virtually non-existent. In the dominions, by contrast, many private stations had been established, some had successfully developed significant local audiences, and most were run on a commercial basis. Some public involvement in broadcasting had been established in New Zealand and Australia, and in Canada the CNR operated network broadcasts on something approximating a public-service basis. However, only in Britain had a publicly owned and administered network of transmitters been established to provide a unified 'national' service that could be picked up by the vast majority of the population. On the other hand, in Britain there was none of the diversity that characterized broadcasting in the dominions. The BBC had a domestic monopoly of radio, breached only during the early 1930s by European stations transmitting across the English Channel.

[50] *Report of the Royal Commission on Radio Broadcasting* (Ottawa, 1929). Reith to Aird [n.d.], LAC, RG33-14, 2/227-14-1. W. H. Clark to Lord Passfield, 19 March 1930, UKNA, DO35/228/2. Mary Vipond, *Listening In: The First Decade of Canadian Broadcasting, 1922–1932* (Montreal, 1992), 217–24.

How would the blank spaces on the empire's broadcasting canvas be filled? And would the varied elements of the picture ever form a harmonious whole? Did the BBC have a coordinating role to play? Much depended upon future developments in the rapidly changing world of radio, and these remained hard to predict. There were signs in Europe and the US that broadcasting would eventually become predominantly 'local' or 'national', with private or public services generally respecting established political boundaries in their search for listeners. However, there were also reminders that radio waves could cross borders with ease: American stations were popular in Canada, and European broadcasters found listeners in Britain. Could wireless also transcend the empire's own internal boundaries and divisions?

Transmissions across national boundaries were limited by the fact that most broadcasters operated on the medium waveband. Their signals could generally reach listeners at best a few hundred miles distant from the transmitter (further at night). However, signals on other wavebands could be picked up by more remote receivers. Long waves were initially the favoured distance carrier. The British GPO opened a long-wave radiotelegraph station at Rugby, Warwickshire, in 1926. This used as much power as a small town. A year earlier the BBC had opened its own long-wave station at Daventry, Northamptonshire, to broadcast to parts of the country inadequately served by medium-wave transmitters. Daventry could also be picked up in parts of Europe. However, it was soon discovered that short-wave transmitters, which required much less power to run, could be a cheap and effective alternative. Marconi's imperial chain of wireless telegraphy stations operated on short wave, as did its trans-Atlantic radiotelephone service, which opened in January 1927. Following the creation of Imperial and International Communications Ltd, the GPO took over the UK-based elements of the radiotelephone system, which was expanded to connect Britain with key parts of the empire and wider world.[51]

Short-wave signals reach distant receivers by bouncing between the ground and ionized layers in the upper atmosphere (the ionosphere). They are thus subject to an initial 'skip' distance, and cannot normally be picked up by receivers located within several hundred miles of the point of transmission. This could be an attractive characteristic: using medium waves for home listeners, and short waves for overseas listeners, it was possible to target domestic and external audiences separately. However, short-wave broadcasting also had its limitations. Contemporaries observed that short waves travelled in an unpredictable fashion. The height and density of the ionosphere is influenced by the relative position of the sun and the earth. Changes in the ionosphere thus occur according to complex and shifting seasonal and geographic patterns, and with the transition from day to night. In the 1920s and early 1930s engineers had few solutions to the resulting transmission

[51] Baker, *Marconi Company*, 212–33. Daniel R. Headrick, 'Shortwave Radio and its Impact on International Telecommunications between the Wars', *History and Technology*, 1/11 (1994), 21–32. Press cutting, E. V. Appleton, 'Cable and Wireless', UKNA, DO35/198/2. 'The First Decade', *World Radio*, 11 November 1932. 'Imperial Conference—Inter-departmental Committee on Economic Questions—Empire Telephony—Brief for Information of Ministers—I.C.(E.C.) Br.2/1929', 20 December 1929, UKNA, CO323/1060/12.

problems. Signals crossing the day/night boundary were particularly seriously disrupted, affecting transmissions from Britain to Australia and New Zealand. Similarly, short-wave signals were disrupted in an unpredictable fashion when they travelled across the polar zone, separating Britain and Canada. Increasing the operational power of transmitters did little to overcome such problems, and repeated changes of frequency and broadcast times brought only limited improvements, while inconveniencing listeners. Moreover, in this period only those who possessed special and expensive receiving sets could tune in to short-wave broadcasts themselves and become 'direct listeners'. For the majority, possessing standard receiving sets, short-wave signals could only be heard indirectly. They first had to be picked up by a local radio station that had access to a short-wave receiver, and then retransmitted by that station on medium wave, as part of its own programme schedule: 'rebroadcasting'. Local collaborators were thus necessary if short-wave broadcasts were to reach a wide audience.

How would the BBC use short wave? Initially, in the Sykes Report and in Reith's early musings on the subject, 'imperial and international broadcasting services' were seen as connected aspects of the same general sphere of external operations, helping to restore order and stability to the international as well as the domestic scene in the wake of the First World War.[52] The quasi-biblical motto of the BBC's coat of arms, 'Nation Shall Speak Peace Unto Nation', reflected the Christian idealism that underpinned Reith's view of the international role of radio:

> [Broadcasting] cannot in the nature of things be restricted by frontiers, natural or otherwise, [it] should link up the earth and promote a spirit of world citizenship. Sooner or later people in all parts will be in touch with each other, and by music, speeches, statements of policy and in many other ways Broadcasting will play its part in the establishment of world unity…
>
> The structure of civilization is at present insecure; another catastrophe might encompass the ruin of all. International hatred and suspicion must be exorcised. The printed word has failed…It is left for Broadcasting to effect something which no other agency has done in the past, something which may not be far from saving civilisation.[53]

Reith was also enthusiastic about most things imperial. He liked telling any Canadians he met about the year his grandfather had spent as general manager of Canada's Grand Trunk Railway. At times, he dreamed of being sent to East Africa as a colonial governor, or to India as viceroy. However, his views on the empire were conventional at best, and sometimes ill-informed and outdated.[54] He had little time for contemporary critics of Britain's role overseas, regarding the empire as natural and positive, bringing benefits for Britain, the dominions and colonies, and the wider world. In Reith's opinion, the empire still needed a firm hand from

[52] *Broadcasting Committee: Report*, 6. Reith, *Broadcast over Britain*, 221–2.

[53] Reith, 'Memorandum of Information', 6, WAC, R4/27/1.

[54] Reith diaries, 3 November 1925, WAC, S60/5/2/1; 17 January 1928, 20 January and 5 February 1929, S60/5/2/2. John M. MacKenzie, 'In Touch with the Infinite: the BBC and the Empire, 1923–53', in MacKenzie (ed.), *Imperialism and Popular Culture* (Manchester, 1986), 186.

Britain to guide its affairs: he thought symptoms of imperial decline a consequence of the introduction of the 'democratic method' at home.

> England has obviously given up all idea of governing India... It would be interesting to trace the development of the democratic inferiority complex in Imperial affairs, from quite small beginnings, just a few men here and there beginning to yield, and far too much lip service prematurely to democratic methods.

Reith hoped it would still be possible 'to make the ties more definite [through] a completely new system of Empire coordination'.[55]

Reith's desire for such 'coordination' was a response to the ambiguities that characterized the relationship between Britain and the dominions during the 1920s. After the First World War, and particularly at the Imperial Conferences of 1923 and 1926, policymakers in Britain and the dominions agreed to abandon the idea of unitary foreign and defence policies for the entire empire. Each country would henceforth develop its own individual external policies. The 1926 Balfour Report on Inter-Imperial Relations recognized the equal constitutional status of Britain and the dominions. Any vestigial ability that Britain had to impose unanimity was now lost. Yet the Balfour Report also acknowledged that the dominions voluntarily remained members of the empire, or as it was increasingly called, the British Commonwealth. Contemporaries did not expect that Commonwealth countries would each go their own way. Instead, through discussion and cooperation, they would ensure that their policies remained compatible.[56] Rather than signal the end of empire, the report unleashed 'a torrent of idealism about the Commonwealth'.[57]

British constructive imperialists and their allies in the dominions now hoped that, with contentious issues of constitutional status settled, it would be possible to establish closer practical, demographic, and economic connections. Reinvigorated bonds of sentiment and interest would ensure that the member states of the Commonwealth travelled together, in the same direction. Yet these hopes were not easily realized. Between 1919 and 1922 the Overseas Settlement Committee operated a free passage scheme to help settle returning British servicemen and women in the dominions. However, only 86,027 migrants took advantage of the scheme, instead of the anticipated 405,000. A lack of enthusiasm for the scheme in the dominions was part of the problem: Canada and New Zealand proved unwilling to admit new, potentially impoverished, migrants at a time of considerable economic dislocation; similar worries in Australia were compounded by clashes between federal and state governments concerning responsibility for settlement schemes; and in South Africa British soldier settlement was hobbled by conflict between English and Afrikaans speakers.[58] The results of the more ambitious 1922 Empire Settlement

[55] Reith diaries, 5 July 1934, WAC, S60/5/4/1 and 10 July 1936, S60/5/4/4.

[56] John Darwin, *The Empire Project: The Rise and Fall of the British World-System, 1830–1970* (Cambridge, 2009), 393–407.

[57] Ronald Hyam, *Britain's Declining Empire: The Road to Decolonisation, 1918–1968* (Cambridge, 2006), 71.

[58] Kent Fedorowich, *Unfit for Heroes: Reconstruction and Soldier Settlement in the Empire Between the Wars* (Manchester, 1995).

Act likewise 'fell far below intentions'.[59] Tensions between British and dominion authorities similarly reduced the effectiveness of the Empire Marketing Board (EMB), founded in 1926 to fund research into the production and marketing of British and empire produce and to run publicity campaigns to encourage sales of such produce in the UK. Some Canadian policymakers feared that the EMB might undermine recently established dominion autonomy. Although the board purported to represent the entire empire, it was based in London, funded by the UK taxpayer, and projected ideas about imperial economic complementarity that, to some extent, presented the dominions as mere suppliers of raw materials and markets for British industry.[60] The BBC's plans for broadcasting to the British world reflected a somewhat analogous, centralizing perspective, and met with similar resistance from the dominions, particularly Canada.

In seeking further to define the BBC's imperial role during the 1920s and 1930s, and to differentiate it from broader international responsibilities, BBC officers and British civil servants, and some of their counterparts overseas, spoke and wrote of 'empire broadcasting'. However, the exact meaning of this phrase was neither self-evident nor uncontested. If the BBC followed the same principles that had begun to emerge in its dealings with broadcasters in the US and Europe, then empire broadcasting might involve a measure of mutual cooperation among broadcasting authorities in Britain, the colonies, India, and the dominions, with programmes flowing freely and reciprocally among them. However, unsurprisingly, given Reith's desire to bring order and discipline to the empire, and his belief in the virtues of well-run monopolies, this was not the BBC's understanding of the term. Rather, for the BBC, empire broadcasting at root meant one thing: the corporation broadcasting from Britain to the empire.

Certainly, this also reflected practical concerns. Short-wave reception and transmission facilities were initially limited in the colonies and dominions, and in many parts of the empire there was no effective 'national' broadcasting authority (public or private) with which the BBC could work. Nevertheless, even later, when a greater measure of collaboration and reciprocity became possible, limits still continued to be imposed by an enduring, if often unstated, assumption that empire broadcasting was something the BBC did from Britain. At its core, the debate about the meaning of empire broadcasting reflected different perspectives on the questions of where power and authority for broadcasting within the British Empire would lie, and whether that power and authority would be unitary or multiple, centralized or devolved. Some of the most pronounced critics of the Reithian centralizing model of empire broadcasting would accuse the BBC of perpetuating 'imperialism', a term which had a specific meaning in this context. Cooperation among the empire's broadcasting authorities on a basis of Britannic equality was regarded by many as a good thing. BBC dominance and centralization in London was seen as 'imperialism' and provoked resistance.

[59] Stephen Constantine, 'Migrants and Settlers', in Judith M. Brown and Wm. Roger Louis (eds.), *OHBE*, iv, *The Twentieth Century* (Oxford, 1999), 173.

[60] Stephen Constantine, 'Anglo-Canadian Relations, the Empire Marketing Board and Canadian National Autonomy between the Wars', *JICH*, 21/2 (May 1993), 357–84.

Despite publicly proclaiming its imperial role from an early stage, in practice the BBC initially paid more attention to establishing links with broadcasters in the US and Europe. It organized its first exchange of programme material with an American station in 1923, and with a European station a year later. In 1925 it played a key role in the founding of the *Union Internationale de Radiophonie*, established to help coordinate broadcasting affairs across Europe.[61] The colonies and dominions were also a priority, but BBC officers were reluctant to act precipitately, for many policy issues needed first to be resolved. Short-wave broadcasts to the empire would inevitably find listeners elsewhere, particularly in the US, and would have to be devised with this in mind. Few non-white listeners within the empire could be expected to understand the English language or enjoy European music, and many would have neither the means nor the inclination to tune in to programmes from Britain in the first place. Technical constraints provided an added argument against premature action, and a convenient excuse to avoid public discussion of the wider policy problems. It could plausibly be argued that short-wave technology was not yet sufficiently advanced to guarantee a reliable service worthy of the BBC's reputation.[62]

However, in 1927 the activities of short-wave broadcasters in Holland and the US and of experimenters in Britain spurred the BBC to begin short-wave transmissions, using a station at Chelmsford, Essex, leased from Marconi. Even though this could be presented as an important new medium of imperial communication, the UK Treasury refused to provide a grant-in-aid, obliging the BBC to fund the experiment from UK listener licence fees.[63] The BBC claimed that UK listeners had to be provided with some sort of return for this investment, in the form of incoming programme material. Little could yet be expected from the colonies and dominions, so the experiment came to focus on exchanging programmes with stations in the US.[64]

Nevertheless, some programmes from Chelmsford were received successfully in the dominions, and the BBC gained an impression of potential direct listening and rebroadcasting patterns should a full short-wave service be established. Several stations in Australia and New Zealand carried an Armistice Day programme from Chelmsford, and the CNR rebroadcast BBC coverage of other special events and ceremonies. Significantly, experimental short-wave transmissions from the dominions were also reaching Britain. The CNR had organized a broadcast to Britain and Ireland in 1925 by boosting the power of its station at Moncton, New Brunswick. In 1927 it helped organize a programme to mark the diamond jubilee of Canadian confederation: Canadian Marconi specially modified its radiotelephone transmitter to allow this to be broadcast internationally. That same year AWA arranged its first short-wave 'Empire Broadcast' from Sydney. There was a strong possibility

[61] Briggs, *History of Broadcasting in the UK*, i, 282–94.

[62] C. F. Atkinson to C. D. Carpendale, *c.* May 1927, WAC, E4/1.

[63] For more information see WAC, E4/1 and Gerard Mansell, *Let Truth Be Told: 50 Years of BBC External Broadcasting* (London, 1982), 1–19.

[64] Carpendale to L. S. Amery, 21 December 1928, UKNA, CO323/1060/12. 'Empire Broadcasting', *World Radio*, 21 October 1932.

that empire broadcasting might be something done from the dominions, as well as from Britain.[65]

Gerald Beadle, BBC station director in Belfast, suggested it might thus be desirable to combine British and dominion efforts in a single scheme. Beadle argued that empire broadcasting offered a means to combat two threats: the worldwide spread of the 'Bolshevic [*sic*] disease'; and the 'national spirit' developing in the dominions (Beadle had spent time working in Natal, and was presumably aware of the growing spirit of Afrikaner nationalism in other parts of South Africa). However, Beadle argued that in the new order ushered in by the Balfour Report, the BBC could only expect the dominions to share the cost of empire broadcasting if they were given a role in the actual running of short-wave services. Beadle thus suggested the creation of a separate empire broadcasting authority, run and financed cooperatively by the BBC and dominion broadcasters in principle, even if dominated by the BBC in practice.[66]

Caution meanwhile continued to be urged by BBC Chief Engineer Peter Eckersley and by Foreign Director C. F. Atkinson. Eckersley argued that as reliable short-wave reception overseas could not be guaranteed, the BBC should modestly aim only at 'lonely' expatriate listeners in the tropical colonies who possessed their own short-wave sets but had 'nothing else' to listen to. It would be years, Eckersley predicted, before short-wave technology had advanced sufficiently to allow satisfactory rebroadcasting by stations in the dominions. Atkinson argued that there was little point in broadcasting to the dominions anyway, because dominion listeners only wanted light entertainment and dance music, which their local stations could easily provide. If the BBC broadcast its normal type of programmes to the dominions, it would be accused of insensitivity to local tastes. If it provided more lowbrow material, then it would compromise its Reithian mission of cultural uplift at home, by encouraging UK listeners to demand the same popular fare. The only thing that dominion listeners wanted from Britain was 'actuality' coverage of important events.[67]

However, a staff change shifted plans onto a different track; arguably, the wrong one. In 1929 Eckersley was named in a divorce case, and was dismissed by Reith for straying from 'the paths of righteousness'.[68] His replacement as chief engineer was Noel Ashbridge, who wanted to end the BBC's reliance on the Marconi transmitter at Chelmsford, and build a short-wave station of his own. Atkinson's modest proposals did not justify such a major capital project. So, in new recommendations prepared jointly by Atkinson, Ashbridge, and Rear-Admiral C. D. Carpendale (Reith's deputy), and presented to the GPO in December 1929, a more expansive

[65] 'London Successfully Re-broadcast by 3YA', New Zealand *Radio Record*, 18 November 1927. Given, 'Another Kind of Empire', 42–5. Weir, *Struggle for National Broadcasting*, 21, 42, and 53–60.

[66] Gerald Beadle, 'Empire Broadcasting', 12 May 1927, WAC, E4/1.

[67] P. Eckersley to Reith, 6 December 1928 and Atkinson, 'Memorandum on Empire and World Broadcasting', June 1929, WAC, E4/2. The requirements of white expatriates had also been highlighted at a Colonial Office Conference held in London in 1927. See 'C.O. 15th Meeting (1927)', 20 May 1927, WAC, E4/1. For a sense of the scope and diversity of the expatriate audience, see Robert Bickers (ed.), *Settlers and Expatriates: Britons over the Seas* (Oxford, 2010).

[68] Boyle, *Only the Wind*, 230.

plan was set out. The BBC would provide a daily short-wave service to the dominions as well as to the colonies. Atkinson's and Beadle's earlier arguments were omitted. The BBC now presented itself as a central broadcasting authority for the entire empire.[69]

The proposed short-wave service would target the 'white population [of territories] under the British flag'.[70] While US listeners might be able to eavesdrop on the empire service, officially they would be served separately, by exchanges of programmes between the BBC and US broadcasters using the radiotelephone system. Non-white inhabitants of Britain's formal empire were similarly excluded from the target audience. It was argued that 'European-type' programmes would not appeal to 'the natives', who would eventually be provided with more appropriate fare by local broadcasting services. Listeners in Britain's informal empire (in Egypt for example) were also to be ignored.

The BBC's proposals acknowledged some of the problems to be faced in broadcasting even to this narrowly defined target audience. Reception quality in India, Australia, New Zealand, and South Africa left 'a good deal to be desired'. Moreover, the requirements of white listeners in the various parts of the empire were not uniform. In the tropical colonies and India (where broadcasting structures were largely undeveloped) expatriate direct listeners with their own short-wave receivers would, it was anticipated, turn to the BBC's empire service for a well-rounded schedule of programmes, including up-to-date news and material taken live from the BBC's normal domestic services. In South Africa, where local broadcasting was deemed to be in an 'unsatisfactory condition', it was similarly assumed that listeners could be treated as overseas fragments of the home audience. But in Canada, Australia, and New Zealand, the situation seemed more complicated. It was anticipated that there would be some direct short-wave listening in these places, but that most people would continue to tune in to local stations broadcasting on medium wave. BBC programmes would only be widely available in the dominions if those local stations rebroadcast the short-wave transmissions as part of their medium-wave schedules. Perhaps public broadcasters would be willing to do this, out of 'a desire for home contacts'? Although public broadcasting remained in what could at best be described as an embryonic state in the dominions, the BBC's proposals hardly mentioned private stations as potential collaborators. Meanwhile, it was acknowledged that dominion listeners would not necessarily want the same sort of programmes that the BBC broadcast to domestic and expatriate listeners. Time-zone differences and problems with recording technologies and repeat performance rights would anyway make it difficult to reuse domestic programmes in transmissions to the dominions. The BBC would thus have to produce programmes especially for dominion listeners, outside normal BBC operating hours. This would be expensive.

[69] N. Ashbridge to Carpendale, 22 October 1929 and Carpendale to Reith, 5 November 1929, WAC, E4/2.

[70] The following three paragraphs are based on 'The British Broadcasting Corporation—Empire Broadcasting', November 1929, UKNA, DO35/198/2.

How would the empire service be funded? The idea of operating the empire service as a commercial concern, perhaps supported by advertising for branded British goods sold overseas, does not seem to have been contemplated seriously. Broadcasts might promote British trade indirectly, but overt sponsorship was ruled out. Thus, from the outset, the domestic principles of non-commercial, public-service operation were extended to external broadcasting. Yet the BBC maintained that UK licence-fee payers should not be asked to fund the construction and operation of the proposed short-wave station. While some reciprocal programmes might be received from the dominions, it was argued that these would hardly compensate UK listeners for the expense of the outgoing service. Perhaps the UK Treasury might cover the costs? This would mean direct state funding, a new departure for the BBC. Would this undermine the corporation's autonomy? Or perhaps the colonies and dominions would make financial contributions to the running of the service, either through grants-in-aid or by passing on a share of their own listener licence fees? If so, how would the colonies and dominions divide the bill?

British civil servants in both the Colonial Office and the Dominions Office were enthusiastic about the general idea of BBC short-wave broadcasting. An empire service might overcome the isolation of colonial administrators, maintain 'sentimental ties' with the dominions, and project Britain overseas. In 'an age of vigorously competing national cultures', it was thought unwise to 'allow the British case to go by default': US dominance of the cinema was 'a sufficient warning of this truth'. However, both departments worried that the BBC's proposals were too ambitious. Civil servants echoed some of Atkinson's earlier concerns. White expatriates in the colonies might want a full broadcast service from Britain, but colonial governments could not justify using revenues contributed by Africans and Asians to help the BBC pay for it. Meanwhile, dominion listeners only wanted occasional programmes from the BBC, covering special events, and would not pay for a full-scale empire service aimed primarily at whites in the colonies. Indeed, the requirements of dominion listeners seemed closer to those of audiences in the US, and might be served in a similar fashion, by periodic link-ups with local stations and networks using the radiotelephone system. The Colonial Office and Dominions Office thus agreed that a modest scheme aimed mainly at white expatriates in the dependent colonies, and funded largely from UK licence fees or tax revenues, was preferable to the BBC's proposals.[71]

Reith was not happy to be overruled by civil servants, but agreed to have estimates prepared for a limited service aimed at expatriates in the colonies, incorporating programmes from the BBC's domestic services, gramophone records, and one or two daily news bulletins.[72] This more modest 'scheme B' was received enthusiastically by white colonial administrators when they visited London for a Colonial Office Conference in the summer of 1930. They agreed that a five-shilling

[71] Minute [by Sir C. Davis, 20 December 1929], UKNA, CO323/1060/13. Minutes by P. H. Morris and Sir Samuel Wilson, [*c*.21 January 1930], CO323/1103/1. Minute by G. L. M. Clauson, 21 January 1930, CO323/1103/1. Minute by Hankinson, 17 February 1930 and minute for Sir E. J. Harding, 26 February 1930, DO35/198/2.

[72] Minutes, 'Meeting to Discuss the Scheme for an Empire Broadcasting Service', 11 March 1930, UKNA, CO323/1103/1.

levy on colonial listener licences might allow a small financial contribution to be made to the running costs of the service.[73]

However, given the world economic crisis, the UK Treasury refused to make up the shortfall until the proposals had been put to the Imperial Conference of the governments of the UK, India, and the dominions, held in London that October.[74] At the conference, a committee on communications duly considered both scheme B and the BBC's original proposals ('scheme A', now offered at a somewhat lower cost).[75] Neither won unanimous support. Scheme B hardly concerned the dominions, and while some interest was expressed in scheme A, it was noted that broadcasting in the dominions was not generating enough revenue to fund local stations satisfactorily, yet alone to contribute to the cost of empire broadcasting.[76] To the annoyance of the Colonial and Dominions Offices, the UK Treasury meanwhile diverged from the agreed British negotiating position, indicating that it would not contribute under any circumstances.[77] Eventually, the chair of the conference's communications committee, Francis Brennan (Australia's Attorney General), ruled that further consultation with broadcasting authorities in the dominions was necessary before any decision could be made.[78] Reith was once again disgusted by what he perceived as the inadequacies of the civil servants and politicians that he had to deal with. He judged the meeting 'quite fatuous', and Brennan 'no use'.[79]

However, Brennan was better informed than Reith allowed. Brennan had presumably seen the Australian brief on the BBC's proposals, written by H. P. Brown of the PMG's Department, which echoed the earlier objections of Atkinson and the Dominions and Colonial Offices. Brown argued that Australian audiences did not need a full broadcast service from Britain. They wanted only commentaries on cricket and rugby test matches, and 'outside broadcast' coverage of major ceremonies and events. These could be fed to Australia via radiotelephone, without the need to construct either a BBC short-wave transmitter or additional Australian receiving facilities.[80] Similar views were expressed more openly by one of the Canadian delegates at the conference, Lt. Col. W. A. Steel of the communications branch of the Canadian Department of National Defence. The Canadian delegation was not averse to imperial cooperation in principle: it strongly supported a reduction in postage rates within the empire, for example.[81] As Steel himself argued: 'The

[73] 'Colonial Office Conference 1930—Extract from Summary of Proceedings', UKNA, CO323/1103/2.

[74] Unattrib. minute, 7 September 1932, UKNA, DO35/198/2.

[75] 'EE(30)46 Committee on Communications—First Report—Empire Broadcasting', 29 October 1930, UKNA, CAB32/82.

[76] 'EE(C)(30) Committee on Communications—Notes of the Second Meeting', 9 October 1930, UKNA, CAB32/95.

[77] Minutes by Vernon and Harding, 24 October 1930, UKNA, CO323/1103/2.

[78] 'EE(C)(30) Committee on Communications—Notes of the Third Meeting', 13 October 1930, UKNA, CAB32/95.

[79] Reith diaries, 13 October 1930, WAC, S60/5/3/1.

[80] 'Notes for Prime Minister on Empire Broadcasting', 3 July 1930, National Archives of Australia, Canberra, A461/10, C422/1/6, 97518.

[81] 'EE(C)(30) Committee on Communications—Notes of the First Meeting', 7 October 1930, UKNA, CAB32/95.

value of broadcasting as a means of tying together the various parts of the Empire is unquestioned.' However, he doubted the practicability of the BBC proposals. There was not enough money in Canadian broadcasting to contribute anything to an imperial project, and neither had a public authority been created that would be able to rebroadcast BBC programmes in Canada. Private stations would not carry them, because the popularity of programmes from Britain could not be guaranteed, and if audiences were lost then advertising revenues would suffer. The railway broadcasting networks could help, but it might first be necessary to construct and finance a suitable short-wave receiving station. Moreover, in Canada as in Australia, Steel claimed that listeners wanted only a limited, bespoke service from Britain, providing coverage of 'important events and particular programmes prepared with the direct intention of popularizing Empire broadcasting'. These could be fed to the railway company networks or to individual stations via the existing radiotelephone service, without the need to finance a new BBC transmitter.[82]

The report of the conference's communications committee thus endorsed the general principle of empire broadcasting as a means of 'strengthening the ties between the various parts of the British Commonwealth', but suggested that the BBC consult with broadcasters and government departments in the dominions before specific action was taken. In the meantime, it recommended 'reciprocal broadcasting of programmes and events of special interest' by radiotelephone.[83] The heads of delegations approved the report without further detailed discussion. Given the turmoil overtaking the world economy, when compared with pressing issues such as tariff policy, broadcasting was simply not a priority for those attending the conference.[84]

Indeed, not only did the UK Treasury remain unwilling to support scheme B, but in August 1931 it even asked the BBC to make a 'voluntary contribution' to the beleaguered national exchequer. Reith agreed to give £200,000 and announced that, in the same spirit, the BBC would itself pay for the construction of a short-wave station and the running of an empire service.[85] Work began on the installation of two new transmitters at the BBC's Daventry site, linked by landlines to BBC production facilities elsewhere. In the interim, the BBC continued to provide a service from Chelmsford, including three daily news bulletins compiled from Reuters news. As Reuters retained its copyright over this news, the bulletins could not be rebroadcast by stations overseas without special permission. Reuters feared that overseas newspapers that subscribed to its service would protest if the bulletins were made available for use by local medium-wave stations. This embargo was not deemed to be a major problem, however, as the BBC service was aimed at expatriate direct listeners in the colonies, not at rebroadcasters in the dominions.[86]

[82] 'EE(C)(30)5 The Position of the Dominion of Canada with Reference to Empire Broadcasting—Statement by Canadian Representative', UKNA, CAB32/95.

[83] 'EE(30)46—Committee on Communications—First Report', 29 October 1930, UKNA, CAB32/82.

[84] Imperial Conference 1930, Meetings of Prime Ministers and Heads of Delegations, UKNA, CAB32/79.

[85] Briggs, *History of Broadcasting*, ii, 342–53.

[86] P. Cunliffe-Lister to colonial governors, 10 February 1932, UKNA, DO35/198/2.

Rumours that the BBC was about to become the 'British Empire Broadcasting Corporation' were clearly exaggerated, but a new BBC Empire Department was established to coordinate the short-wave service, issue publications for overseas markets (notably the BBC journal for overseas listeners, *World Radio*), and despatch recordings of BBC programmes to broadcasters overseas.[87] The department was placed under the direction of Cecil G. Graves, the nephew and heir of Viscount Grey of Fallodon and a former soldier and intelligence officer.[88] J. Beresford Clark was appointed empire programme director, Graves's number two.[89] Graves had formerly been assistant director of programmes, Clark principal assistant to the BBC's North Regional Director. Both were capable administrators, but neither had detailed, up-to-date imperial experience. Graves later admitted that 'my knowledge of the British Empire...was shockingly small'.[90] Thus, initially at least, neither of them would question the broader BBC assumption that empire broadcasting was something that should be done almost exclusively from Britain, by the corporation. Subsequently, they followed a steep learning curve, but in the meantime they were frequently obliged to turn to the Colonial Office and Dominions Office and to the dominion high commissions in London for guidance. Indeed, relations between the Empire Department and officialdom would remain close and generally cordial, even deferential.

CONCLUSIONS

In July 1930 *The Times* claimed that 'It has often been said that the gift of wireless was vouchsafed to humanity for the special benefit of the British Empire.'[91] Yet, a few months later, the Imperial Conference indicated that much of the potential value of radio as a means of imperial mass communication had yet to be realized, and that unleashing it would not be a straightforward task. Britishness remained a potent force to conjure with in most parts of the British world, and many contemporaries believed that broadcasting would work to strengthen complementary Britannic and national identities. Quite how this would be achieved, however, remained unclear. During the 1920s the BBC had focused on putting its own house in order, and in developing links with broadcasters in Europe and the US. Little sustained thought had been given to the empire. In the meantime, in the dominions, broadcasting had tended to follow the US model, with significant adaptations to suit local circumstance. Private interests had become entrenched, and public involvement remained limited. Moreover, just when the BBC belatedly began to project its influence overseas, and to seek a coordinating role for

[87] Press cutting, Garry Allighan, 'Will it Become the B.E.B.C.?', *Evening Standard* (London), 11 July 1932, UKNA, DO35/198/2. Atkinson to Colonial Office, 18 August 1932, CO323/1198/2.

[88] 'Obituary—Sir Cecil Graves', *The Times* (London), 14 January 1957. Reith to Harding, 12 September 1932, UKNA, DO35/198/2.

[89] 'Introducing J. B. Clark', *BBC Empire Broadcasting*, 16 December 1932.

[90] Sir Cecil Graves, 'Pioneer Days', *London Calling*, 10 December 1953.

[91] 'Empire Broadcasting', *The Times* (London), 4 July 1930.

broadcasting in the empire, the world economic depression intervened. This reduced the funds available for developing radio in the colonies, India, and the dominions, and for establishing an empire service in the UK. The BBC had decided that empire broadcasting was to be run on a non-commercial basis, like broadcasting in the UK. However, the state subsidies that had supported other imperial communication schemes in the past were not forthcoming and, unsurprisingly, the dominions could not be persuaded to fund a service primarily aimed at the colonies.

The BBC's proposals were not sunk by 'nationalist' resistance to the basic idea of pan-empire broadcasting. At the Imperial Conference, the delegates had generally agreed that improved cooperation among the empire's broadcasting authorities would be welcome, and that some material from the BBC was needed in the dominions. Rather, the problem stemmed from the weaknesses of the BBC's specific proposals. The idea of a single service for 'lonely listeners' in India and the colonies, and rebroadcasters in the dominions, was obviously delusional even at this early stage; a product of Ashbridge's desire to build his own BBC short-wave station, and the vain hope that the dominions could be persuaded to foot the bill. Rebroadcasters in the dominions seemed only to want bespoke services from Britain, tailored to their specific interests. Moreover, broadcasters such as the CNR in Canada and AWA in Australia expected empire broadcasting to be a reciprocal activity, involving a two-way flow of programmes. The BBC's proposals of 1929 were hardly a convincing 'Britannic solution' to the question of how to use broadcasting to promote the unity of the British world.

2

Discord, 1932–35

During the 1920s debates about broadcasting and empire had largely revolved around a very basic question: what sort of institutions should be given responsibility for managing radio services, within and across the empire's internal boundaries? To some extent, this question was answered during the early 1930s. As new public broadcasting authorities were established in the dominions, institutional structures in the different parts of the British world seemed to converge, and the threat posed to the British Broadcasting Corporation (BBC) model by American-inspired commercial alternatives receded somewhat. Decisions were made in Canada, Australia, and New Zealand that fundamentally shaped and potentially improved opportunities for imperial broadcasting collaboration, as significant steps were taken towards establishing the principle of public ownership and operation of radio. These decisions reflected dissatisfaction with existing domestic arrangements, and were also the product of a growing familiarity with the British model (to some degree encouraged by the BBC itself) and a modest cross-fertilization of ideas about broadcasting among the dominions.

However, if the public broadcasting authorities established in the dominions in this period bore some sort of resemblance to the BBC, nowhere was the likeness close. This in itself hampered cooperation. Public broadcasters working in very different circumstances, and according to various constitutional arrangements, often had divergent requirements which BBC officers found difficult to understand and accommodate. BBC empire-building created significant complications; in particular, the corporation's attempts to extend the domestically generated principles of monopoly and public service overseas to its operations in places where those principles were by no means familiar or widely accepted, proved problematic. Moreover, broadcasting authorities in the dominions showed little subservience to the BBC. They were guided by their own agendas, understandings of local tastes and needs, and ideas about empire broadcasting. Their officers often wanted collaboration to be a mutual and reciprocal activity. They thus sometimes resisted, and often resented, the centralizing BBC approach.

These tensions became particularly apparent after 1932, when debates about broadcasting and empire ceased to be just about institutional structures, and broadened out to encompass programmes as well. From 1932 the BBC's new short-wave Empire Service targeted white listeners in the tropical colonies, India, and around the British world. The BBC also began to issue series of recorded programmes (known as 'transcriptions') to supplement the flow of material from Daventry. Rather than passively receive what they were given, public broadcasters in the dominions sought actively to shape the programmes the BBC provided, and the terms on which they were made available. In Australia, while the public broadcasting authority rebroadcast BBC Empire Service programmes and purchased BBC transcriptions, it constantly pressed the BBC to modify the nature and scope of those services. In Canada the public authority frequently refused to rebroadcast the Empire Service and rejected BBC transcriptions altogether. Only in New Zealand were collaborators relatively undemanding.

During the 1930s the global economic depression exerted a pervasive influence on the BBC's plans for empire broadcasting. To some extent, the effects were obvious and direct: in those years it was difficult to fund national, let alone pan-empire, radio services. However, the slump also influenced empire broadcasting more subtly, by reshaping the underlying imperial relationship between Britain and the dominions, in a complex and somewhat contradictory fashion. In some ways, the depression weakened the bonds of empire. As international trade declined, foreign markets were closed off by tariff barriers, the British economy contracted, and Britain's overseas imperial influence was correspondingly reduced. Yet the depression also encouraged the closer unity of the British world, fulfilling the hopes of constructive imperialists. World prices for primary products fell sharply. The dominions were key exporters of such goods, and became increasingly reliant upon remaining opportunities in the British market. The 'sterling bloc' came to offer a trading zone protected from currency-value fluctuations, and in 1932 Britain imposed tariffs of its own on imports, but granted valuable concessions to empire countries, agreed at that year's Imperial Economic Conference in Ottawa. The constitutional autonomy of the settler states had been confirmed during the 1920s, and the 1931 Statute of Westminster set the seal on the new constitutional consensus. Yet renewed imperial economic complementarity seemed to render these constitutional changes somewhat irrelevant.

The increased coherence of the British world was of course ephemeral, and comforting only because of the formidable nature of wider international challenges. The turn to protectionism and imperial preference probably did the British economy more harm than good. Moreover, underlying tensions within the British world still periodically bubbled up to the surface. At the Ottawa conference, dominion negotiators bargained hard, and infuriated their UK counterparts by refusing British manufacturers unrestricted reciprocal entry to their markets. Later, when Sir Otto Niemeyer of the Bank of England visited Australia, his deflationary recommendations were accepted by the government, but provoked widespread hostility against 'financial imperialism'. New Zealand faced a fiscal crisis in 1931–2, and had to be rescued through the intervention of the Bank of England: subsequently, however, the New Zealand government took steps to secure greater financial independence. In a rather different sphere, the English cricket team's infamous 1932/3 'bodyline' tour of Australia called into question some of the most basic shared values upon which the Britannic connection was assumed to rest.[1]

[1] John Darwin, *The Empire Project: The Rise and Fall of the British World-System, 1830–1970* (Cambridge, 2009), 431–51. Ric Sissons and Brian Stoddart, *Cricket and Empire: The 1932–33 Bodyline Tour of Australia* (London, 1984). Simon Boyce, '"In Spite of Tooley Street, Montagu Norman, and the Reserve Bank's Governor": Recolonization or the Eclipse of Colonial Financial Ties with Britain in the 1930s?', *New Zealand Journal of History*, 39/1 (April 2005), 75–92.

Enduring divisions between French and English Canadians, and between Afrikaners and English-speaking South Africans, further complicated the picture. Nevertheless, it was by no means clear to contemporaries that the overall trend was towards imperial disintegration: to some, it seemed quite the opposite. The backdrop to the BBC's early experiments with empire broadcasting was thus confused, and confusing.

PUBLIC BROADCASTING FOR THE DOMINIONS?

During the early 1930s public broadcasting became a more pervasive presence across the British world. In New Zealand and Australia, as part of attempts to regulate and improve broadcasting, private companies had already been contracted by the state to provide 'public service' programmes and reach audiences neglected by commercial broadcasters. However, criticism of the standard of programming provided by these companies was widespread. In New Zealand the Radio Broadcasting Company (RBC) was attacked for its over-reliance on classical music (audiences wanted greater variety) and local talent (which was deemed inadequate in terms of both quantity and quality). In October 1930 the New Zealand postmaster-general announced that the RBC's contract would not be renewed.[2] The Australian Broadcasting Company had meanwhile been starved of revenue, and some believed that it had failed to use properly what funds it did have. The election of a federal Labor government under J. H. Scullin raised the promise of reform. Labor believed that its views were under-represented in the newspapers and in the programmes broadcast by private radio stations, many of which were owned by newspaper companies. In the past Labor had experimented with municipally run broadcasting, and during the 1930 Imperial Conference Scullin met with John Reith further to discuss radio.[3]

In Australia responsibility for drawing up new broadcasting legislation was delegated to H. P. Brown of the PMG's Department. As Brown's thinking developed, he told Reith that he planned to introduce a broadcasting commission 'very similar in character' to the BBC, but partly supported by on-air advertising or sponsorship, and with his own department retaining responsibility for technical operations.[4] Reith conceded that these significant modifications might be necessary to make public broadcasting work in Australia, but warned that the division of technical and programming responsibilities had not proved a success in Germany. He hinted that he could travel to Australia 'at your request to give expert advice about

[2] *New Zealand Parliamentary Debates* (*NZPD*), 226: C. L. Carr, 3 October 1930, 392–4; J. B. Donald, 3 October 1930, 399.

[3] K. S. Inglis, *This is the ABC: The Australian Broadcasting Commission, 1932–1983* (Carlton, Vic., 1983), 12–17. Michael S. Counihan, 'The Construction of Australian Broadcasting: Aspects of Australian Radio in the 1920s', MA thesis (Monash University, 1981), 156–61, 227–9, 243–7. Murray Goot, 'Radio LANG' in Heather Radi and Peter Spearritt (eds), *Jack Lang* (Neutral Bay, NSW, 1977). John Reith diaries, 29 October 1930, BBC Written Archives Centre (WAC), S60/5/3/1.

[4] H. P. Brown to Reith, 17 November 1931 and Brown to Reith, 1 December 1931, WAC, E1/341/1.

your new Broadcasting constitution'. Brown replied that 'some day it may be possible to arrange a visit', but did not issue an invitation.[5] He had either made up his mind about the nature of the new legislation already, and thought a visit by Reith unnecessary, or (perhaps recalling Niemeyer's visit) believed that direct British intervention, particularly by someone of Reith's status and combative personality, might damage the prospects for public broadcasting in Australia. Others were more open in their avowal of the merits of the UK approach. In January 1932 the Australian postmaster-general met a delegation of university professors, educationalists, and representatives of listeners' associations and various women's, voluntary, and trade groups, who together advocated the introduction of 'the British system of broadcasting control' and emphasized 'the educational value of broadcasting, as a medium for the development of culture and public opinion... and the need for freedom from sectional or political bias'.[6]

Cultivated settler elites had in the past often looked to British cultural imports as a means of smoothing the rough edges of frontier and small-town life, and of asserting their own cultural authority. In Canada they mobilized with particular efficacy in the broadcasting debate, invigorated by the new language of national responsibility that had emerged after the First World War. Their vehicle was the Canadian Radio League (CRL), driven by Graham Spry and Alan Plaunt. Both men were young, well educated, and well connected. Spry was a Rhodes scholar, and Plaunt had also studied at Oxford University, and was wealthy enough to help bankroll the CRL. Like some other Canadian intellectual reformers of the time, and like Reith in Britain, they were high modernists, and hoped that broadcasting could improve the working of democracy and encourage national unity if properly controlled and placed under the guidance of public-spirited experts. Spry believed that Britishness had little appeal for many French Canadians or recent migrants from Europe, and should no longer form the basic defining feature of Canadian identity. However, he retained a strong personal sense of connection with Britain (he spent much of his life there), and saw cooperation with the UK as crucial to Canada's international role. Plaunt's views evolved differently, but in the early 1930s neither could have been called 'anti-British'. Both were willing to work closely, if covertly, with the BBC.[7]

The existence of the CRL on the ground as potential collaborators gave the BBC an opportunity to intervene in the Canadian broadcasting debate that had not presented itself in either Australia or New Zealand. Canada was also closer and easier to communicate with and visit in a period when long-distance commercial air travel was still in its infancy. The fact that Canada was the dominion most

[5] Reith to Brown, 22 December 1931, ibid.; Reith to Brown, 27 January 1932 and Brown to Reith, 28 January 1932, E1/341/2.

[6] 'Broadcasting', *Sydney Morning Herald*, 21 January 1932. Notes re: visit of delegation to Postmaster-General, 20 January 1932, National Archives of Australia (NAA), Melbourne, MP341/1, 1932/3887, 349572.

[7] Michael Nolan, *Foundations: Alan Plaunt and the Early Days of CBC Radio* (Toronto, 1986). Doug Owram, *The Government Generation: Canadian Intellectuals and the State, 1900–1945* (Toronto, 1986), esp. 135–67. David James Smith, 'Intellectual Activist: Graham Spry, a Biography' (D.Phil. thesis, York University, Ont., 2002), esp. 178–9 and 247–50.

susceptible to the rival influence of US broadcasting models and networks provided an added incentive for BBC involvement. The CRL was soon in touch with William Ewart Gladstone Murray, the BBC's Canadian-born director of publicity and, like Spry, a Rhodes scholar. Murray invited the Canadian Prime Minister, Bennett, to visit the BBC while he was in London for the 1930 Imperial Conference. Although he failed to acknowledge Murray's invitation, on his return to Canada he announced plans for a radio bill.[8]

Both the BBC and the CRL were aware that their relationship needed to be constructed and utilized with care, in order to avoid the accusation that supporters of public broadcasting wanted 'to bring Canada under the influence of the B.B.C. and of Britain'.[9] The CRL also faced claims that the BBC was unpopular with listeners and subservient to the government, and that the UK model was generally unsuited to Canada's dispersed and provincially divided population. Commercial interests resumed the offensive soon after the Imperial Conference, as the Canadian Pacific Railway (CPR) sought to establish itself as the major private broadcasting network, and to restrict the Canadian National Railways (CNR) to ownership of a stunted public network. In support of this scheme, J. Murray Gibbon, the chief public relations officer of the CPR, published a damning critique of the BBC, and alleged that British-style public broadcasting would merely drown out US programmes that were better suited to the 'North American mentality'. Spry quickly demolished Gibbon's claims in a published counterblast, and charged the CPR with seeking a highly profitable monopoly of radio advertising, which would saddle the CNR and thus the taxpayer with the bill for public broadcasting, and prevent the erection of any effective barrier against the continued dominance of American programmes and advertisements on Canadian stations. With Spry's support, Gladstone Murray meanwhile confronted CPR officials in London who, after hinting that heavy-handed BBC protests would not be well received in Canada, eventually disavowed Gibbon's article. Without seeming too intimately connected with the BBC, the CRL had drawn on British support and played up the danger of US influence in order to neutralize the arguments of powerful domestic opponents.[10]

Spry and Murray also exploited the Canadian prime minister's known enthusiasm for the British connection. For Christmas 1931, the BBC planned an ambitious transnational network programme. It would organize a sequence of contributions by radiotelephone from around the empire, add its own material, and then broadcast the composite programme on short wave. Stations around the empire would be able to rebroadcast the programme on medium wave for their own listeners. While the CNR agreed to help, the Canadian arrangements depended on landlines owned by American companies, which refused to cooperate. When the

[8] For more details, see Simon J. Potter, 'Britishness, the BBC, and the Birth of Canadian Public Broadcasting, 1928–1936' in Gene Allen and Daniel J. Robinson (eds), *Communicating in Canada's Past: Essays in Media History* (Toronto, 2009).

[9] R. E. L. Wellington to G. Spry, 7 October 1930, Library of the University of British Columbia Special Collections Division (UBCSCD), Alan Plaunt fonds, 11/14.

[10] Potter, 'Britishness', 86–8.

BBC thus announced the cancellation of the entire project, Bennett intervened personally and the landline companies offered to reverse their decision. However, the BBC claimed it was too late to reinstate the broadcast.[11] Both the CRL and the BBC were in fact pleased with this dramatic outcome, for it seemed to demonstrate the need for the establishment of a Canadian public broadcasting authority capable of taking charge of such events. Spry later recalled with especial satisfaction the impact that the affair had on Bennett, 'the great imperialist'.[12] Bennett appointed a Special Commons Committee on Radio Broadcasting and the CRL brought Murray to Ottawa to give evidence. Careful to avoid charges of BBC interference, Murray stressed that he was there to provide information only, 'and in no sense to suggest that what is happening outside Canada should necessarily be a criterion of what Canada should do'. Nevertheless, Murray outlined the UK approach in some detail, and emphasized the importance of establishing a public authority under remote state control and with a domestic broadcasting monopoly.[13]

During the early 1930s policymakers in the dominions were thus urged to adopt aspects of the BBC model. However, the reforms that were eventually introduced corresponded only loosely with those that had been advocated. In New Zealand, with the days of the RBC numbered, Post and Telegraph (P&T) Department officials mixed together elements of the British and Australian models. New legislation, passed on 11 November 1931, handed programming responsibilities for A-class stations over to a government-appointed three-person New Zealand Broadcasting Board (NZBB), to be funded by listener licence fees, and assisted in its executive functions by a general manager. Some features of the BBC model were thus introduced, and in parliament the postmaster-general and others spoke approvingly of the BBC and Canada's Aird Report.[14]

Yet the new structure differed in key respects from that advocated by either Reith or Aird. As in Australia, technical responsibilities were separated out and entrusted to the postal authorities. There was still no national networking, and neither was the NZBB to enjoy a domestic monopoly: private B-class stations would continue to operate, and were now authorized to accept some commercial sponsorship. State control of the NZBB was meanwhile far from remote. Not only would the P&T Department run the technical side of operations, but a senior postal employee, E. C. Hands, was appointed as the NZBB's first general manager. Controversial broadcasting, now permitted in Britain, remained banned.[15] Yet the

[11] E. AustinWeir, 'Inter-empire Christmas Broadcast Proposed for Christmas Day 1931', Library and Archives Canada (LAC), E. Austin Weir papers, M-715. Weir, *The Struggle for National Broadcasting in Canada* (Toronto, 1965), 141. See also 'Memorandum on the Cancellation of the Empire Christmas Broadcast', Plaunt fonds, 12/10.

[12] Typescript MSS, unpublished Spry memoirs, ch. 4, LAC, Graham Spry papers, 84/13.

[13] [W. E. G. Murray], 'Radio Broadcasting: Memorandum Submitted as Basis of Evidence, Ottawa 1932', University of Toronto Archives and Records Management, Vincent Massey papers, 59/3.

[14] *NZPD*, 230: 2 November 1931, 655–8; J. Parr, 6 November 1931, 802; C. J. Carrington, 6 November 1931, 807. As in Canada and Australia, during the debate some also questioned whether the BBC was doing a good job of entertaining UK listeners: E. J. Howard, 2 November 1931, 671.

[15] Patrick Day, *A History of Broadcasting in New Zealand*, i, *The Radio Years* (Auckland, 1994), 107–8, 136–53.

main thrust of criticism of the new arrangement was not that state control was excessive, but rather that it was too limited. Michael Joseph Savage, soon to become leader of the Labour Party, argued that the P&T Department should have been assigned programming as well as technical responsibilities.[16]

When public broadcasting was introduced in Canada, similarly drastic divergences from the BBC model occurred. This was despite the recommendations of the Aird Report, the CRL, and Gladstone Murray. Based on the findings of the 1932 inquiry, a Canadian Radio Broadcasting Commission (CRBC) was established. The commissioners would be appointed, like the BBC's governors, by the government of the day. Unlike the BBC governors, however, the CRBC commissioners would work full time, without a BBC-style chief executive officer. State control would be more direct than in Britain, as the federal government would retain significant authority over finances and appointments. Funding would come from advertising as well as listener licence fees. Crucially, there would be no domestic broadcasting monopoly for the CRBC. Private stations would remain, and the CRBC's relationship with them would be complex: it would regulate them and allow them to affiliate with its network (no private networks were to be permitted), but its own public stations would also have to compete with them. Indeed, the CRBC was arguably intended to help private broadcasters compete with the US radio industry, rather than to entrench the principle of public involvement. By carrying well-resourced CRBC network programming, it was hoped that private stations would be able to win over listeners from US stations. The CRBC was, in a way, a means indirectly to subsidize the private stations, providing money to make better Canadian programmes which they could then broadcast.[17] Ultimately, the CRBC looked more like a state-run version of an American network than a variant of the BBC model.

Nevertheless, the foundation of the CRBC was accompanied by plenty of stirring national and Britannic rhetoric, often deployed in tandem. The Special Commons Committee on Radio Broadcasting had argued that radio could act as 'one of the most efficient mediums for developing a greater National and Empire consciousness within the Dominion and the British Commonwealth of Nations'.[18] Bennett asserted that only public broadcasting could meet Canada's 'national requirements and empire obligations', and presented the CRBC as 'a dependable link in a chain of empire communication by which we may be more closely united one with the other'.[19] He told Spry that 'you have saved Canada for the British Commonwealth'.[20] Plaunt thought it no coincidence that

> the bill to create a unified Canadian system and another—the final link—in an all Empire chain, was passed on Empire Day, just before the Imperial Economic Conference, just before, it may be, the beginnings of the 'Fourth British Empire'. The CRBC

[16] *NZPD*, 230, 2 November 1931, 659–62.

[17] Mary Vipond, *Listening In: the First Decade of Canadian Broadcasting, 1922–1932* (Montreal, 1992), 264–80.

[18] Quoted in Frank W. Peers, *The Politics of Canadian Broadcasting, 1920–1951* (Toronto, 1969), 96.

[19] Quoted in Weir, *Struggle*, 133 and Vipond, *Listening In*, 270.

[20] Quoted in Vipond, *Listening In*, 268.

> may well be indeed one of the links, to quote Sir Wilfrid Laurier, 'light as air, yet strong as the bonds of steel' that will bind the new empire.[21]

The CRL suggested that Gladstone Murray should be appointed chairman of the CRBC, but Murray was unwilling to accept the substantial pay cut this would have entailed. Instead, Bennett selected Hector Charlesworth, a music and drama critic, and editor of the Conservative-leaning Toronto journal *Saturday Night*. The vice-chairman was Thomas Maher, a Francophone Conservative from Quebec who had failed to gain a seat in the 1930 elections, and who was also director of a private radio station. The third commissioner, Lt. Col. W. A. Steel, had been the Canadian delegate at the Imperial Conference who helped sink the BBC's empire broadcasting proposals.[22] Reith was disgusted by rumours that it was Steel who had persuaded Bennett to make the CRBC 'more or less a Government Department' instead of an autonomous public corporation.[23] Reith openly described the Canadian departure from the BBC principle of remote state control as 'surprising and, to many, disquieting'.[24]

From the outset the CRBC was subject to considerable criticism in Canada. The commission purchased the CNR's radio stations, but a shortage of funds meant than plans to build additional high-powered transmitters were not implemented, and the CRBC found itself limited to providing private stations with a meagre daily schedule of networked programmes, accompanied by an increasing amount of advertising. This was hardly the bulwark against US broadcasting that had been envisaged. Moreover, the CRBC's handling of Canadian linguistic divisions proved problematic. Rather than confine broadcasting in French to Quebec, the CRBC carried French programmes on its network in English-speaking areas too, arousing significant protests. Some of the CRBC's decisions as a regulator of the private stations seemed partisan, and were also criticized.[25]

Early in 1933 Bennett decided to bring Gladstone Murray over from Britain at the federal government's expense (Murray reported that Bennett made 'stirring appeals to my patriotism and my imperialism') to investigate how the CRBC's difficulties might be overcome. In his interim report to the Canadian House of Commons, Murray stressed once again the importance of the British principle of remote state control. The government's direct authority over CRBC revenue, appointments, and policy should, he argued, be diluted. This led to heated debate in Parliament, and claims that a visitor from Britain had no business getting involved in Canadian affairs. Murray's final report was, in the short term at least, largely ignored.[26]

It was only in Australia that a public authority even vaguely resembling the BBC was introduced at this time. Brown's broadcasting bill survived the fall of Scullin's

[21] Plaunt, 'Some Notes on the Radio League', 25 May 1932, Plaunt fonds, 12/3.

[22] Knowlton Nash, *The Microphone Wars: A History of Triumph and Betrayal at the CBC* (Toronto, 1994), 90–3.

[23] Reith diaries, 11 November 1932, WAC, S60/5/3/2.

[24] Press cutting, Reith, 'The Future of British Broadcasting', *Daily Telegraph* (London), 15 November 1932, National Library of Australia, Herbert and Ivy Brookes papers, MS1924, 26, f. 1273.

[25] Nash, *Microphone Wars*, 94–116.

[26] Potter, 'Britishness', 90–1.

Labor government, and on 10 March 1932 was introduced to parliament with only minor revisions by the new United Australia Party government, headed by Joseph Lyons, a former member of Scullin's cabinet. The postmaster-general publicly praised the BBC, the NZBB, and the Canadian Aird Report, and claimed that the bill 'mimicked the British system as closely as Australian conditions will permit'. However, others judged the new proposals 'an exceptionally pale shadow' or a 'parody' of the UK model.[27] The Australian Broadcasting Company's programme responsibilities, assets, and share of the listener licence fee were to be transferred to a commission made up of five part-time members, with a general manager, but the PMG's Department was to retain control of the technical side of broadcasting. The department's authority in other areas was diluted as the legislation was revised, but the commission would still require ministerial approval for large items of expenditure, long leases of property, and decisions regarding the location of studios. Moreover, the minister retained the power to direct the commission to broadcast or censor material, according to his own conception of the public interest. Was this remote or direct state control? Much would depend upon interpretation and usage.[28]

There were other divergences from the British model. The commission was not to enjoy the security from sudden legislative reform conferred upon the BBC by its ten-year renewable royal charter. As in New Zealand and Canada, the new Australian broadcasting authority was established by an act of parliament, which could be amended at any time. The Australian commissioners were to be paid much less than the BBC's governors, and there was to be greater regional devolution of authority, to offices of the commission in the different Australian states. Moreover, as in the other dominions, the new commission was denied a domestic broadcasting monopoly. Powerful newspaper interests with radio ramifications lobbied to ensure the survival of the B-class stations. During the parliamentary debates over the new legislation there was renewed criticism of the BBC's monopoly in Britain and the resulting prevalence of 'highbrow' programming.[29] However, in a way, defenders of private broadcasting brought the commission more closely in line with BBC operations: under pressure, the new legislation was modified to prevent the commission accepting commercial sponsorship, a privilege to be reserved for private broadcasters.[30]

The Australian Broadcasting Commission (ABC) transmitted its first programme on 1 July 1932. While the commissioners had political sympathies with the government that appointed them, they did seem more likely to pursue Reithian goals than their NZBB and CRBC counterparts. The first chairman of the ABC was Charles Lloyd Jones, and the first vice-chairman was Herbert Brookes: both

[27] *Commonwealth Parliamentary Debates* (*CPD*), 133: 9 March 1932, 840–1; 10 March 1932, 958, 976.

[28] Counihan, 'Construction of Australian Broadcasting', 247–55.

[29] Amalgamated Wireless (Australasia) Ltd (AWA) arranged for Peter Eckersley (the former BBC chief engineer) to visit from Britain to speak in favour of an AWA technical monopoly, with competitive private provision of programming. See Brown to Reith, 6 July 1932, WAC, E1/341/2.

[30] Inglis, *This is the ABC*, 18–26.

were businessmen with artistic interests and a history of public service. They were joined by Prof. R. S. Wallace, the vice-chancellor of the University of Sydney; Mrs E. M. R. Couchman, president of the Australian Women's National League and a member of several other voluntary bodies, some organized on an empire-wide basis; and R. B. Orchard, a former actor and parliamentarian.[31] In a press statement issued after the commission's first meeting, Jones claimed that Lyons had instructed him to 'walk in the footsteps of the B.B.C. and fall in behind Britain'.[32] Writing to Reith, Jones referred to the BBC as 'the parent Company', and in an inaugural radiotelephone message he stated that:

> My Commission is mindful of the fact that your Corporation has given a wonderful lead to all that is best in broadcasting throughout the world, and it is our wish to emulate your example, draw upon your experience, and cooperate in every way possible for the benefit of the [Australian] Commonwealth and the Empire... my Commission is anxious to do all in its power to foster the development of this means of welding the Empire into a still-closer relationship, and looks forward to the opportunities of broadcasting to the people of Australia the addresses given over your system by the many eminent authorities who are leaders of the world's thought.[33]

Jones also struck a decidedly Reithian note when he claimed that the ABC was 'under a moral obligation to realise the taste and improve the culture of the community, to spread knowledge, encourage education, and foster the best ideals of our Christian civilisation', to balance 'national service' and 'national entertainment'.[34]

By the end of 1932 public broadcasting authorities were thus operational in Canada, Australia, and New Zealand. They were the product of domestic concerns, and of the activities of domestic lobbyists, civil servants, and politicians. Overseas models, and particularly the UK approach, had been evaluated and discussed in detail. In the case of Canada, BBC officers had made direct and sometimes controversial contributions to debate. Ideas about Britannic identity had been deployed alongside those concerning Canadian, Australian, and New Zealand nationhood and national requirements. However, only in Australia was anything vaguely resembling the BBC introduced, and even here marked divergences were apparent. The degree to which ideas about public service broadcasting had been transformed as they were adapted to domestic requirements reflected the primacy of local agendas. No matter how marked the rhetoric of Britishness, there was no strict adherence to the UK approach.

To justify public funding, the new broadcasting authorities would have to discharge functions that private operators could or would not. Listeners would have to be educated and informed, as well as entertained, not an easy balance to strike.

[31] Ibid., 20.

[32] *The Argus* (Melbourne), 28 May 1932, quoted in Alan Thomas, *Broadcast and be Damned: The ABC's First Two Decades* (Carlton, Vic., 1980), 19.

[33] C. L. Jones to Reith, 22 March 1933, WAC, E1/315/2. 'The Empire and its Broadcasting', *World Radio*, 15 July 1932.

[34] *The Argus* (Melbourne), 5 May 1933. For wider debates, see Lesley Johnson, '"Sing 'Em Muck, Clara": Highbrow versus Lowbrow on Early Australian Radio', *Meanjin*, 41/2 (June 1982), 210–22.

Many listeners might not want to hear too many highbrow programmes, or indeed any at all. If listeners in Britain were dissatisfied with what the BBC broadcast, they might be able to pick up a continental European station instead, but otherwise their only option was to switch off. Public broadcasters in the dominions, however, faced competition from private rivals who often focused on material designed to be popular with as many listeners as possible. If public stations carried too much highbrow material, then listeners would simply change stations, and much of the justification for public funding would be lost. From the outset, Hector Charlesworth stressed that the CRBC would make radio a 'cultural force' as well as a means of entertainment, but also referred to complaints that BBC programmes were too 'high-brow': he presumably meant to safeguard the CRBC against similar accusations.[35]

Moreover, public broadcasters had to strike a workable balance between local content and imported material. It might seem culturally desirable, and good for public relations, to sponsor Canadian, Australian, or New Zealand talent, but to do so was expensive. Moreover, local artists might not be as polished as the products of artistic centres in London, Paris, or New York. Poor performances might have an adverse impact on the prestige of the broadcasting authority. Yet if public broadcasters brought in too much material from abroad, then they might be accused of failing to serve national interests, of perpetuating stultifying influences from Old World sources, or even of foisting alien and incomprehensible material on dominion audiences. To further complicate matters, decisions about balancing domestic and imported material were often inextricably linked with debates about the respective merits of high and popular culture.

Dominion public broadcasting authorities approached the BBC's empire broadcasting efforts with another significant question in mind: who was going to speak to whom? The Australian postmaster-general seemed worse than naïve when he enthused that empire broadcasting would allow

> naked blacks to listen-in in the jungle to the world's best operas. We may also reach the period when brown-skinned Indians will be able to dance to one of England's best orchestras, and when fur-clad Canadians in distant snowbound outposts may listen to a description of the running of the English Derby... It should also help to deepen our Empire spirit considerably if we, through the wireless, can listen to the greatest British artists, speakers and lecturers, who participate in broadcast programmes.[36]

The caricatures of Indians and Africans (and Canadians) were atavistic and offensive, as was the assumption that all the good music, actualities, and information would come from the centre, from England. Others believed that, if radio was to strengthen the Britannic connection, then public broadcasting authorities in Britain and the dominions would need to work together in a way that mirrored the

[35] 'Personnel of the Canadian Radio Broadcasting Committee', *World Radio*, 21 October 1932. Nash, *Microphone Wars*, 90–3. For more on Charlesworth's views about culture and broadcasting, see Mary Vipond, 'Cultural Authority and Canadian Public Broadcasting in the 1930s: Hector Charlesworth and the CRBC', *Journal of Canadian Studies*, 42/1 (winter, 2008), 59–82.

[36] *CPD*, 133, 10 March 1932, 845.

broader constitutional pattern of voluntary inter-war cooperation (Plaunt's 'Fourth British Empire'). Radio would thus not just be the voice of Britain, it would also provide the dominions with a more audible presence overseas. For the former Australian Prime Minister William Hughes, radio meant that 'Our voice can now be heard in every other dominion, and we, in turn, can hear theirs.'[37] The problem was that the BBC remained largely oblivious to the emergence of a 'Fourth British Empire'. In the BBC's Empire Department a centralizing 'imperialism' remained dominant.

REACTIONS TO THE BBC EMPIRE SERVICE

While the BBC was installing its new short-wave transmitters at Daventry, in Ottawa the 1932 Imperial Economic Conference was reiterating the need for reciprocity: 'closer liaison between the responsible bodies controlling radio in all parts of the Commonwealth' was needed in order to secure an 'interchange' of programmes and facilities.[38] The BBC took little notice. Under pressure from Noel Ashbridge, and despite protests from C. F. Atkinson, the idea of a modest short-wave service primarily for expatriate direct listeners in the dependent colonies was again dropped in favour of the original, more ambitious BBC scheme, whereby short wave would also serve rebroadcasters in the dominions. The BBC would go 'the whole hog'.[39] The proceedings of the 1930 Imperial Conference were thus set aside, and the recommendations of the 1932 Imperial Economic Conference ignored, as the BBC dusted off its own plans from December 1929.

The new BBC Empire Service was to broadcast for ten hours a day, with two hours of programming designed to reach five different 'zones' in sequence at peak local listening times: essentially, the Pacific, including Australia and New Zealand; the Far East; the Mediterranean, Middle East, and East, Central, and South Africa; West Africa and the South Atlantic; and North America, including Canada and the British West Indies.[40] Given the size, diverse conditions, and multiple time zones included within each target area, and the fact that Daventry was equipped with only two transmitters, serious compromises would have to be made. There was little spare capacity to allow for special broadcasts for particular groups of listeners within each zone, or to deal with the potentially divergent requirements of direct listeners and rebroadcasters. The modest nature of the transmission equipment installed at Daventry reflected the scarcity of available funding, diverted from the BBC's share of UK listener licence fees. The BBC had applied the domestically generated non-commercial public-service model to its external broadcasting, but without finding any adequate source of revenue, at home or overseas, to support it.

[37] Ibid., 958.

[38] 'IEC(32)182 CD(5-F)—Imperial Economic Conference—1932—Report of Sub-Committee on Films and Radio to Committee on Methods of Economic Co-operation', Hocken Library, William Downie Stewart papers, MS-0985-020/022.

[39] Minutes, 'Empire Broadcasting', 31 March 1932, WAC, E4/6.

[40] *The Empire Broadcasting Service* (BBC, 1933), 4–7, 18.

Resources for programming were thus similarly limited. At the very beginning, the Empire Department's programme budget was a miserly £10 per week. This had only grown to £200 per week by the end of 1933.[41] For some zones, material from the BBC's domestic services could be reused at minimal cost, but what would overseas audiences make of programmes prepared with UK listeners in mind, or even of BBC programmes designed specifically for the Empire Service? The BBC was keen to emphasize that, while 'the "home" standards governing tone and material will be scrupulously adhered to', the Empire Service would be sensitive to differences in local taste and habits.[42] Yet the Reithian philosophy dictated that audience preferences were mutable. In Britain, the BBC was seeking to modify and raise tastes and expectations. Would it attempt to 'uplift' audiences in the dominions and colonies, too, and adopt a civilizing mission overseas?

The creation of the CRBC, ABC, and NZBB had potentially removed one of the obstacles to the success of empire broadcasting. These national authorities might rebroadcast BBC programmes as a public service. Yet this could not be guaranteed. The Empire Service was designed to reach audiences in each zone at peak listening times, primarily due to the perceived requirements of direct listeners. Rebroadcasters were not permitted, for copyright and performing rights reasons, to record programmes for use in later time slots. Would public broadcasters in the dominions accept these inflexible conditions, and turn over their peak audiences to unproven programmes provided by another organization? If they did, and the programmes were unpopular, they might lose both prestige and listeners. Would the BBC be willing to respond to any demands they might have in terms of the scope and content of the service, given the general Reithian unwillingness to cater to what audiences thought they wanted, and the fact that the Empire Service was to be conducted without any dominion financial contributions? And would public broadcasters in the dominions expect the BBC to rebroadcast their programmes to UK listeners in return? The BBC had not seemed very enthusiastic about this prospect in the past.

A more fundamental concern was whether the Empire Service could actually be received in the dominions at a quality suitable for rebroadcast. A *World Radio* correspondent described the experience of listening to the Empire Service in New Zealand:

> There is a constant heavy hum, like the sound of immense dynamos on full load, and a rhythmic surge as the radio waves reach us. Then, to thrill us, from this noisy background, come music, military words of command, or the National Anthem. In spirit we are participating in a typical British ceremony, so that our loneliness in the great wastes of the South Pacific does not feel so great. Radio has linked us instantaneously with the heart of the Empire. We have not heard the whole programme, but our patriotic imagination has filled in the blanks.[43]

[41] Gerard Mansell, *Let Truth Be Told: 50 Years of BBC External Broadcasting* (London, 1982), 20–6.

[42] *Empire Broadcasting Service*, 17–19.

[43] [Charles E. Wheeler], 'Empire Broadcasting in the Pacific', *World Radio*, 4 November 1932.

This was being diplomatic: as the Dominions Office was well aware, not everyone agreed that sentimental attachment to Britain was enough to compensate for appalling sound quality.[44] The problem derived partly from the unpredictable nature of short-wave propagation, but also reflected the limitations of the transmitters at Daventry. Contemporaries often remarked that American, French, and German short-wave transmissions were received much more clearly in Australia and New Zealand.[45] Would stations in the dominions be able and willing to rebroadcast Daventry's inferior and unpredictable signal?

A worrying number of questions thus remained when the BBC Empire Service began broadcasting in October 1932. The BBC sought answers by issuing a listener questionnaire and sending a new employee, Malcolm Frost, on a world tour.[46] However, characteristically, no provision was made for systematic, continuous audience research. The BBC was thus largely dependent upon the dominion broadcasting authorities for feedback on its programmes. Yet these authorities themselves undertook little in the way of formal listener research. If the BBC and dominion broadcasting authorities were to disagree as to what sorts of programmes the Empire Service should provide, the relative merits of conflicting opinions could not be judged against any concrete or convincing evidence of audience response. Disagreements would thus be difficult to resolve.

Initially, dominion broadcasting authorities seemed to react favourably to the Empire Service, reflecting the novelty of the transmissions, and the special effort devoted by the BBC to a 1932 Christmas Day transnational network programme. Listeners were taken on a tour westward round the empire, with messages of greetings from Dublin, *S.S. Majestic* in mid-Atlantic, Halifax, Montreal, Toronto, Winnipeg, Vancouver, Wellington, Sydney, Melbourne, Brisbane, the *S.S. Empress of Britain* at Port Said, and finally Cape Town and Gibraltar.[47] The availability of transmitters and radiotelephone links determined the itinerary: the result was a focus on the most economically developed parts of the empire, the British world, and on Britannic identities. The programme closed with a royal Christmas message, written by Rudyard Kipling, and read by King George V. The King's speech offered solace in hard economic times, and appealed explicitly to families listening in their homes as well as those in outposts of empire 'so cut off by the snows, the desert or the sea, that only voices out of the air can reach them'. Kipling's words emphasized the imperial potential of the medium: it was 'a good omen that Wireless should have reached its present perfection at a time when the Empire has been linked in closer union . . . it offers us immense possibilities to make that union closer still'.[48]

[44] Minute by Holmes, 29 October 1932, United Kingdom National Archives (UKNA), DO35/198/2.

[45] E.g. M. A. Frost to C. G. Graves, 17 June 1933, WAC, E4/50.

[46] 'The B.B.C. Empire Tour', *World Radio*, 25 November 1932.

[47] R. H. Roberts, 'The Inter-Empire Broadcast Christmas 1932' [*c.*1932/3], Weir papers, M-548.

[48] Tom Fleming (ed.), *Voices out of the Air: The Royal Christmas Broadcasts, 1932–1981* (London, 1981), 11.

Reith thought that this prestige programme 'evoked more widespread appreciation than any we have ever transmitted', and the general manager of the ABC congratulated the BBC on a 'magnificent success'.[49] Given the problems in Canada that had led to the cancellation of the 1931 Christmas broadcast, the CRBC was particularly keen to rebroadcast the 1932 programme, and indeed made it the occasion for its inaugural network transmission. With the help of the CNR, the CRBC relayed the empire Christmas broadcast to forty-two stations across Canada.[50] Charlesworth judged it 'the finest transatlantic broadcast from the point of view of reception and coverage that has occurred since radio was invented': the CRBC's controller of programmes thought that it had awakened 'many memories in those from the old land and strange feelings of affection for the country of their forefathers among a vast number of native born Canadians'.[51]

However, problems with the Empire Service also became apparent from an early stage. Arriving in South Africa, Frost found reception quality to be extremely poor. The inaugural programme was only half-intelligible at Cape Town, and virtually unintelligible at Johannesburg and Durban. Music programmes were ruined. The African Broadcasting Company took a radiotelephone feed of the Christmas programme, rather than rebroadcast Daventry.[52] In an attempt to create support for rebroadcasting and for a South African financial contribution to the Empire Service, Frost played up the 'reciprocal element': he arranged for a programme to be produced live from the top of Table Mountain and fed via radiotelephone to the BBC for rebroadcast in Britain and overseas. Yet Afrikaner hostility, poor reception quality, and the fact that the African Broadcasting Company did not want local broadcasting revenues to be diverted to the BBC, made the prospect of financial contributions from South Africa seem remote.[53] Meanwhile, from Australia, Brown had written to Reith expressing the hope that Britain and Australia could 'develop a closer and more equitable relationship' in the field of short-wave broadcasting.[54] Did Brown mean future financial contributions, a reciprocal flow of programmes, or some other form of 'equitable' cooperation? S. M. Bruce, Australia's Resident Minister in London, was more forthright, arguing that the reciprocal exchange of programmes was crucial.[55]

The BBC could seem insensitive to the requirements of dominion public broadcasting authorities. It granted all stations everywhere blanket permission to rebroadcast Empire Service programmes, without charge. Only rebroadcasting of Reuters copyright news and certain programmes involving special performing rights issues

[49] Reith to Jones, 3 February 1933, WAC, E1/315/2. H. P. Williams to N. Ashbridge, 26 December 1932, E1/315/1.

[50] H. Charlesworth, 'Interim Report—Canadian Radio Broadcasting Commission—1st November 1932 to 10th January 1933', LAC, René Landry papers, 1.

[51] Charlesworth to W. A. Steel, 27 December 1932, WAC, E1/508. Weir to CRBC network stations, 20 January 1933, Weir papers, M-548.

[52] Frost to [Graves], 11 January 1933, WAC, E4/49.

[53] Frost to Graves, 13 January 1933, WAC, E4/45. Frost to [Graves], 24 January 1933, E4/49. Mansell, *Let Truth Be Told* 30.

[54] Brown to Reith, 17 November 1931, WAC, E1/341/1.

[55] S. M. Bruce, 'Pioneers of Empire and the New Service', *World Radio*, 21 October 1932.

was restricted. This policy reflected the BBC's hope that 'maximum generosity' would increase the chances of financial contributions from the dominions and colonies.[56] However, the ABC and the NZBB wanted exclusive rights to rebroadcast the Empire Service in Australia and New Zealand respectively. Both were seeking to establish themselves in the face of private competition, and to justify their position as publicly funded authorities providing a different type of programming to that available on B-class stations. When the BBC refused to grant exclusive rights, the ABC claimed that the decision damaged 'the prestige of the National Service in Australia', and led to BBC programmes being associated with inappropriate advertising (one Sydney B-class station allegedly broadcast Empire Service programmes in a slot sponsored by 'the Specialty Blouse Stores'). The ABC hinted that it might drop Empire Service rebroadcasts altogether, rather than share BBC material with the B-class stations. The NZBB similarly maintained that exclusive rights to Empire Service programmes would help it combat the allure of the 'popular entertainment' broadcast by the private stations. However, the BBC refused to budge.[57]

The ABC also pressed for Empire Service transmissions to the Pacific to be timed according to its own requirements, arguing that it knew better than the BBC what was required. When one of the ABC's commissioners visited London, he urged the Director of the BBC's Empire Department C. G. Graves not to broadcast to Australia in peak listening periods as the ABC was producing high-quality programmes of its own for these slots. Graves seemed to assume that Australians lived ' "at the back of beyond", and that the Empire programmes will catch on because we have none of our own!'[58] H. P. Williams, the ABC's general manager, similarly urged Graves to target Australia in off-peak periods, when the ABC could rebroadcast poor-quality short-wave signals and risk fewer complaints from listeners.[59] The limited flexibility of the Daventry transmitters, and the BBC's reluctance to abandon direct listeners in other parts of the Pacific zone, made it difficult to respond to these demands.[60] In 1933 a new ABC general manager, W. T. Conder, thus reported that the ABC was rebroadcasting virtually nothing from the Pacific transmissions. He suggested to Frost that the BBC close down the Empire Service, and instead distribute recorded programmes on disc free of charge.[61] It was only when the BBC agreed to the ABC's scheduling demands, and moved its transmissions that June to target Australia before the peak listening period, that the ABC started systematically rebroadcasting the Empire Service.[62]

If reception conditions for the Empire Service were poor in Australia, they were truly awful in New Zealand and the rest of the Pacific zone. Frost believed that, as

56 Graves to Frost, 2 and 4 January 1933, WAC, E4/45.

57 Jones to Reith, 23 September 1932, WAC, E1/315/1. Reith to Jones, 3 February 1933, E1/315/2. E. C. Hands to C. F. Atkinson, 12 December 1932 and Graves to NZBB, 14 and 30 December 1932, E1/1091/1.

58 R. S. Wallace to Jones, [*c.* late 1932], NAA, Sydney, SP1558/2, box 76, 'Appointment London Rep—Inward and Outward Corres., 1932–1946'.

59 Williams to Graves, 17 January 1933, WAC, E1/315/2.

60 Graves to Reith and Ashbridge, 23 February 1933, WAC, E1/315/2.

61 W. T. Conder to Graves, 13 May 1933, WAC, E1/315/2. Frost to Graves, 20 June 1933, E4/47.

62 Conder to Graves, 28 June 1933, WAC, E1/315/2.

in South Africa, this 'complete failure' ruled out any financial contribution from the New Zealand government.[63] In June 1933 only between 5 and 20 per cent of a speech by the King carried by the Empire Service was deemed intelligible.[64] Music was entirely spoiled. After the initial novelty of rebroadcasting wore off, the use of Empire Service material by the NZBB declined.[65]

What types of programmes did the BBC think its listeners in Australia and New Zealand needed? More importantly, what sort of material would local rebroadcasters accept? At the outset, the BBC expected that in general 'the most popular type of programmes will be those that reflect everyday life in this country—items, in fact, of the kind that are best illustrated by the broadcasting of Big Ben'.[66] C. H. Chomley, editor of the *British Australian and New Zealander*, a London periodical, agreed. Australians and New Zealanders were 'ninety-eight per cent of British blood', and 'the more typically British' the Empire Service was, the more favourably it would be regarded, particularly as a riposte to American dominance of the cinema. However, Chomley also hinted that Australian and New Zealand listeners would not necessarily want exactly what was served up to UK audiences: what they needed above all from the Empire Service was news, sports commentaries, and serious talks.[67] Conder argued that only such material would appeal to native-born Australians, and that given poor reception quality only the best drama and light entertainment could hope to compete with local efforts.[68]

For British expatriates in Africa and Asia, and for recent British migrants to the dominions, existing sentimental links with 'home' might be sufficient motivation for direct listening.[69] However, if the BBC wanted its programmes rebroadcast in Australia and New Zealand, something carefully tailored to match the more particular needs of listeners and rebroadcasters was required. Yet, initially, the Empire Department simply did not have access to the necessary funds, and was even obliged to use commercial gramophone records to fill airtime. This seemed pointless: such recordings were already available in the dominions, and could be broadcast locally at much higher quality than from Daventry.[70]

[63] Frost to Graves, 12 June 1933, E4/49. Frost to Graves, 17 June 1933, E4/50.

[64] Frost to Hands, 15 June 1933 and Hands to Frost, 28 June 1933, WAC, E5/53.

[65] *Second Annual Report of the New Zealand Broadcasting Board* (Wellington, 1934), 2. *Third Annual Report of the New Zealand Broadcasting Board* (Wellington, 1935), 2.

[66] *Empire Broadcasting Service*, 26.

[67] C. H. Chomley, 'The Empire Station and Australasia', *World Radio*, 11 November 1932. Similarly, see Brown to Reith, 17 November 1931, WAC, E1/341/1.

[68] Conder to Sir Ian Fraser, 27 November 1934, NAA, Sydney, SP1558/2, box 81, 'BBC Programmes for ABC—Shortwave broadcast from Daventry—Part 1—1932–38'.

[69] Emma Robertson, '"I Get a Real Kick out of Big Ben": BBC Versions of Britishness on the Empire and General Overseas Service, 1932–1948', *Historical Journal of Film, Radio and Television*, 28/4 (October 2008), 459–73.

[70] Sir E. J. Harding to Graves, 10 April 1933 and New Zealand P&T Department, 'Empire Broadcasting Service', *c.* May 1933, UKNA, DO35/198/2. Only in Western Australia could Empire Service listeners hear live music, in transmissions aimed at the Far East, produced during normal BBC working hours. See Basil Kirke, 'Observations on British Empire Programmes—Indian Zone', *c.* January 1933, WAC, E1/315/2.

During 1933 the Empire Department's programme budget was increased somewhat, and over the next two years it began to offer commentaries on major sporting and ceremonial events: rugby matches, cricket test matches (with accompanying talks), the English Football Association Cup Final, the Davis Cup and the Wimbledon Tennis Championship, the Derby and Grand National horse races, and the Oxford and Cambridge Boat Race; royal weddings and celebrations; speeches by well-known British politicians and public figures; and special programmes to mark Empire Day, Armistice Day, and Christmas Day. Particular emphasis was placed on events deemed of special interest to dominion listeners: memorial ceremonies to mark Anzac Day; celebrations on Australia Day, Dominion Day in New Zealand, and Union Day in South Africa; speeches by the dominion high commissioners and by dominion politicians visiting Britain; and coverage, for example, of the start of the 1934 London–Melbourne air race.[71]

The Empire Service's emphasis on ceremonial and sport reflected the extension overseas of methods already deployed by the BBC to stimulate national unity at home, and also by public broadcasters in the dominions.[72] The CRBC argued that there was a 'widespread and common interest on the part of Canadian people' in 'outstanding events of national, Empire, or international importance': 'a broadcast which simultaneously brings [listeners] into intimate touch with the event is accordingly a major unifying influence'.[73] For the BBC Empire Service, royal events offered a particularly important opportunity to build a sense of Britannic community. The crown remained the key constitutional connection between Britain and the dominions, and the royal family represented in human form more abstract ideas about a Britannic heritage and set of shared values. Broadcast speeches by members of the royal family, and royal occasions more generally, became a regular feature of Empire Service programming. The King's 1932 Christmas message to the empire set a precedent, and thereafter the monarch's voice was used each Christmas to establish an annual moment of imperial communion: 'by a million firesides his words echoed . . . throughout the British world'.[74] According to Reith, the King's Christmas message demonstrated that

> broadcasting is unique not only in its range but in its intimacy . . . For the British peoples, at a moment when national life in the constituent parts in the Empire is in a mood of vigorous expansion and independence, this new invention is the most timely of scientific gifts.

[71] *Second Annual Report NZBB*, 2. *Third Annual Report NZBB*, 2. *Fourth Annual Report of the New Zealand Broadcasting Board* (Wellington, 1936), 2. *Second Annual Report of the Australian Broadcasting Commission* (Sydney, 1934), 10–11. *Third Annual Report of the Australian Broadcasting Commission* (Sydney, 1935), 20–1. WAC, E1/315/2 and E1/315/3.

[72] Mark Pegg, *Broadcasting and Society, 1918–1939* (London, 1983), 191–220. David Cardiff and Paddy Scannell, 'Broadcasting and National Unity', in James Curran, Anthony Smith, and Pauline Wingate (eds), *Impacts and Influences: Essays on Media Power in the Twentieth Century* (London and New York, 1987), 153–60. Mike Huggins, 'BBC Radio and Sport 1922–39', *Contemporary British History*, 4/21 (November 2007), 491–515.

[73] *Annual Report of the Canadian Radio Broadcasting Commission for the Fiscal Year ended March 31, 1935* (Ottawa, 1936), 13.

[74] 'The World's Biggest Hook Up', *CHNS Year Book 1933* (Halifax, NS, 1933).

The *Ottawa Citizen* similarly argued that the King's speech gave the Christmas broadcasts a 'final touch of emotional significance'.[75]

Such claims about the intimate, emotional appeal of radio, and its special ability to unite the far-flung 'British peoples', foreshadowed arguments formulated thirty years later by the Canadian theorist of communication, Marshall McLuhan. McLuhan's writings were both modish and oracular, and some have not aged well. Few would now be comfortable with the easy distinction he drew between 'natives' and 'civilised man'. Yet this categorization was central to his arguments about radio. For McLuhan, new technologies often submerged 'natives with floods of concepts for which nothing has prepared them'. But radio worked differently, bringing 'the oral and tribal ear-culture to the literate West'. Radio was the new 'tribal drum'. Britain and America had, McLuhan claimed, been sufficiently inoculated by literacy to resist the worst effects of the new medium. However, 'the more earthy and less visual European cultures were not immune to radio', which made it the medium of choice for fascists during the 1930s; a way to combine the appeal of the intimate and the communal, to reawaken savage tendencies in supposedly civilized peoples. McLuhan's is, at best, a dubious and radically simplified account of Europe's descent into the collective nightmare of the 1930s. Yet if there is something to his claim that radio revived 'the ancient experience of kinship webs', we might argue that far from being oblivious or immune to this phenomenon, broadcasters in the British world consciously hoped to use radio in precisely this way, to reinvigorate the bonds of Britannic sentiment.[76]

However, if we do accept this argument, then we also need to note that funding restrictions and technical limitations continued to undermine radio's ability to act as the British world's tribal drum. In 1934 and 1935 the ABC could, for example, offer only 'synthetic' coverage of Australia's cricket performances in England: in Australian studios ABC commentators produced dramatized accounts of test matches, using reports sent from the grounds by telegraph and cable.[77] In New Zealand during the month of June 1936 only one Empire Service programme was rebroadcast by NZBB stations. This was a speech given by Stanley Baldwin on the death of King George V: reception was only 'fairly good'.[78] Better results were obtained when the BBC provided coverage of the New Zealand runner Jack Lovelock's triumph at the 1936 Berlin Olympics. Harold Abrahams, the BBC's commentator and a close friend of Lovelock, provided a euphoric description that departed markedly from the BBC's usual tone: 'Come on Jack!'; 'Lovelock leads!'; 'My God he's done it!' However, a subsequent attempt to provide a talk by Lovelock from Britain for rebroadcasting in New Zealand was again blighted by poor reception.[79]

[75] Reith, 'The King's Broadcast', in *King George's Jubilee Trust: Official Programme of the Jubilee Procession*, LAC, W. E. G. Murray fonds, 1, 'BBC Broadcasts—Royalty'. CRBC booklet, *Canada Celebrates Christmas 1935*, Landry papers, 1.

[76] Marshall McLuhan, *Understanding Media: The Extensions of Man* (London, 1987 [1964]), 16–17, 50, 297–307.

[77] *Second Annual Report ABC*, 10–11. *Third Annual Report ABC*, 20–1.

[78] Hands to Ashbridge, 23 July 1936, WAC, E1/1097/2.

[79] 'Lovelock wins 1500 metres in Berlin, 1936', Radio New Zealand Sound Archive, T6238A. Hands to Ashbridge, 21 September 1936, WAC, E1/1093.

The focus on ceremony and sport, and the ban on rebroadcasting Reuters news, also meant that radio often failed to transmit detailed knowledge of political debate across the internal boundaries of the British world during this critical period in world affairs. Some speeches by prominent politicians were broadcast, but other voices were seldom allowed to tackle topics relating to British or international current affairs. The restrictions on controversial broadcasting had been lifted in Britain in 1928, but the Empire Service continued to shy away from politics. Partly, this reflected a judgement about what listeners in the dominions wanted. Although Conder claimed that Australian audiences would appreciate BBC commentaries about international affairs, Frost thought dominion listeners largely uninterested in the outside world.[80] Reluctance to cover domestic and international political controversies also reflected a fear among BBC officers that, due to the corporation's status as a national public body, broadcast news and comment might be interpreted overseas as a reflection of British official policy. Even the BBC's home listeners were thus provided with minimal coverage of foreign affairs, for fear that broadcasts would be overheard by overseas listeners, or relayed to them second-hand.[81] In external broadcasting, ceremonial and sports could be covered without much risk of controversy, the 'bodyline' tour notwithstanding.

If the response to the Empire Service among rebroadcasters in Australia, New Zealand, and South Africa was lukewarm, in Canada it seemed openly hostile. At the 1930 Imperial Conference, Steel had questioned most of the assumptions upon which the BBC was to base the Empire Service. When he visited London early in 1933, in his new capacity as a CRBC commissioner, the BBC wooed him, and transmission times to Canada were adapted in line with his suggestions. Some special programmes for Canadian listeners were also provided, such as a Dominion Day broadcast that included a message from the Canadian high commissioner in London, and coverage of a dinner at Grosvenor House attended by Bennett and the Prince of Wales.[82] However, according to Frost, Gladstone Murray's subsequent visit to Canada to advise on the reform of the CRBC generated 'considerable resentment on the part of the members and staff of the Commission who all suffer from an acute inferiority complex and are only too ready to react to the slightest suggestion of criticism or interference on our part'.[83] Not all of the slights were imagined. Murray claimed to have told Bennett that the BBC would not rebroadcast reciprocal Canadian programmes in Britain until the CRBC achieved 'some semblance of organisation and some reasonable guarantee of creditable production'.[84]

During Murray's visit Steel reiterated the points he had made in 1930: Canada needed a modest service of special programmes from Britain, transmitted by

[80] A. Mason to Graves, 27 August 1933, WAC, E1/315/2. Frost to Graves, 11 January 1933, E4/49.

[81] Siân Nicholas, '*War Report* (BBC 1944–5) and the Birth of the BBC War Correspondent' in Mark Connelly and David Welch (eds), *War and the Media: Reportage and Propaganda, 1900–2003* (London and New York, 2005), 142.

[82] Ashbridge to Steel, 27 January 1933; Steel to Ashbridge, 16 February 1933; Graves to Weir, 26 May 1933, all in WAC, E1/508.

[83] Frost, 'Empire Broadcasting in Canada', [*c.* March 1934], WAC, E1/492.

[84] Murray to Reith, 30 January 1934, WAC, E4/48.

radiotelephone, rather than any comprehensive, regular service from Daventry. Graves did agree to experiment with the radiotelephone, but Steel then lodged a further request that the BBC send him recordings of programmes on Blattnerphone tapes. The CRBC had invested heavily in this new, somewhat unreliable, recording technology.[85] Graves refused, claiming that the BBC did not have the necessary funds. Privately, he feared that complying with Steel's request would undermine both the short-wave Empire Service and BBC plans to sell recorded programmes on disc.[86]

Steel's fellow commissioner, Thomas Maher, meanwhile urged on the BBC his own pet sub-imperial project, which he thought 'very much worth while for the British Empire': the BBC should distribute European programmes in North America, and the CRBC distribute North American programmes in Europe. Later, the ABC would ask the BBC to relay American programmes over the Empire Service. The BBC refused to fall in with such requests: the Empire Service was a tool of imperial communication, not a means to serve the dominions with non-British programmes, no matter how good those programmes might be. When the German short-wave station at Zeesen agreed to relay American programmes to the ABC, Clark of the BBC Empire Department commented rather pityingly that 'to secure good-will Germany will do anything'. Frost similarly noted that the Germans would 'put themselves to any amount of trouble to assist people overseas'.[87] This certainly did not seem to be the policy of the BBC, to the increasing chagrin of the CRBC's commissioners. In January 1934 following Graves's refusal to comply with Steel's requests for regular radiotelephone relays or Blattnerphone recordings, the CRBC ceased rebroadcasting BBC programmes on its network.[88] By chance, Frost had planned a visit to Canada: when he met the CRBC commissioners, Steel became 'abusive', and claimed that a patronizing BBC had flagrantly ignored the wishes of the CRBC and the dominion delegates at the 1930 Imperial Conference, and had foisted inappropriate programmes on the dominions.[89]

Reith already had a poor opinion of the CRBC, which he believed was failing to protect Canadian listeners from domination by US stations. His mood was scarcely improved when, during a brief visit to Ottawa, he was kept waiting outside Charlesworth's office door, seemingly because the CRBC chairman did not know who Reith was.[90] Murray agreed that the CRBC was 'hopelessly incompetent'.[91] Frost

[85] Weir to BBC, 27 May 1933; memorandum by Graves, 9 June 1933; Steel to Murray, 26 August 1933, all in WAC, E1/508. Mary Vipond, 'The Canadian Radio Broadcasting Commission in the 1930s: How Canada's First Public Broadcaster Negotiated "Britishness"' in Phillip Buckner and R. Douglas Francis (eds), *Canada and the British World: Culture, Migration, and Identity* (Vancouver, 2006), 274–5.

[86] Memorandum by Graves, 28 August 1933 and BBC to Steel, 12 September 1933, WAC, E1/508.

[87] T. Maher to BBC, [15 October 1933] and BBC to Maher, 14 November 1933, WAC, E1/522/1. C. Moses to Clark, 12 November 1936 and Clark to Graves, 1 December 1936, WAC, E2/162. Frost to C. A. L. Cliffe, 4 June 1937, WAC, E1/353.

[88] Graves to A. G. C. Dawnay, 29 January 1934, WAC, E4/50.

[89] Frost to Graves, 20 January 1934, WAC, E4/50.

[90] Reith, 'Visit to Canada and the United States of America, November 1933', 25 November 1933, WAC, E15/178. Alan Thomas interview with Charles A. Bowman, 18 February 1960, UBCSCD, SP FC 3803 U54 N. 3:6.

[91] Murray to Reith, 30 January 1934, WAC, E4/48.

thought Charlesworth a poor manager, who habitually drank until his judgement was impaired. Frost also refused to take Maher (a French Canadian) seriously, due to his 'Latin temperament': 'his capacity for choosing female artistes, sometimes with complete disregard for their broadcasting ability, has given rise to considerable comment'. Steel neither drank nor smoked, was an acknowledged technical expert, and had Bennett's ear, yet he was 'obviously extremely impulsive' and sometimes indulged in 'considerable rudeness even at the expense of his personal dignity'.[92] On leaving Canada, Frost lamented that 'I don't think I have ever had a more unpleasant body of people to deal with.' He reckoned the CRBC would not last long.[93]

Frost also thought that the CRBC had made a mistake in adopting 'American principles of production and presentation' instead of British ones. No doubt this reflected a certain complacent confidence in the virtues of the BBC approach, and the limited time that Frost had available to gain an insight into broadcasting conditions in Canada. Nevertheless, he did appreciate one of the basic dilemmas facing broadcasters in Canada. Frost believed that instead of offering Canadian listeners something different, the CRBC, with its much slimmer resource base, had mistakenly entered into a hopeless contest by trying to produce programmes similar to those made by the powerful American networks. The CRBC's under-funded offerings were bound to seem second-rate. 'If only they had adopted the British style, the Canadian listener would at least have been assured of an alternative type of programme.' Frost also acknowledged that, as American networks were willing to sell their programmes in Canada at a loss in order to strengthen their position in the market, the BBC's attitude had seemed haughty in comparison. If the BBC wanted the CRBC to rebroadcast the Empire Service, Frost argued, programmes would have to be tailor-made, with more attention paid to local tastes and broadcasting techniques, to compensate for poor reception quality. To generate goodwill, the BBC would also have to accept reciprocal Canadian programmes for rebroadcast in Britain.[94]

Graves was willing to take some reciprocal material (perhaps one CRBC programme per month) and to produce more programmes aimed at Canada. But he maintained that the CRBC would have to pay in full for anything produced for its exclusive use.[95] Infuriated, Steel again suspended rebroadcasting of the Empire Service, even refusing to carry a BBC programme specially produced to mark the 400th anniversary of Cartier's landing at Quebec. The BBC lodged a complaint with the Canadian government, but intervention was not forthcoming.[96] Relations

[92] Frost to Graves, 20 January 1934, WAC, E4/50.

[93] Frost to V. H. Goldsmith, 30 December 1933 and Frost to R. Jardine Brown, 15 February 1934, WAC, E4/48.

[94] Frost, 'Empire Broadcasting in Canada', [*c.* March 1934], WAC, E1/492. See also Frost to Graves, 20 January 1934, E4/50.

[95] Graves to Dawnay, 29 January 1934, WAC, E4/50. Reith to Charlesworth, 8 February 1934, E1/522/1.

[96] Steel to Frost, 1 March 1934; Murray to Reith, 12 March 1934; Graves to C. D. Carpendale, 20 September 1934, all in WAC, E1/522/1. Graves to Charlesworth, 18 May 1934; Steel to Graves, 2 June 1934; Murray to Lt. Col. G. P. Vanier, 19 June 1934; Graves to Steel, 20 June 1934, all in E1/508.

were temporarily re-established between the BBC and the CRBC in early 1935, but broke down once again when Steel was informed that the CRBC could not record Empire Service programmes for delayed rebroadcast due to performing rights issues.[97] Partly due to this dysfunctional relationship, Canadian audiences seldom had the chance to hear the Empire Service. Direct listening was rare, as few Canadians possessed short-wave receiving sets, and private stations were seldom willing or able to rebroadcast Daventry outside CRBC network hours.[98]

The BBC did take some reciprocal programming from the dominions in this period, often relating to events of special interest to listeners in the UK or other parts of the empire. In October 1933, for example, the ABC arranged a radiotelephone feed of a talk by the Australian aviator Sir Charles Kingsford Smith, who had completed a seven-day solo flight from England to Western Australia.[99] More generally, however, BBC officers worried that programme standards in the dominions were low, and that material touching on controversial political subjects might raise complex issues of editorial responsibility. The BBC domestic and empire services thus normally confined dominion contributions to commemorative and ceremonial occasions. At such times, it was thought, the stimulus provided to Britannic sentiment would perhaps compensate for and justify inferior production or reception standards.

Christmas Day provided one such opportunity to arrange non-controversial transnational networked programmes, on an occasion when goodwill could largely be assumed. Seeking to exploit the ability of radio to address its audience on an intimate, personal level, in order better to harness and consolidate the bonds of imperial sentiment, the BBC moved away from using dry and formal messages from dignitaries, and instead attempted to give a sense of how 'ordinary' people celebrated Christmas around the empire and expressed their loyalty to the Crown.[100] Much emphasis was placed on the power of live actualities to create a genuine sense of interconnection. In 1933, an ambitious and successful attempt was made to depict Christmas celebrations from evocative places around the UK, with outside broadcasts from a coal mine in South Wales, a lighthouse on the English coast, and a Birmingham children's hospital. In 1934 the BBC encouraged overseas contributors to arrange similar live actualities. However, this was not an easy request for some of the participating broadcasting authorities, which were only just beginning to experiment with such techniques and to acquire appropriate equipment and expertise. The ABC planned a complex contribution featuring a toll-keeper on the recently opened Sydney Harbour Bridge, a lifesaver at Bondi Beach, an attendant in the Melbourne Botanical Gardens, a Queensland cattle drover, a glimpse of a service in Adelaide Cathedral, a worker on the Trans-Australia Railway, and a

[97] Murray to Plaunt, 9 February 1935, Plaunt fonds, 3/14. E. Bushnell to Murray, 30 January 1935, 12 February 1935, LAC, RG41, 390/20–9. Clark to Steel, 16 October 1935 and Steel to Clark, 13 November 1935, WAC, E1/522/2.

[98] Frost, 'Empire Broadcasting in Canada', [*c.* March 1934], WAC, E1/492.

[99] Graves to Conder, 18 October 1933, WAC, E1/315/2.

[100] Paddy Scannell and David Cardiff, *A Social History of British Broadcasting*, i, *1922–1939: Serving the Nation* (Oxford, 1991), 286.

Hobart fisherman.[101] However, listeners complained when these segments were passed off as live actualities: due to time zone differences, it was obvious that they had been recorded in advance (families were clearly not still enjoying the sun at Bondi at 1 a.m., Sydney time).[102] As Paddy Scannell has noted, 'sincerity' was a crucial feature of broadcasting, increasingly prized by producers and listeners alike. Simulation could ruin the impact of any radio performance.[103]

For the 1935 Christmas broadcast, the BBC did not encourage overseas broadcasters to try anything so ambitious. Instead, picking up on a phrase from the King's message of the previous year, and on the well-worn image of an imperial family of peoples, the BBC selected 'This Great Family' as the theme for the programme, which would feature 'genuine actualities' from 'typical homes of different kinds' in the UK, India, and the dominions.[104] After the broadcast, one small newspaper in Nova Scotia proclaimed that nothing could have been 'more binding to the Empire', but the Canadian high commissioner in London thought the CRBC's contribution 'hopelessly dull'.[105]

Empire Day provided another occasion when the BBC could overcome its reservations about standards: dominion contributions could hardly be any worse than the tedious UK ceremonies and speeches, and the 'unimaginative flag-waving' that the BBC found itself obliged to broadcast.[106] To make Empire Day more interesting, BBC officers began to ask dominion broadcasting authorities to contribute entire programmes, for UK and empire listeners. The ABC's production for 1934, *Australia will be Here* (a variation on the title of a song of the First World War, *Australia will be There*), started with the chimes of the General Post Office clock in Sydney, followed by a laughing kookaburra, a message of goodwill to the empire, and some conventionally romanticized settler boosterism: 'three dramatic cameos' representing the discovery of gold, the birth of Merino sheep farming, and the beginnings of wheat-growing.[107] The following year, a temporary *rapprochement* between the BBC and the CRBC made possible an Empire Day programme from Canada. The CRBC was given responsibility for scripting and producing the programme, and used it as an opportunity to emphasize past and present Canadian loyalty to the empire. In addition to a message from the Canadian prime minister, the programme included dramatized historical accounts of the arrival of the United Empire Loyalists at Fort Howe in 1783, and of the Canadian Clementina Fessenden's contribution to the establishment of Empire Day itself. As Mary Vipond

[101] Clark to Conder, 16 August 1934 and Conder to Graves, 16 October 1934, NAA, Sydney, SP1558/2, box 81, 'ABC Progs for BBC part 1—1933–39'.

[102] Conder to H. G. Horner, 24 April 1935, NAA, Sydney, SP1558/2, box 68, 'Jubilee George V 1935'. J. C. S. Macgregor to Bushnell, 27 January 1939, WAC, R47/768/1.

[103] Paddy Scannell, *Radio, Television and Modern Life: A Phenomenological Approach* (Oxford, 1996), 58–74.

[104] Clark to A. L. Holman, 13 September 1935, NAA, Sydney, SP1558/2, box 68, 'Christmas Day 1935'.

[105] *Canada Celebrates Christmas 1935*. Massey to Murray, 7 January 1936, Massey papers, 166/8.

[106] Quoted in John M. MacKenzie, 'In Touch with the Infinite: the BBC and the Empire, 1923–53' in MacKenzie (ed.), *Imperialism and Popular Culture* (Manchester, 1986), 173.

[107] 'Empire Day', *The Scotsman* (Edinburgh), 24 May 1934. 'Programmes—Empire Day', *The Times* (London), 24 May 1934.

has argued, the programme mixed an older view of empire, characterized by 'pompous patriotism and sentimentality', with the emerging inter-war ideal of an empire devoted to international peace and cooperation.[108] A dramatized section of the programme, representing the celebration of Empire Day in a contemporary Canadian home, emphasized not only the diasporic links of kinship that personally connected many Canadians with 'British' friends and family around the empire, but also the fact that the empire contained

> British citizens of all races, creeds and color... united in bonds of understanding and singleness of purpose [by] a power of brotherhood and purpose, governed by high ideals and a sincere desire to elevate mankind![109]

This reflected some of the broader ambiguities inherent in British identity: did it mark out an exclusive, 'racial' community, or an inclusive cultural one, capable of uniting people of diverse origins? In mixing old and new attitudes to empire, the CRBC may have judged accurately the diversity of inter-war opinion, in Britain, Canada, and the other dominions, particularly among those outside the intellectual elite. Similarly, a series aired in the CRBC's final months, *Within the Empire*, told 'the story of British subjects the world over', seeking to 'reflect the spirit, ideals and achievements of those who owe allegiance to the British Crown [and] stimulate a closer bond between the "one great family" which comprises the commonwealth of nations'.[110] The CRBC may have played a national, or even nationalist, role, but to the end it also remained happy to beat the tribal drum of Britishness.

REACTIONS TO BBC TRANSCRIPTIONS

At the root of the problems facing the BBC Empire Service in the early 1930s was the simple fact that rebroadcasters in the dominions did not want or need a full service from Britain. As one British civil servant had pithily remarked at the outset: 'It is really [more] that we [Britain] want to give the programme than that they [the dominions] want to receive it!'[111] The lack of financial support from either the British or dominion governments made things worse. BBC officers could not afford to produce the kinds of programmes that dominion rebroadcasters did want, even if they had the inclination to tailor their offerings to dominion requirements. Potentially, the situation was different with transcriptions (recorded programmes on disc). There were clear signs that dominion broadcasters wanted the BBC to provide certain types of high-quality recorded programmes. These might feature artists or deploy expertise not available locally, and balance the flow of recorded

[108] Vipond, 'The CRBC in the 1930s', 280–2. See also MacKenzie, 'In Touch with the Infinite', 186–7.

[109] 'Canadian Radio Broadcasting Commission—Empire Day Announcements', [1935], WAC, E1/493/1.

[110] CRBC press release, '"Within the Empire" Sweeps Round the World to tell of British People', LAC, RG41, 161/11-1/1.

[111] Minute, 26 February 1930, UKNA, DO35/198/2.

programmes now coming from the US. However, the BBC approached transcriptions with the same non-commercial, public-service principles in mind that had so influenced its short-wave broadcasting efforts. It soon became apparent that the BBC was reluctant to allow dominion broadcasters to dictate what sort of transcriptions it should provide. It denied that they were customers, and claimed instead that BBC transcriptions were provided as a public service. Subscribers to the service were meant to adjust their expectations accordingly. Thus, even when BBC officers had the resources to cater to the demands coming from dominion public broadcasters, they did not necessarily want to do so.

As with short-wave broadcasting, the BBC hesitated before entering the field of transcriptions. By 1931, American producers were already supplying recorded programmes to broadcasters in the US and overseas. In New Zealand the RBC had been criticized for importing this 'cheap American rubbish'. Complaints about the inferior standards of American popular culture were tinged with ideas about race. Reflecting the perceived influence of African American idiom on US transcriptions, and thus potentially on New Zealand popular culture, one parliamentarian singled out for opprobrium the broadcasting of 'American slang terms such as "Say bo," "gol darn," "O.K. baby," and "ma lil yello baby." Is it not fine to teach our children such rubbish?'[112] Another complained about 'second-class American jazz bands and third-grade American comedy': 'I think that New Zealand broadcasting should be essentially British, and British people should not [be], and are not, satisfied with entertainment by American performers.'[113] The New Zealand prime minister immediately cabled the BBC asking if recorded programmes could be provided to assist the new NZBB, which lodged a similar request with the BBC as soon as it had held its first meeting.[114] Meanwhile, in Australia, Brown argued that transcriptions, not short wave, were the future for empire broadcasting.[115]

Before joining the BBC's Empire Department, Malcolm Frost had been the managing director of Colonial Radio Programmes, a company established in 1931 to sell recordings of 'all-British' programmes on disc to colonial, dominion, and European stations. Frost had visited the US, Australia, New Zealand, South Africa, India, and Ceylon, and noted that American transcriptions enjoyed unchallenged dominance of British Empire markets. Britain had to compete: the real long-term issue at stake, he argued, was television. Production costs in television, once established, would be far greater than in sound. Individual broadcasters would find it hard to finance the production of programmes so, as in cinema, some sort of 'world wide distribution of programmes' would be required. 'America is out to control the entertainment industry of the world and with the advent of television . . . the American film industry will transfer their present activities to broadcasting.' If

[112] *NZPD*, 230, 2 November 1931, 667. Patrick Day, 'American Popular Culture and New Zealand Broadcasting: the Reception of Early Radio Serials', *Journal of Popular Culture*, 30/1 (summer 1996), 203–14.

[113] *NZPD*, 230, 6 November 1931, 805–7.

[114] G. W. Forbes to BBC, 6 November 1931, WAC, E5/16/1. Hands to Reith, 25 January 1932, E5/53.

[115] Atkinson, note of interview with Frost, 23 November 1931, WAC, E5/16/1.

Britain did not establish an overseas presence with radio transcriptions, it would inevitably lose the coming television battle.[116]

Senior BBC officers were reluctant to act. Selling transcriptions overseas might involve a degree of commercial competition with US and other producers, and a need to cater to audience demands, that short-wave had not required. Nevertheless, in July 1932 the BBC announced it would produce its own transcriptions for sale overseas, and took Frost into its employ.[117]

In response to earlier arguments about the need for characteristically 'British' material, an attempt was made to render the BBC transcriptions 'peculiarly national in character'. This was to be achieved by emphasizing the UK's regional diversity, reflecting one of the BBC's key strategies for projecting Britishness at home and overseas. The first series included programmes of traditional music from the Isle of Man ('listeners will find familiar echoes of its English, Scotch and Irish links'), England (*Cakes and Ale*), and Scotland. Frost had a Manx background, and Reith (a Scot) himself wrote the outline for the Scottish programme. Such material might not represent the high-cultural fruits of metropolitan British civilization, but, deriving from folk culture rather than the 'popular' mass cultures of the modern Americanized world, it still seemed to offer a means of discharging the BBC's Britannic duty. The series also included an historical drama about 'a typical Yorkshire manufacturing family', a biographical radio-play (*Christopher Wren*), a story for children (*Robin Hood and the Sorrowful Knight*), a tale told by the popular, pseudonymous broadcaster A. J. Alan (*My Adventure at Chislehurst*), and a programme of semi-topical commentaries (*It Happened Yesterday*). Initially, the BBC planned to avoid supplying popular entertainment programmes altogether, so as not to antagonize commercial broadcasters and gramophone companies. However, it was decided to include several musical comedy and variety offerings, drawing on popular British styles of humour. In one programme, the well-known comedian Tommy Handley thus introduced various 'typically British' music hall acts.[118]

By producing transcriptions, the BBC thus aimed to project British culture as well as to entertain listeners. Although money was to change hands, the Empire Department still tried to present the new initiative as a non-commercial public service. Revenue would cover costs, but not produce a profit. Users were asked to 'subscribe' in advance to entire batches of programmes (called 'series', despite the fact that each batch generally contained different types of programme, produced as stand-alone efforts), rather than to select and buy individual programmes. This would provide the stable, guaranteed revenues required to undertake production without imposing a significant financial burden on the BBC. It was also hoped that operating the service on this basis would minimize expectations that the BBC would follow the dictates of the market. Frost took sample discs for the first series with him on his tour, but, for future series, overseas broadcasters would be asked

[116] Frost to Goldsmith, 25 April 1932, WAC, E5/16/2.

[117] Transcript of Frost's BBC oral history interview, [*c.*1979], WAC, R143/46/1.

[118] Jardine Brown, note of conversation with Atkinson, 19 May 1932, WAC, E5/16/3. BBC brochure, 'Empire Recorded Programmes', November 1932, UKNA, DO35/198/2.

to make a commitment in advance, without the chance to audition material. Subscribers would, it was hoped, accept whatever the BBC provided, as BBC programmes would be self-evidently superior to local efforts. Frost was instructed not to 'give the impression that we are only going to give what we think fit', but his superiors thought that subscribers would nevertheless fall in with the BBC's choices.[119]

The African Broadcasting Company and the NZBB both snapped up the rights to the first series of transcriptions.[120] This was despite indications that New Zealand listeners wanted more lowbrow and entertaining programmes than the RBC had previously provided: the BBC transcriptions were not particularly helpful in this regard, and some were criticized for their 'educational' tone. Nevertheless, in general the series was acclaimed as 'a milestone in radio', and the NZBB was eager to buy more, particularly given the poor sound quality of Empire Service rebroadcasts.[121] Subsequently, it subscribed to every series of transcriptions that the BBC offered.[122]

However, the cost to subscribers of the transcriptions was higher than initially planned, because the CRBC refused to buy them, and others thus had to make up the shortfall. Given the increased cost of the recordings, the BBC's claims that it was running a non-commercial service began to seem somewhat implausible, and subscribers felt justified in urging their own particular requirements upon the BBC. The NZBB had to pay five times more per hour for the second series than it had anticipated. It thus asked if the third series could include more frequent appearances by famous speakers and performers, and more entertainment.[123]

ABC officers were meanwhile outraged when Frost offered them only 'first performance' rights to the first series, while selling the right to broadcast subsequent performances to private Australian stations. Playing on the BBC's argument that the transcriptions were provided as a non-commercial, imperially minded public service, Jones told Reith that if the BBC wanted to encourage 'the spreading of knowledge and the advancement of culture in a growing and not unimportant section of the British race', then in future it should sell transcriptions exclusively to the ABC. Graves lamented that 'these Australian people are very touchy', but reluctantly fell in with Jones's demands.[124]

Conder, the ABC's general manager, also argued that in future the ABC should not be obliged to subscribe to entire series of transcriptions, but should be allowed to pick what it wanted, leaving the rejects for the private stations. Graves thought the private stations would not tolerate this. He also feared that, without a guaranteed

[119] 'Notes with Regard to your World Tour', n.d., WAC, E4/49.

[120] Frost to Graves, 11 January 1933, WAC, E4/49. Record of telephone conversation between Graves and Frost, 23 May 1933, E4/46.

[121] Ian K. Mackay, *Broadcasting in New Zealand* (Wellington, 1953), 43. Hands to Atkinson, 4 October 1932, Archives New Zealand (ANZ), AAFK 890/19d. Frost to B. B. Chapman, 15 June 1933 and press cutting from Wellington *Dominion*, 5 June 1933, WAC, E4/46.

[122] Hands to Graves, 13 January 1934 and NZBB to BBC, 24 June 1935, WAC, E5/53.

[123] Hands to Clark, 24 July 1935, WAC, E5/53.

[124] Jones to Reith, 19 July 1933; Graves to Reith, 19 October 1933; Graves to Dawnay, 28 November 1933; Graves to Conder, 6 December 1933, all in WAC, E5/5/1.

Australian subscription, the entire transcription project would collapse for want of funds. He thus accepted an alternative proposal, that the ABC buy the entire second series, on condition that the BBC consult with ABC staff in London on the planned schedule of programmes, and consider any suggested modifications.[125] Demand from subscribers for a say in what was being produced was clearly growing.

Frost duly met with Arthur Mason (a long-expatriated Australian journalist and musician who had been appointed the ABC's London agent) and Ewart Chapple (the ABC's controller of programmes for New South Wales, who was visiting Britain at the time). Of the twenty-six hours of material planned for inclusion in the series, Frost was advised that only two programmes seemed unsuitable. One was a debate about the League of Nations: the speakers (Sir Norman Angell and Sir Charles Petrie) did not possess sufficient 'publicity value' in Australia. The other was a play about the First World War: Frost was told that this subject had now fallen out of favour with Australian audiences. Frost thought Mason out of touch with Australian conditions, and felt the BBC could placate the ABC with a minor gesture. Graves agreed to drop the play, but decided to keep the League of Nations programme, on the grounds that, although the ABC was paying the largest single subscription to the service, it should not be allowed to dictate what was included.[126]

Like the first series, the second contained a mix of light and traditional music, variety, and drama. The emphasis on the regional diversity of the UK was maintained, with a Scottish programme and *Irish Bulbuls* ('a light presentation of true Irish brogue and music') presumably drawing on material from Northern Ireland. More uplifting programmes were also provided, including a number of fifteen-minute talks on *The Causes of War*, with speakers including Dean Inge, Sir Norman Angell, Lord Beaverbrook, Aldous Huxley, Winston Churchill, G. D. H. Cole, and the psychoanalyst R. Money-Kyrle. These proved a great success with the NZBB.[127] Conder, however, was not satisfied with the discs when they began to reach Australia. He protested at the inclusion of an Armistice Day programme, emphasizing that the ABC only wanted the BBC to provide material 'which we are incapable of producing in this country'.[128] Conder also objected to the BBC's refusal to clear world performing rights for music and other incidental material included in the transcriptions: the BBC maintained that local rights could be dealt with easily and more cheaply by individual subscribers, but this meant that the recordings were accompanied by hidden costs to the end users.[129] Again, the BBC emphasized that this was not a commercial service: it was not a 'vendor', but was rather the 'co-ordinator' of the scheme. The implication was that purchasers of the

[125] Conder to Graves, 15 February 1934; Jones to Graves, 14 March 1934; Graves to Conder, 28 March and 19 April 1934; Conder to Graves, 24 April 1934, all in WAC, E5/5/2.

[126] Frost to Graves, 1 June 1934 and Graves to Frost, 2 June 1934, WAC, E5/5/2.

[127] BBC to NZBB, 28 September 1934, WAC E5/53. Clark to Mason, 2 May 1935, E5/5/3. Hands to Clark, 23 May 1935, ANZ, AAFK 890/20b.

[128] Conder to Graves, 30 October 1934, WAC, E5/5/2.

[129] Conder to Frost, 4 December 1934 and Clark to Mason, 12 December 1934, WAC, E5/5/2.

transcriptions should be less demanding.[130] However, the ABC was not so easily satisfied.

In June 1935 the BBC offered a third series, comprising fifteen hours of programming. General entertainment material was promised, with a smattering of talks. We know little about the precise content of the series, but a few select titles perhaps suggest the general tone and character: *Devonshire Tea*; *Victorian Melodies*; *Gert and Daisy Take a Zoo 'Oliday* (featuring the pioneering comedy act devised by Elsie and Doris Waters, in which cockney accents were presented realistically rather than in caricature); *Echoes of Ulster*; and *Ceremony of Guard Mounting*.[131] The ABC agreed to subscribe, at the urging of T. W. Bearup, ABC manager for Victoria. Bearup argued that the transcriptions were providing light musical performances by artists not available on commercial recordings, and had acted as 'a healthy antidote' to the American transcriptions carried by private stations. Drama productions had provided ABC producers with 'an insight into the methods used and the standard of proficiency attained by overseas broadcasters', and topical programmes and talks had 'enabled listeners to hear at first hand the point of view of many eminent people on questions of world-wide interest'.[132] Nevertheless, other ABC officers remained highly critical of the BBC's approach to making and selling transcriptions, as would soon become apparent.

CONCLUSIONS

The early 1930s could be described as an experimental period for empire broadcasting, or, less charitably, as a time when radio failed to provide effective channels of imperial mass communication. Officers at the BBC had some success in beating the tribal drum of Britishness, particularly with Christmas Day and Empire Day broadcasts that combined the appeal of royal and 'ordinary' voices to harness the particular strengths of radio as an intimate mass medium. However, they had little knowledge of reception conditions or listener tastes in the dominions, and did not establish good working relationships with their counterparts in public broadcasting authorities overseas.

These difficulties partly reflected the fact that public broadcasting was yet to be established in a stable form in the dominions. The BBC did not always know who its collaborators were, whether they knew what they were doing, or whether they would be around for long. Hesitant attempts to develop alternative relationships with private stations (for example through the proposed sale of transcriptions in Australia and Canada) met with little success.

[130] Clark to Holman, 10 October 1935, WAC, E5/5/3.

[131] D. A. Stride to Hands, 12 February, 2 April, and 25 June 1936, WAC, E5/53. Stride to Hands, 9 July and 16 October 1936, WAC, E5/54.

[132] BBC to ABC, 21 June 1935 and Holman to BBC, 2 July 1935, WAC, E5/5/3. Bearup to ABC commissioners, 28 June 1935 and Clark to Conder, 5 July 1935, NAA, Sydney, SP1558/2, box 80, 'BBC Recorded Programs 2 1935–40'.

A shortage of funds in Britain and the dominions, in the context of a continuing global economic depression, exacerbated these problems. Without a state subsidy, the Empire Service lacked the resources fully to meet the requirements of rebroadcasters when they did become apparent. Public radio authorities in the dominions similarly lacked the means to produce the sorts of reciprocal programmes that the BBC wanted, or to transmit them to the BBC without recourse to the expensive radiotelephone system. Meanwhile, with a limited number of subscribers, BBC transcriptions had to be sold at a high price, even if the Empire Department denied that they were a commercial, market-oriented product. The BBC's determination to operate overseas on the same public-service principles that informed its activities at home caused some serious problems.

As a result, BBC empire broadcasting achieved few of its goals in these years. While BBC transcriptions were partly intended to counter 'Americanization' in the dominions, and were explicitly 'British' in nature, they were not sold at all in Canada, where US cultural influence was greatest. Elsewhere, they provided only a weak inoculation of BBC culture. The second and third series of transcriptions each ran over a full year, but comprised only twenty-six and fifteen hours of programming respectively. Graves doubted whether they were worth the effort. Meanwhile, the Empire Service was not widely rebroadcast. Without sufficient funding, the BBC could neither run enough transmitters nor produce enough programmes to suit the divergent needs of direct listeners in the colonies and rebroadcasters in the dominions. Even Ashbridge, the champion of the 'whole hog' approach, came to recognize this.[133] But if it could not serve both colonies and dominions, then which should it prioritize? As Graves acknowledged, the BBC seemed to have 'veered from one extreme to the other' in trying to answer this question, and had as a result satisfied no-one.[134]

The BBC Empire Department did have some success in exploiting similarities between UK, Australian, and New Zealand tastes in sport, humour, and popular entertainment in this period. However, few Canadian listeners shared this common ground. Indeed, UK and Canadian tastes and programming styles were so different that the CRBC was reluctant to carry either Empire Service programmes or BBC transcriptions, for fear of compromising its own ability to counter US influence. If UK programmes proved unpopular, the CRBC would lose listeners to American stations, or to Canadian private stations carrying US material. Frost hoped that the CRBC would draw on British material to provide a distinctive alternative to US programming styles, but to Charlesworth, Maher, and Steel, this did not seem practicable. Though geographically closer to Britain, Canadian listeners thus heard even less BBC programming than did their contemporaries in Australia and New Zealand.

The failures of empire broadcasting in the early 1930s reflected technical and financial limitations, and cultural and geopolitical problems, compounded by unhelpful attitudes and organizational structures. The BBC's approach to empire

[133] Ashbridge to Reith, 2 May 1934, WAC, E4/7.
[134] Graves to Ashbridge, 3 March 1933, WAC, E4/6.

broadcasting was very much a centralized and centralizing one, involving the organization of all activities from London. Directives and programmes flowed outwards from the empire's media hub, with little reciprocity. Broadcasters in the dominions resented the BBC's inflexible and superior attitude, while BBC officers often thought their counterparts in the dominions (especially in Canada) unreasonable, ill-informed, and overly sensitive to criticism. Frost wrote of the constant need to 'make allowances for the inferiority complex of Dominion broadcasting organizations'.[135]

This left the BBC Empire Service vulnerable to charges of 'imperialism'. Run from London, its officers tended to view the exercise from the perspective of the centre and to expect dominion broadcasters to conform to British ideas about good programming standards and technical practice. The BBC financed the service itself, and thus often felt justified in making key decisions without consulting others. Unsurprisingly, Gerald Beadle's earlier suggestion for a more cooperative, less centralized form of empire broadcasting, in which costs and responsibilities would be shared, found their echo during this period. In 1934 L. R. C. Macfarlane, a member of the NZBB, visited Britain and proposed the creation of an 'Overseas Empire Central Broadcasting Bureau', a cooperative organization that would take over many of the functions of the centralized BBC Empire Department. Managed and funded by the broadcasting authorities of Britain, the dominions, and the colonies, the proposed bureau would be staffed by people from all around the empire. This would encourage broadcasting personnel to circulate continuously, and develop mutual knowledge and understanding. Such a body might be 'a real partnership within the English speaking world', running an empire broadcasting service, and eventually assuming responsibilities for imperial news and television services.[136] Nothing came of Macfarlane's proposals, but the ideas behind them would resurface repeatedly, in various guises, over the years ahead.

[135] Frost to Graves, 26 January 1934, WAC, E1/522/1.

[136] Frost to Atkinson, 7 September 1934, WAC, E5/53. L. R. C. Macfarlane, 'Overseas Empire Central Broadcasting Bureau', [*c.*1934], E1/1091/1.

3

Integration, 1935–39

As an institution, the British Broadcasting Corporation (BBC) had been designed to meet the domestic requirements of UK policymakers. However, the decisions that created and confirmed its public monopoly of broadcasting in the UK exerted a largely unintended, but highly significant, influence over the empire's wider broadcasting landscape. Private commercial broadcasting was left largely untouched as a field for imperial cooperation. Entrepreneurs in the dominions were denied UK partnership or inspiration even if they would have welcomed it, and thus inevitably looked to the US. Sometimes the BBC even actively obstructed attempts by private stations in the dominions to develop broadcasting links with the UK. When, for example, an Australian entrepreneur tried to use UK General Post Office (GPO) radiotelephone facilities to send programme material from Britain direct to private stations in Australia, the BBC persuaded the GPO not to cooperate, claiming that the corporation derived from its domestic monopoly 'the right to handle all programmes performed in this country, including those for radiation overseas'.[1]

The extension of the BBC's monopoly from domestic to external broadcasting meant that collaboration was largely confined to the interaction of public authorities around the British world. These authorities had been established in Canada, Australia, and New Zealand in the early 1930s, but not as clones of the BBC. They developed along markedly divergent lines, pursuing their own agendas that reflected their particular requirements and circumstances. They often lacked the resources required to act as effective imperial collaborators. Moreover, the centralizing attitude of the BBC, which viewed empire broadcasting as an activity largely to be carried out in and coordinated from Britain, soured relations, particularly with the Canadian Radio Broadcasting Commission (CRBC), but also to some extent with the Australian Broadcasting Commission (ABC).

During the four years that preceded the outbreak of the Second World War some significant changes occurred. First, the BBC began to play a more direct role in supporting the establishment and consolidation of public broadcasting authorities around the empire (and in Europe). Between 1933 and 1939 the BBC sent staff to Egypt, Palestine, Newfoundland, Jamaica, India, Canada, and South Africa, as well as Holland, Denmark, and Belgium.[2] To some extent, broadcasting structures

[1] J. B. Clark to C. G. Graves, 11 and 21 November 1935; 'Statement to [BBC] Board of Governors', *c.* 25 November 1935; S. Tallents to Reith, *c.* January 1936, all in BBC Written Archives Centre (WAC), E1/358.

[2] 'B.B.C. Staff Loaned or Transferred to Overseas Broadcasting Concerns', [*c.*1938], WAC, E4/51.

in Britain and the dominions continued to converge, facilitating collaboration. Second, broadcasting officers from India, the colonies, and the dominions also began to travel overseas more frequently, meeting each other, establishing contacts, and improving working relations. Third, the BBC put more money into empire broadcasting, started to target particular audiences in ways that catered to their perceived tastes and requirements, and tried to include a greater amount of reciprocal programming from the dominions on its home and overseas services. Fourth, although tensions remained, the approach of war encouraged many broadcasting officers to recognize the urgent necessity of collaboration.

BROADCASTING IN INDIA AND THE DEPENDENT COLONIES

Previously neglected, by the mid-1930s the need for the development of broadcasting in India and the dependent colonies had come to seem urgent. Prior to its collapse in 1930, the Indian Broadcasting Company had only established low-power stations at Bombay and Calcutta, serving a few thousand licensed listeners. The two stations were subsequently run, with few improvements, by the government of India. However, with the opening of the BBC's Empire Service, the number of licensed listeners in India doubled, providing a motive and resources for local expansion. Following discussions between John Reith and the India Office, in 1935 Lionel Fielden, Special Assistant to the BBC's Director of Talks, was seconded to the government of India as controller of broadcasting. With technical assistance from two senior BBC engineers, Fielden developed India's broadcasting infrastructure under the aegis of a rebranded state body, All India Radio (AIR). Although still badly under-resourced, AIR worked to establish countrywide coverage, using a mixture of medium- and short-wave transmitters. Daventry and other European short-wave stations were rebroadcast, but AIR also produced programmes of its own in English and vernacular Indian languages. Broadcasting began to be seen as something for Indians as well as for white expatriates. Indeed, by 1939 the majority of licence-holders were non-European, Indian music and talks in Indian languages predominated, and experiments had begun with group listening using communal village sets.[3]

The BBC also began to renew its appeals for support for its overseas activities from the British state, particularly for the development of broadcasting in the tropical colonies and in Britain's informal empire and League of Nations mandates. In these places, as in India, listening to the Empire Service was restricted to those with access to expensive short-wave receivers. If more people were to hear BBC programmes, then other means to generate audiences for the Empire Service

[3] 'B.B.C. Staff Loaned or Transferred to Overseas Broadcasting Concerns'. H. R. Luthra, *Indian Broadcasting* (New Delhi, 1986), 66–70, 82–8. Partha Sarathi Gupta, *Radio and the Raj, 1921–47* (Calcutta, 1995), 22–9. Lionel Fielden, *The Natural Bent* (London, 1960), 149–216. 'Broadcasting—the Community Set', *The Times* (London), 23 March 1937. *Report on the Progress of Broadcasting in India* (Delhi, 1940).

and BBC transcriptions would have to be provided. Early in 1934 the BBC asked the Colonial Office to help fund the construction of medium-wave stations in the dependent colonies to facilitate rebroadcasting, and rediffusion services to bring the Empire Service and transcriptions to the homes of urban subscribers (rediffusion had already been introduced at Port Stanley in the Falkland Islands and Freetown, Sierra Leone). The Colonial Office was sympathetic but pleaded poverty: it did however agree to relay information about rebroadcasting and rediffusion to the colonial governments.[4] The Colonial and Dominions Offices also sent out circulars asking for information about the size of audiences for foreign short-wave services.[5]

Crucially, however, despite this new interest in broadcasting in the dependent colonies, there was little attempt to introduce the BBC model of remote state control. As in Britain, competition in the sphere of private broadcasting was to be restricted, but instead of a BBC-style public broadcasting authority, two alternatives seemed more attractive to contemporary policymakers: monopolistic private broadcasting, under strict state regulation, or else a monopoly under direct state control.

The first option had been pursued in the early days of broadcasting in India, when Marconi had owned a majority share in the Indian Broadcasting Company. It was also followed in Egypt, a former British protectorate which, while recognized as a 'sovereign independent' country in 1922, remained under British control in certain key respects, most notably as an imperial military base and communications hub.[6] In 1931 the Egyptian government closed down existing private stations, and the following year signed a ten-year renewable contract with Marconi to construct and operate a radio station. Although Marconi had withdrawn from broadcasting in the UK, it remained a key supplier of equipment to the BBC and other broadcasters around the world, and maintained overseas affiliates that continued to operate radio and radiotelephone transmission facilities of their own. To run the new Egyptian State Broadcasting Service (ESBS) Marconi received 60 per cent of the listener licence fees collected by the state. From May 1934, the ESBS transmitted programmes in English, French, and Arabic. Within five years, the number of licensed sets in Egypt had risen to an impressive total of more than 86,000, comparable to the number issued in India, and reflecting Egypt's sizeable, cosmopolitan, and relatively prosperous urban community. The ESBS was not just for expatriates: many Arabs listened communally to sets installed in cafés. Nevertheless, the broadcaster was still clearly part of the superstructure of British informal empire. Many ESBS employees were British or European,

[4] Minutes of meeting of 12 February 1934, United Kingdom National Archives (UKNA), CO323/1277/7.

[5] Circular to colonial governors, 8 May 1935, UKNA, CO323/1338/6. Circular to colonial governors, 14 November 1935, enc. 'BBC—Introductory Memorandum on Broadcasting and the Colonial Empire', 25 October 1935, DO35/199/1.

[6] Glen Balfour-Paul, 'Britain's Informal Empire in the Middle East' in Judith M. Brown and Wm. Roger Louis (eds), *Oxford History of the British Empire*, iv, *The Twentieth Century*, 499.

the Cairo studio was built in mock-Tudor style, with a self-consciously 'English' interior, and the BBC hovered as a background presence.[7]

By early 1936 private organizations were running stations in Kenya, British Guiana, Malaya, Mauritius, Newfoundland, Fiji, and Papua; and rediffusion services in Malta, Gibraltar, Bridgetown (Barbados), and Port-of-Spain (Trinidad).[8] Colonial states supervised their operations and, to a degree, restricted their freedom to exploit radio's commercial possibilities. This reflected the broader inter-war tendency of colonial governments to allow entrepeneurs considerable scope (thus minimizing state intervention and expenditure) while regulating them in an attempt to protect the perceived interests of the colony's inhabitants.[9]

With the creation of AIR, India presented an alternative model, marked by a conscious turn towards direct state control. A similar path was chosen for the British Mandate of Palestine, where the Palestine Broadcasting Service (PBS) was created in the same year as AIR, and headed by a succession of seconded BBC officers. BBC West Regional Programme Director R. A. Rendall was appointed as the first manager of the PBS. He was succeeded in Palestine by Stephen Fry, formerly of BBC Outside Broadcasts, and later by Crawford McNair, formerly the deputy conductor of the BBC Northern Orchestra. As in the dependent colonies, PBS audiences were initially concentrated in urban areas. Only 5,900 radio listener licences had been issued in Palestine by mid-1935, mostly to expatriates and Jewish settlers. But, as in Egypt, communal listening in cafés was widespread and, as in India, the PBS began to provide experimental village listening facilities.[10] Direct state control of radio was similarly established in Southern Rhodesia, Ceylon, and Hong Kong, and state-controlled rediffusion services operated at Port Stanley, Freetown, Accra (Gold Coast), and Lagos (Nigeria).[11]

In early 1936 the Colonial Office established a Committee on Broadcasting Services in the Colonies, chaired by the Earl of Plymouth, and including four senior BBC representatives. Its report, issued the following year, recommended direct control of broadcasting by the colonial state, rather than either the licensing of private companies or the establishment of autonomous public broadcasting authorities along BBC lines. Modest systems would develop slowly (perhaps rediffusion first, then medium-wave broadcasting) and thus, it was argued, could be managed easily by colonial officials. Direct government control would guarantee that broadcasting would support rather than undermine the colonial state. The state could restrict rebroadcasting and rediffusion to British programmes. As most

[7] Douglas A. Boyd, *Broadcasting in the Arab World: A Survey of the Electronic Media in the Middle East* (Ames, Iowa, 1999, 3rd edn), 17–19. 'B.B.C. Staff Loaned or Transferred to Overseas Broadcasting Concerns'.

[8] 'An Analysis of Empire Broadcasting', January 1936 and memorandum by E. B. Bowyer, 27 January 1936, WAC, E4/9.

[9] Michael Havinden and David Meredith, *Colonialism and Development: Britain and its Tropical Colonies, 1850–1960* (London, 1993), 140–83.

[10] 'B.B.C. Staff Loaned or Transferred to Overseas Broadcasting Concerns'. Andrea L. Stanton, 'A Little Radio is a Dangerous Thing: State Broadcasting in Mandate Palestine, 1936–1949' (Ph.D. thesis, Columbia University, 2007).

[11] 'An Analysis of Empire Broadcasting', January 1936, WAC, E4/9.

colonial subjects could not afford short-wave receiving sets of their own, they would thus be isolated from 'objectionable wireless propaganda' from other countries. Meanwhile, communal listening, 'in circumstances psychologically favourable', especially to coverage of royal and other ceremonial occasions, might impress upon Britain's colonial subjects 'a more vivid realization of their connection with the Empire'. Programmes could be provided for non-European audiences as well as expatriates. Radio would thus promote 'both local and Imperial interests', acting as a tool of 'advanced administration' and basic education, particularly in the fields of public health and agriculture. All this justified the expenditure of colonial state revenues on local broadcasting.[12]

The BBC had been shaped according to the perceived characteristics and requirements of British parliamentary democracy. It would have been surprising if, during the 1930s, contemporaries had thought that the UK's broadcasting system could be exported wholesale to the dependent colonies, for few aimed to replicate there the broader set of British political institutions of which it was a part. According to the Simon Commission's report of 1930 on constitutional reform in India, the parliamentary system fitted Britain 'like a well-worn garment, but it does not follow that it will suit everybody'. Only after the Second World War did such attitudes alter, to match the rapidly changing political realities of decolonization. Until then, colonial government remained direct and paternalistic, and it seemed that broadcasting needed to follow a corresponding pattern. The BBC model was perceived as something for the distant future.[13]

BROADCASTING IN THE DOMINIONS

The BBC model was, however, deemed fit for export to the 'British' societies of the dominions. Indeed, in the eyes of senior BBC officers, the more closely public broadcasting authorities in the dominions resembled the BBC, the better. This was made clear when, towards the end of 1934, Reith spent over a month in South Africa, advising on broadcasting reform. In asking Reith to act in this capacity, the South African Union government probably knew in advance what he would recommend. Reith certainly did. Before he left England, he ascertained that

[12] *Interim Report of a Committee on Broadcasting Services in the Colonies* (1937, Colonial No. 139). Subsequently, the committee met on an occasional basis, to discuss broadcasting in different territories. See the committee's minutes, and working papers on Malaya, Mauritius, and British Guiana (WAC, E2/127/1) and *Broadcasting Services in the Colonies: First Supplement to the Interim Report of the Committee* (1939). To compare British colonial broadcasting policy with that of other European imperial powers, see for example Rebecca P. Scales, 'Subversive Sound: Transnational Radio, Arabic Recordings, and the Dangers of Listening in French Colonial Algeria, 1934–1939', *Comparative Studies in Society and History*, 52/2 (2010), 384–417; Rudolf Mrázek, *Engineers of Happy Land: Technology and Nationalism in a Colony* (Princeton, NJ, 2002); and Marcus Power, 'Aqui Lourenço Marques!! Radio Colonization and Cultural Identity in Colonial Mozambique, 1932–74', *Journal of Historical Geography*, 26/4 (2000), 605–28.

[13] A. F. Madden, '"Not for Export": the Westminster Model of Government and British Colonial Practice', in Norman Hillmer and Philip Wigley (eds), *The First British Commonwealth: Essays in Honour of Nicholas Mansergh* (London, 1980), quote at 20.

Schlesinger would be ready to place the African Broadcasting Company's private monopoly in public hands. He also asked the Dominions Office whether a royal charter could be used to create a South African equivalent of the BBC.[14]

In his subsequent report for the Union government, Reith argued that the African Broadcasting Company should be replaced by a public broadcasting monopoly under remote state control, capable of establishing a national network. Departures from the UK approach would be necessary. Given the size of the territory to be covered, direct state subsidies were required. Bilingual broadcasting would also be necessary (Reith recommended that Afrikaans and English programmes be carried side by side on the same network; however, as in Canada, two separate language services on different networks were ultimately created). Reflecting the BBC's growing emphasis on the importance of broadcasting to non-whites, Reith also recommended that services be introduced for 'Asiatics', and that programmes and communal listening facilities be provided for Africans in urban and rural areas. However, in other respects, Reith's recommendations derived directly from the BBC model. The proposed South African authority would be directed by a board of part-time governors, selected on non-partisan grounds, enjoying freedom from interference by politicians, and devolving responsibility for day-to-day operations to a chief executive officer. As in Britain, national networking would be accompanied by some regional programme variations, and controversial, religious, and schools broadcasting would all be developed. Programme policy would be 'framed in the assurance that a supply of good things well presented will create a demand for them'.[15] Although Schlesinger complained that Reith's report imposed a UK model unsuited to South African conditions, and although the report also raised difficult questions about funding and language policy, it was accepted by the South African parliament in 1936 as the basis for new legislation with only minor modifications. The assets of the African Broadcasting Company were taken over by a new public authority, the South African Broadcasting Corporation (SABC). The first network link, between Grahamstown and Johannesburg, was established in 1938.[16]

Reith's South African report became a work of reference for those seeking to adapt the BBC model to circumstances in the other dominions. It was certainly used by members of the Canadian Radio League, now resuscitated by Plaunt, as they urged the Canadian government to replace the CRBC with a public broadcasting authority under remote state control. Felix Greene (the BBC's new North American representative, and a cousin of the novelist Graham Greene) likewise

[14] J. C. W. Reith, *Into the Wind* (London, 1949), 196–205. Sir John Reith diaries, 31 July 1934, WAC, S60/5/4/1. The Dominions Office recommended an Act of the Union parliament instead of a charter. See Sir E. J. Harding to Reith, 22 August 1934, UKNA, DO35/201/6.

[15] *Union of South Africa—Department of Posts and Telegraphs—Report on Broadcasting Policy and Development by Sir J. C. W. Reith* (Pretoria, 1935). Graham Hayman and Ruth Tomaselli, 'Ideology and Technology in the Growth of South African Broadcasting, 1924–1971', 31–2, in Ruth Tomaselli, Keyan Tomaselli, and Johan Muller (eds), *Broadcasting in South Africa* (Bellville, 1989).

[16] W. H. Clark to J. H. Thomas, 22 March 1935, and Liesching to Thomas, 8 May 1935, both in UKNA, DO35/201/6.

drew on Reith's South African report when he advised the Canadian government on proposed broadcasting legislation.[17]

The creation of a new Canadian broadcasting authority, the Canadian Broadcasting Corporation (CBC) in 1936 was welcomed by the BBC. Like the BBC, the CBC was placed under remote state control, directed by a board of part-time governors, and its day-to-day affairs were administered by a chief executive officer (the general manager). However, significant divergences from the British approach were also apparent. Crucially, the CBC would not enjoy a domestic broadcasting monopoly. Private stations remained, and although some of them would be affiliated with the CBC's network, that network would be partially reliant for its funding on sales of advertising time. In Canada public broadcasting thus remained quasi-commercial.

Paradoxically, the subsequent appointment of the BBC's Gladstone Murray as general manager of the CBC offered further evidence of the limits of BBC influence over the Canadian broadcasting debate. By 1936 Murray had fallen out badly with his senior colleagues in Britain, who suspected him of having a drink problem, misusing his expense account and leaking confidential BBC information to the press. That April Murray was effectively sacked, but Reith agreed not to make his dismissal public, to avoid prejudicing his chances of the CBC job. In Canada Plaunt, largely unaware of Murray's fall from grace in London, worked to bring him to Ottawa: this involved convincing Mackenzie King and others that Murray was 'truly Canadian' and not an 'imperialist'. Murray, the former Rhodes scholar, meanwhile continued to urge on Reith and others at the BBC that his move to Canada would allow him to 'fulfil the intention of the Founder of the Rhodes Trust in terms of service to my own country in the British Commonwealth'.[18]

When Murray was subsequently offered the Canadian job, and accepted it, Felix Greene (who had opposed his candidacy) worried that the appointment would scupper any chance of future cooperation between the BBC and the CBC. Reith was not overly concerned, as he thought Murray would not last long.[19] However, to Greene's surprise, Murray proved willing to forgive and forget, and even to treat Greene (who he knew had intrigued against him) as a confidant and ally. From the outset Murray complained to Greene of the 'spurious nationalism' of the Canadians he was dealing with: 'A very different country, he said, to the one he left years ago.'[20] Although the BBC turned down Murray's request to have Greene seconded to the CBC for a few months to assist him, Greene did visit Canada on a regular

[17] Simon J. Potter, 'Britishness, the BBC, and the Birth of Canadian Public Broadcasting, 1928–1936' in Gene Allen and Daniel Robinson (eds), *Communicating in Canada's Past: Essays in Media History* (Toronto, 2009), 91–2. On Greene, see Jeremy Lewis, *Shades of Greene: One Generation of an English Family* (London, 2010), 181–93.

[18] Potter, 'Britishness', 93–7. 'Statement of Major W. E. Gladstone Murray to Board of Governors and Director-General', 29 April 1936, WAC, L2/154/1.

[19] F. Greene to Graves, 6 October 1936, WAC, E1/113/2. Reith diary, 23 September 1936, S60/5/4/4.

[20] Greene to Graves, 24 November 1936, WAC, E1/113/2.

basis to gather information and offer advice. However, he remained concerned, and repelled, by Murray's continued and very public heavy drinking.[21]

Murray seemed to make a conspicuous success of his early years as general manager, working with the governors' support to establish the CBC's national network, and to restrict private stations to localized 'community' broadcasting. The number and transmitting power of the CBC's stations was increased, as were network programme hours.[22] More generally, Murray attempted to bring broadcasting more closely in line with the British approach. Soon after his arrival in Canada, he distributed copies of the *Listener* (the BBC's weekly journal for the 'intelligent listener') to selected 'leaders of Canadian thought', to generate support in Canada for 'the serious side' of broadcasting.[23] In an article published in the *The Times*, he linked the CBC's uplifting public broadcasting ethos directly with that of the BBC, and with a broader imperial British ideal: 'The public service tradition of broadcasting, pioneered and firmly founded in the United Kingdom, spreads across the world, one hopes as a new safeguard of the civilization from which it emerged.'[24] While out of necessity the CBC carried a substantial amount of American commercial programming, Clark of the BBC's Empire Department thought that the CBC's long-term policy 'seems now to be firmly based on the public service conception of broadcasting', and that its strong regulatory influence over private stations might usefully be copied in Australia.[25] ABC officers had in fact already noted with interest the Canadian idea of giving the public authority a measure of control over programming on private stations.[26] In Canada, however, at least one critic of the CBC suggested that inspiration should flow in the other direction: broadcasting in Canada should copy the more clear-cut division of responsibility that characterized the Australian 'dual system', with a non-commercial public authority running the 'Cultural Network', leaving private stations free to compete with one another in the field of entertainment.[27]

How far did the ABC see itself as providing Australia's 'Cultural Network'? The ABC's Chairman, Charles Lloyd Jones, had already adopted a Reithian tone. His successor, William Cleary, chairman from 1934 to 1945, similarly argued that the ABC should help overcome the influence in Australia of 'the Philistines' and of American 'canned culture'. Only public broadcasting organizations 'devoted to national service and not to profit can properly be entrusted with the preservation

[21] Greene to Graves, 18 November 1936, WAC, E1/113/2. Greene to Graves, 4 January and 24 September 1937, E1/113/3.

[22] Knowlton Nash, *The Microphone Wars: A History of Triumph and Betrayal at the CBC* (Toronto, 1994), 141–7.

[23] Murray to W. L. M. King, 11 May 1937, Library and Archives Canada (LAC), William Lyon MacKenzie King papers, MG26JI, C-3728/239/205824.

[24] Murray, 'Broadcasting: Satisfying a Scattered Public', *The Times* (London), 15 May 1939.

[25] Clark to Graves, 16 November 1937, WAC, E1/491. Clark to Reith, 3 May 1938, WAC, E1/490.

[26] T. W. Bearup to C. Moses, 11 June 1936, National Archives of Australia (NAA), Sydney, SP1558/2, box 83, 'London Representative, 1938–59'.

[27] Press cutting, H. H. Stallsworthy, 'What Price Radio?', *Canadian Business*, March 1938, Library of the University of British Columbia Special Collections Division, Alan Plaunt fonds, 4/10.

and cultivation of those "finer things of life"'.[28] Cleary questioned the commitment of his general manager, Conder, to this uplifting vision of broadcasting, but it was allegations of financial impropriety that ultimately led to Conder's dismissal. Cleary filled the vacancy with his protégé, Charles Moses, a British-born former soldier, car salesman, and ABC announcer and sports commentator.[29] Moses, while not intellectually inclined, knew how to navigate the complexities of his new position. When asked whether he had any political leanings, he replied: 'None whatever... beyond the fact that I am proud to be British.'[30] While a supporter of the broadcasting of 'serious' music, Moses (with Cleary's backing) also introduced more light talks and dance music, to compete with the B-class stations. In the UK BBC policies were being modified in a similar way at the same time: both organizations began to offer 'light refreshment' as an enticement to listeners to consume 'more nutritious fare'.[31]

Reith and his senior colleagues at the BBC were well aware of the difficulties created for the ABC by commercial competition, and urged Australian policymakers to bring private broadcasters to heel. Australian Prime Minister Joseph Lyons suggested that Reith visit in an advisory capacity, but Attorney General Robert Menzies was less encouraging, claiming that the popularity of the B-class stations among listeners, and their political influence, was enough to render any crackdown impossible.[32] Towards the end of 1935 the Lyons government did seek to regulate the network-building activities of Amalgamated Wireless (Australasia) Ltd and Keith Murdoch's Herald media group. However, the commercial stations and their supporters fought back, calling for the postmaster-general's regulatory powers to be transferred to an independent radio commission (in the US, such a body had done little to limit entrepreneurship). As rumours circulated concerning Reith's possible visit to Australia, supporters of commercial broadcasting complained that any BBC adviser would be bound to be 'very much biased in favour of the British system of broadcasting'. Murdoch meanwhile sought to persuade Lyons that US-style private networks would mean better-quality programmes for B-class stations.[33] The government caved in, and watered down the regulations. Reith judged the situation 'impossible' and 'hopeless': 'I asked [Menzies] if they were ever going to put things right, & he said "No, we haven't the guts."'[34]

[28] Alan Thomas, *Broadcast and be Damned: The ABC's First Two Decades* (Carlton, Vic., 1980), 48. 'Why National Broadcasting? ABC Chairman's Talk at Legacy Club Luncheon', 8 August 1940, National Library of Australia (NLA), William James Cleary papers, MS5539, 'Speeches as Chairman ABC'.

[29] Thomas, *Broadcast and be Damned*, 50–2.

[30] Press cutting, 'Charles Moses, General Manager of the A.B.C. at Home', *Radio Pictorial of Australia*, 1 December 1935, Cleary papers, MS5539, 'News Cuttings—Australian Broadcasting Commission'.

[31] K. S. Inglis, *This is the ABC: The Australian Broadcasting Commission, 1932–1983* (Carlton Vic., 1983), 48.

[32] Reith diaries, 6 June and 15 July 1935, WAC, S60/5/4/2.

[33] Bridget Griffen-Foley, *Changing Stations: The Story of Australian Commercial Radio* (Sydney, 2009), 22–4. M. A. Frost, 'Broadcasting in Australia—Debate in the House of Representatives', 17 February 1936, WAC, E1/390. K. Murdoch to J. Lyons, 25 November 1935, NLA, Sir Keith Murdoch papers, MS2823, 1/1/4.

[34] Reith diaries, 11 June 1936, S60/5/4/4.

In New Zealand the Post and Telegraph (P&T) Department, and G. W. Forbes's right-leaning coalition government, seemed to have more guts. When Malcolm Frost visited in 1933, B-class stations were being starved of funds and strictly regulated, and he was told that the P&T Department hoped to eliminate them entirely within the next three years. Frost thought that the New Zealand Broadcasting Board (NZBB) was being conducted 'almost exactly on the same lines' as the BBC, even if programme standards left much to be desired.[35] In London that July, Reith and the BBC chairman discussed broadcasting with J. G. Coates, the New Zealand Minister of Finance.[36] Four months later, the New Zealand postmaster-general banned the sponsorship of programmes. The government had already begun to purchase insolvent B-class stations and close them down, and to transfer their plant to the NZBB. Broadcasting, it was announced, would now be managed as a

> national utility... used for information, educational, and entertainment purposes... In broadcasting we are following a safe guide in pursuing a similar policy to that of the British Broadcasting Corporation. This is looked upon as the best system in the world.[37]

Subsequently, the 1934/5 Broadcasting Amendment Act seemed to bring New Zealand even closer to the UK model. The NZBB gained greater autonomy from the P&T Department, and increased scope for controversial broadcasting. The NZBB was also granted new regulatory powers over the B-class stations, which were henceforth to be limited in number. Clark was hopeful that all this indicated 'a steady trend toward a public service monopoly'. Reith continued to urge Coates to 'get hold' of the B-class stations.[38]

However, Clark's optimism was misplaced, and Reith's warning belated. The B-class stations and their supporters were vocal in opposing attempts to impose the British model on New Zealand broadcasting: they claimed that the country was too small to fund radio through licence fees alone, and that the 'high cultural ideal' of the BBC was 'anathema to a large section of New Zealanders'.[39] The new Labour Prime Minister M. J. Savage, elected in 1936, was sympathetic to these arguments. However, Savage also wanted to use radio to counterbalance what he perceived to be the conservative, anti-Labour bias of the newspaper press. Savage thus moved to permit advertising, in order to fund more popular styles of radio, while at the same time taking all broadcasting under state control: a strange mix of populism and 'kiwi totalitarianism'.[40] Reith thought this 'deplorable': he sent the NZBB chairman a copy of his South African report, and also of the recently

[35] Frost to Graves, 17 June 1933, WAC, E4/50.

[36] Reith diaries, 20 July 1933, WAC, S60/5/3/2.

[37] *New Zealand Parliamentary Debates* (*NZPD*), 237, 10 November 1933, 146–7.

[38] Clark to Reith, 18 May 1935, WAC, E1/1091/1. Reith diaries, *c.*26 June 1935, S60/5/4/2. Patrick Day, *A History of Broadcasting in New Zealand*, i, *The Radio Years* (Auckland, 1994), 183–4, 195–7.

[39] Press cutting, 'Two Stations', *Wellington Post*, 22 November 1933 and unattributed press cutting, 4 December 1933, Alexander Turnbull Library, Ian Keith Mackay papers, box 3, scrapbook.

[40] Barry Gustafson, *From the Cradle to the Grave: A Biography of Michael Joseph Savage* (Auckland, 1986), 141, 193–7. R. J. Gregory, *Politics and Broadcasting: Before and Beyond the NZBC* (Palmerston North, 1985), 18. James Belich, *Paradise Reforged: A History of the New Zealanders from the 1880s to the year 2000* (Auckland, 2001), 295.

published report on British broadcasting of the Ullswater Committee, in the hope that these documents would provide evidence of the importance of remote state control.[41]

Labour abolished the NZBB, and transferred its responsibilities and assets to a new National Broadcasting Service (NBS). The state meanwhile purchased as many of the B-class stations as possible, and subsequently brought them together under the aegis of a National Commercial Broadcasting Service (NCBS). One service would be funded from licence-fee revenue, the other largely through advertisements. Both would be under direct state control, staffed by public servants.[42] In parliament, the new postmaster-general deployed the rhetoric of public service, but rejected the case for remote state control. Broadcasting, he argued, 'should be directly controlled by the Government for, and in the interests of, the people and... used not only for the entertainment of the people, but for enlightening them on matters of public interest and public welfare'. Opponents attacked the authoritarian implications of the legislation, and correctly challenged the postmaster-general's claim that the Ullswater Report had recommended direct state control for Britain. They emphasized that Canada was at that very moment moving closer to the BBC model of autonomy.[43] However, not everyone fetishized the British model. One supporter of the new system thought it a good thing that it was 'not "Daventry" or "Broadcasting House," or the "British Broadcasting Corporation," but New Zealand—"Here"'.[44]

Private broadcasting had thus been effectively eliminated in New Zealand, but public involvement took a very different form than in Britain or any of the other dominions: a duopoly, part-funded by advertising, wholly under direct state control, and still without any unified national networking. Reith did not approve, and told Savage as much when he met him in London in June 1937.[45] Nevertheless, the new NBS collaborated enthusiastically with the BBC, perhaps more effectively even than had the NZBB. This was due in part to the Reithian sympathies of its first director, James Shelley, an Englishman who had come to New Zealand in 1920 as professor of education at Canterbury University College, and had advised Labour on education issues before it came to power. Like Reith, Shelley believed that social progress in a democracy could only be achieved through firm guidance by experts. In broadcasting, this meant the setting of the highest possible standards. Shelley was also a firm believer in the crucial importance to New Zealand of its link with Britain.[46] In his first letter to Reith, Shelley welcomed the prospect of

[41] Reith to H. D. Vickery, 8 January 1936, WAC, E1/1095. Clark to Vickery, 16 March 1936, Archives New Zealand (ANZ), AAFK 890/20c. On the Ullswater Report see Asa Briggs, *The History of Broadcasting in the United Kingdom*, ii, *The Golden Age of Wireless* (Oxford, 1995 [1965]), 461–7.

[42] Day, *History of Broadcasting in New Zealand*, i, 216–21. Only two stations remained in private hands.

[43] *NZPD*, 245, 9 June 1936, 748–71 and 10 June 1936, 779–830.

[44] *NZPD*, 245, 9 June 1936, 762.

[45] Reith to Vickery, 1 July 1936, WAC, E1/1091/1. Reith diaries, 21 June 1937, S60/5/5/1.

[46] Day, *History of Broadcasting in New Zealand*, i, 219–21. Ian Carter, *Gadfly: The Life and Times of James Shelley* (Auckland, 1993), 59–61, 198–206, 231, 274. Gregory, *Politics and Broadcasting*, 27.

future cooperation with and assistance from the BBC, and explained how the Empire Service brought 'thrills to an imaginative Englishman' such as himself.[47]

Savage appointed Colin Scrimgeour Controller of the NCBS. Scrimgeour was a charismatic entrepreneur and socially conscious non-denominational religious broadcaster, the founder and former president of the New Zealand Federation of B-Station Proprietors, and a Labour partisan. He was mandated to focus on light entertainment, and began to promote Australian and American presentation styles and programme formats and buy in material from overseas, particularly from the US and Australia.[48] In 1939 he visited North America to investigate broadcasting first-hand. Scrimgeour argued that, just as the NBS was developing strong links with the ABC and the BBC, so it was appropriate for the NCBS to forge connections with commercial broadcasters in Australia, Canada, and the US.[49] In the 1930s commercial broadcasting clearly had transnational affinities of its own. Due to the BBC's domestic monopoly, they did not encompass the UK.

MOBILITY

If Reith and others sought to export the BBC model to the dominions, then during the mid-1930s their efforts had mixed results. The SABC was closely related to the BBC in terms of its constitution at least, but the ABC and the CBC were more distant cousins, and in New Zealand the government seemed to have left the family fold by deciding to impose direct state control. Yet despite these divergences, during the later 1930s public broadcasters were able to work more effectively with their counterparts around the British world than had earlier been the case. This was partly due to improvements in long-distance communications. An expanded radio telephone system was not just used for broadcasts: although expensive, it could allow discussion of urgent administrative matters. Airmail services were another novelty of the 1930s. People in the world of broadcasting were particularly aware of such innovations, and drew listeners' attention to them: the opening of the airmail service between Britain and Australia in December 1934 was, for example, marked by a special Empire Service broadcast.[50] Broadcasting officers could not yet routinely avail themselves of long-distance air travel, but ocean liners allowed senior administrative and programme staff to undertake long if infrequent overseas visits, and acquire first-hand knowledge of local conditions.

Three ABC commissioners were able to visit Britain on fact-finding missions during the 1930s, as was Professor W. J. Dakin, a member of the ABC's National Talks Advisory Committee, and Professor G. V. Portus, who investigated BBC educational broadcasting, publications, and talks programming on behalf of the ABC. Dakin and Portus both subjected British broadcasting practices to critical

[47] J. Shelley to Reith, 5 October 1936, WAC, E1/1097/2.
[48] Day, *History of Broadcasting in New Zealand*, i, 221–3, 238.
[49] C. Scrimgeour to M. J. Savage, 6 January 1939, ANZ, AAFL 563/4b 1/-/-.
[50] Clark to A. Mason, 6 December 1934, WAC, E1/315/3.

scrutiny, reporting on what could usefully be copied, and also noting mistakes made by the BBC that could be avoided in Australia.[51] Keith Barry meanwhile freelanced for the BBC in London before returning to Australia and joining the ABC. He wrote several articles presenting his observations on broadcasting in Britain, and on the common problems faced by the BBC and ABC.[52] Barry would eventually become ABC controller of programmes: when asked by a newspaper whether the ABC should give listeners what they wanted, he proffered 'the advice of Sir John Reith... "Give them something a little bit better than they think they want."'[53] An even more dedicated supporter of the Reithian philosophy of broadcasting was T. W. Bearup, ABC manager for Victoria. Bearup was Australia's 'high priest of the B.B.C. cult': he printed up Reith's dictum that broadcasting should bring 'to the greatest number of people as much as possible of contentment, of beauty and of wisdom', and hoped that 'many of my [Australian] colleagues would "paste it inside their hats"!'[54] In 1936 Bearup was sent to the US and Britain to arrange programme exchanges and visits to Australia by performing artists, to investigate overseas programming policies and studio designs, and to try 'to dispel the impression—as far as you honestly can!—that here we are a set of provincial boobies without a thought for the outer world except to beg favours from it'.[55] Bearup was amazed at the lavish resources available in the US, but disgusted by the commercialism that drove the American broadcasting industry.[56] His subsequent time in Britain formed the basis of a strong, enduring, and influential personal connection between the ABC and the BBC.

Visitors also came to the BBC from India and South Africa. In 1937 Lionel Fielden visited London on leave: that same year the first batch of Indian students also arrived at the BBC's staff training school. Eight members of AIR's programme staff were trained in Britain before the outbreak of war, including Fielden's protégé Zulfaqar Bokhari.[57] René S. Caprara, the director of the new SABC, also visited the BBC in 1937: in characteristically condescending terms, Reith judged him 'not up to his job by a long way'.[58]

At the CBC, Gladstone Murray worked to create strong personal connections between public broadcasting in Britain and Canada. Murray looked after his

[51] Simon J. Potter, 'The Colonization of the BBC: Diasporic Britons at BBC External Services, *c.*1932–1970' in Marie Gillespie and Alban Webb (eds), *Diasporas and Diplomacy: Cosmopolitan Contact Zones at the BBC World Service* (forthcoming).

[52] Press cutting, *Sunday Telegraph*, 16 November 1960 and press cutting, Keith Barry, 'Parallel Problems of the ABC and BBC', *Wireless Weekly*, 3 August 1934, National Film and Sound Archive, Keith Barry papers.

[53] Press cutting, 'Listeners' Varied Tastes', *Sydney Morning Herald*, 24 June 1937, Barry papers.

[54] Isabel Hodgson to Graves, 18 June 1936, WAC, E1/320. Bearup to J. Darling, 6 July 1965, NLA, Sir James Darling papers, MS7826, 33/3.

[55] Cleary to Bearup, 25 March 1936 and Hodgson to Bearup, 20 April 1936, NAA, Sydney, SP1558/2, box 83, 'London Representative, 1938–59'.

[56] Bearup to Moses, 20 and 26 April 1936, ibid.

[57] Reith diaries, 2 and 12 April and 21 May 1937, WAC, S60/5/5/1. *Report on the Progress of Broadcasting in India* (Delhi, 1940), 139.

[58] R. S. Caprara to C. A. L. Cliffe, 7 April 1937, WAC, E1/11. Reith diaries, 24 June 1937, S60/5/5/1.

friends, and he brought several of them with him over from the UK. Bob Bowman (Charles Bowman's son) had spent several years at the BBC as a sports commentator and news writer: he was now appointed CBC supervisor of special events. Later, Murray found a desk at the CBC for Richard 'Rex' Lambert, whose position at the BBC had become untenable following a high-profile legal case relating to his investigations into the paranormal activities of a talking mongoose.[59] Meanwhile, several CBC governors and officers visited Britain, examining BBC facilities and discussing common concerns.[60]

Murray also arranged an exchange of producers with the BBC. In October 1937 Laurence Gilliam, the producer of most of the BBC's 'round-the-empire' Christmas and royal broadcasts, went to Canada for six months to make feature and drama programmes for both organizations.[61] The following year Lance Sieveking left Britain for his own six-month visit to Canada, to experiment with extended outside broadcasts, and to produce a series of exchange programmes between Britain and Canada, focusing on shared democratic values.[62] The CBC meanwhile sent two of its own producers, George Taggart and Rooney Pelletier, to Britain.[63] Murray judged the experiment a success, and arranged a similar exchange of producers with the ABC.[64] Ernie Bushnell, CBC General Supervisor of Programmes, and one of Murray's trusted lieutenants, himself visited Britain in autumn 1938, travelling widely and meeting most of the BBC's senior officers. Like some previous Australian visitors, Bushnell cast a critical eye over BBC operations, and reported on their strengths and weaknesses. One of the most important lessons that the CBC could learn from the BBC, he felt, was a negative one: the need to entertain and innovate in order to placate the hostility of listeners and newspapers.[65] Dominion visitors to Britain often admired the BBC, but seldom thought it perfect.

THE EMPIRE SERVICE

Travelling in the opposite direction, J. B. Clark (who had succeeded C. G. Graves as Director of the Empire Service) undertook his own tour of the empire in 1937,

[59] Peter Stursberg, *Mister Broadcasting: The Ernie Bushnell Story* (Toronto, 1971), 134. Len Kuffert, 'R. S. Lambert and the Great Heritage: A Briton at the CBC', unpublished paper delivered at the British World Conference, University of Calgary, 12 July 2003.

[60] Reith diaries, 22 September 1937, WAC, S60/5/5/1. R. T. Bowman to J. C. S. Macgregor, 17 March 1938, E1/585/1. Murray to Graves, 16 April 1937; Murray to Macgregor, 11 May 1937; Cliffe to Murray, 1 July 1937; J. C. Stadler to L. W. Hayes, 4 August 1937, all in E1/585/2.

[61] 'B.B.C. Staff Loaned or Transferred to Overseas Broadcasting Concerns'. 'Notes on Canadian Broadcasting Corporation', 22 August 1946, WAC, E1/493/2.

[62] L. Sieveking, 'Some Provisional Notes on Proposed Work during Six Months', October 1938, Plaunt fonds, 16/3.

[63] Bushnell, 'Report on Visit to BBC, 1938', LAC, Ernest L. Bushnell papers, 14, 'CBC—Bushnell appointed as North American correspondent of BBC'.

[64] Murray to F. W. Ogilvie, 21 December 1938, WAC, E1/576/1. *Seventh Annual Report and Balance Sheet of the Australian Broadcasting Commission* (Sydney, 1939), 29–30.

[65] Bushnell, 'Report on Visit to BBC, 1938', Bushnell papers, 14, 'CBC—Bushnell appointed as North American correspondent of BBC'.

to meet overseas broadcasters and gauge listener responses to BBC programmes. The trip also allowed Clark to assess improvements in reception quality following the installation of three new BBC transmitters and a new aerial system at Daventry.[66] Neither direct state subsidies nor colonial or dominion financial contributions had been forthcoming to fund the Empire Service. Instead, to pay for technical and programming improvements, the GPO agreed to pass on a larger proportion of licence-fee revenue to the BBC. British licence-holders thus continued to foot the bill for overseas broadcasts that they never heard.[67]

With the expansion of transmitter facilities at Daventry, Empire Service operations increased from ten to over eighteen hours daily, and the British 'branding' of the service was improved: each transmission now closed with the national anthem; the chimes of Big Ben were broadcast; and listeners were regularly told: 'This is London calling'.[68] In October 1934 a special Empire Orchestra was established, to perform live outside normal working hours, and thus help overcome the problems caused by time zone differences. An Empire Music director was appointed, who worked to increase the number of performances by artists from the dominions (112 in 1936), without compromising the 'musical value' of programmes.[69] An Empire News editor was also taken on, with a staff of three sub-editors, to compile bulletins for empire audiences from agency news.[70] Programme schedules were distributed in weekly airmail pamphlets, and reprinting by overseas newspapers encouraged.[71] Malcolm Frost became Head of the Empire Press Section of the BBC's Information and Publications Branch, and worked to improve publicity for the Empire Service.[72]

In Canada, Murray and Bob Bowman sought to increase rebroadcasting of the Empire Service.[73] CBC schedules were inflexible due to the demands of advertisers and the US networks for regular slots, and the BBC did not permit recording of Empire Service programmes for delayed rebroadcast. Eventually, Murray and Bushnell persuaded the BBC to fall in with the CBC's own scheduling requirements, by placing special 'flagship' programmes for rebroadcasting in Canada in a regular, precisely timed thirty-minute slot.[74] Subsequently, towards the end of 1937 the BBC agreed to modify Empire Service schedules further, to facilitate mid-afternoon rebroadcasting in Canada. This provided the CBC with a cheap way to fill its expanding off-peak network hours.[75]

[66] Eric Foss to Clark, 19 April 1937 and Macgregor to Clark, 28 April 1937, WAC, E4/14.

[67] Asa Briggs, *The BBC: The First Fifty Years* (Oxford, 1985), 141.

[68] *The Empire Broadcasting Service* (London, 1937), 7–8.

[69] Foss to Hands, 5 October 1934, ANZ, AAFK 890/19e. Foss to Clark, 19 April 1937, WAC, E4/14.

[70] Gerard Mansell, *Let Truth Be Told: 50 Years of BBC External Broadcasting* (London, 1982), 29.

[71] Graves to Hands, 17 September 1934, ANZ, AAFK 890/19e. See 'BBC Empire Service—Programmes for All Transmissions', WAC publications collection.

[72] Clark to H. Charlesworth, 11 February 1935, LAC, RG41, 390/20-9/7.

[73] Murray, script for broadcast 'Chatting with Listeners', 21 June 1937, WAC, E1/543.

[74] R. T. Bowman to Macgregor, 16 June 1937, WAC, E1/522/3. Greene to Clark, 4 January 1937, E1/561/1.

[75] Greene to Graves, 17 September 1937, WAC, E1/113/3. Greene to Clark, 12 April 1938, E1/522/4. 'Notes on Canadian Broadcasting Corporation', 22 August 1946, E1/493/2.

Rebroadcasting of the Empire Service in Australia still remained limited, due to poor reception quality, and the fact that many of the programmes were designed for expatriate listeners in the dependent colonies, rather than dominion listeners used to quite different material. The ABC manager for Tasmania described some of them as 'an insult to our intelligence'.[76] It was eventually agreed to move the transmissions aimed at Australia to after the peak listening period, when reception conditions might be better, and when state managers would be willing to chance rebroadcasting an unreliable signal from Daventry. This had the intended results, and Moses subsequently asked for a regular service of special programmes for Australia to be rebroadcast in a *B.B.C. Hour* every Sunday, beginning in April 1938.[77] New Zealand stations meanwhile found that new transmission times and the new equipment at Daventry finally made it possible to rebroadcast speech reliably. NBS and NCBS stations began to carry a fair number of BBC talks, outside broadcasts, and sports commentaries.[78]

The Empire Service continued to focus on serving 'British whites', and to approach controversial broadcasting with caution. Sport and ceremonial remained staples. During the 1938 Australian cricket team's tour of England 'synthetic' coverage, reconstructed in local studios from telegraphic accounts, began to give way to live coverage from the grounds, provided via the Empire Service's new transmitters. The 'miracle of wireless' now made it possible 'for any Australian sitting in his home, to hear the click of the ball against the bat at Lords [*sic*], or the cheers of the crowd at the Oval'.[79] Even if Canadians did not want to hear short-waved cricket (or rugby) commentaries, audiences in Australia and New Zealand did, and all seemed interested in sports such as horse racing, golf, tennis, and boxing. Recent migrants from Britain to the dominions also demanded soccer coverage.[80] Overseas, as at home, Britishness (and, more specifically, Englishness) continued to be imagined and performed through sport.[81]

Royal events still offered a means to beat the Britannic tribal drum. On the death of King George V in January 1936, the CBC cancelled all broadcasts for the day. From then until the funeral, any 'lighter material which might offend public opinion' was omitted. A memorial speech by the British Prime Minister Stanley Baldwin, rebroadcast in the dominions, dwelt upon the King's love for the people

[76] L. R. Thomas to Moses, 6 January 1936 and Moses to Clark, 28 January 1936, NAA, Sydney, SP1558/2, box 81, 'BBC Programmes for ABC—Shortwave Broadcast from Daventry—Part 1—1932–38'.

[77] Clark to Greene, 9 December 1936, WAC, E1/522/3. Moses to Clark, 16 December 1937, NAA, Sydney, SP1558/2, box 81, 'BBC Programmes for ABC—Shortwave broadcast from Daventry—Part 1—1932–38'.

[78] Shelley to Ashbridge, 28 May 1937, WAC, E1/1093. *Annual Report of the National Broadcasting Service for the Twelve Months ended 31st March 1938* (Wellington, 1938), 3. *Annual Report of the National Broadcasting Services for the Twelve Months ending 31st March 1939* (Wellington, 1939), 3, 6.

[79] *Sixth Annual Report and Balance Sheet of the Australian Broadcasting Commission* (Sydney, 1938), 37.

[80] 'Notes on Canadian Broadcasting Corporation', 22 August 1946, WAC, E1/493/2. W. T. Conder to Graves, 8 May 1934, E1/315/3.

[81] Jack Williams, *Cricket and England: A Cultural and Social History of the Inter-war Years* (London, 2003 [1999]), 12–16.

of the empire, 'that family to whom he spoke last Christmas'.[82] Less than a year later, a very different royal occasion, the abdication of King Edward VIII, again demonstrated the ability of short wave to bring events of imperial significance home to distant audiences. The CBC was able to rebroadcast Edward's abdication message, and in Australia the crisis reportedly led to an unprecedented level of short-wave listening.[83]

The new transmitters at Daventry were worked to capacity for the subsequent coronation of King George VI in 1937.[84] The Empire Service carried coverage of the entire two and a half hour ceremony, and of the royal procession, and in the surrounding hours and days also offered a number of special talks, religious services, concerts and music programmes, and drama and variety productions, many of them taken from BBC domestic schedules. Public broadcasters in the dominions rebroadcast much of this: as on subsequent royal occasions, the barriers between UK and dominion listeners were partially dismantled, and audiences were encouraged to imagine themselves as members of a single imperial community, simultaneously partaking of the same programmes.[85] The dominion public broadcasters also prepared their own programmes to mark the coronation, and adapted the overall tone of their schedules to suit this solemn yet joyful British imperial event. During coronation week in New Zealand NBS stations carried only 'music and spoken matter by British composers, authors and/or musicians': particular care was taken to exclude German and American performers. Talks were to be 'British, in the broad sense of the term. They should be popular in subject and treatment, and non-British names should be avoided.'[86] On coronation day itself, each NBS station carried over four and a half hours of rebroadcasts. One NBS announcer claimed that the coronation represented 'the most historic day in the history of a nation... a symbol of our national heritage'.[87] It was the imperial British nation to which he referred, of which New Zealand was imagined to be a part.

Great events like the coronation were one thing, but what of the more day-to-day work of the Empire Service? During the later 1930s increases in the Empire Department's programme budget, and extensions to the operating hours of the Daventry transmitters, allowed for more diverse and impressive schedules, including talks, drama, features, children's programmes, religious broadcasts, and light entertainment.[88] But would rebroadcasters in the dominions be interested in this sort of material if they were already producing similar programmes of their own?

[82] LAC, RG41, 240/11-37-2-1. 'The Memory of King George V Broadcast on 21st January 1936 by the Prime Minister, the Rt. Hon. Stanley Baldwin, M.P.', RNZSA, HMV C 2819–21.

[83] 'Notes on Canadian Broadcasting Corporation', 22 August 1946, WAC, E1/493/2. *Radiola Rambles*, 31 December 1936, WAC, E4/17.

[84] 'Speech by Mr J. B. Clark at Inauguration of New Transmitter and Studio, Colombo, Ceylon', 6 June 1937, WAC, E4/21. For more on the coronation in Britain see Thomas Hajkowski, *The BBC and National Identity in Britain, 1922–53* (Manchester, 2010), 88–93.

[85] 'BBC Overseas Press Bulletin no. 23', 17 February 1937, LAC, RG41, 240/11-37-2. Head Office programme notification, 'Programme for Wednesday, May 12th', 11 March 1937, ANZ, AADL 564/17d 1/6/5. Bearup to Clark, 19 May 1937, WAC, E4/17.

[86] Head Office programme notifications, 8 and 13 April 1937, ANZ, AADL 564/17d 1/6/5.

[87] Script, 'Coronation Programme', 12 May 1937, ANZ, AADL 564/17d 1/6/5.

[88] *The Empire Broadcasting Service* (BBC, 1937, rev. edn), 8–12.

Would the BBC find it difficult, as it had in the past, to reconcile the varying tastes of listeners in the different parts of the British world?

Talks by prominent figures in British political and cultural life proved one of the least problematic elements of the Empire Service's extended schedules. Earlier fears about covering current affairs for overseas audiences had, perhaps, been unfounded. According to Murray, Canadians were eager to hear 'leaders of British thought and affairs', 'descriptions of home events', and commentaries by people like Vernon Bartlett, the former foreign correspondent for *The Times* and well-known broadcaster.[89] In Australia, by 1938 the ABC was regularly rebroadcasting Empire Service talks, with a BBC commentary on international affairs every Thursday evening, and two further talks every Sunday.[90] The NBS also rebroadcast BBC talks and commentaries: tight regulation and ministerial control meant that discussion of controversial topics was otherwise absent from the New Zealand airwaves. Programmes prepared by the BBC could be rebroadcast without the NBS having to take responsibility for their contents. The BBC also attempted to include in the Empire Service talks by visitors from the dominions, particularly in new series such as *Empire Exchange* and *Cards on the Table*.[91] Murray was even able to arrange for the Empire Service to carry reports from a new 'CBC London Correspondent', Graham Spry, who had moved to Britain.[92]

Surprisingly, music could prove more contentious. Poor reception often marred musical performances on short wave. During his world tour Clark found that as a result, among rebroadcasters, 'sopranos especially have many enemies'. In New Zealand reception quality remained poor enough to rule out the rebroadcasting of music entirely. And even when Empire Service listeners could hear music relatively clearly, they did not always like the type of music that was broadcast. Clark noted that in Ceylon, for example, 'crooning of any kind is loathed', but dance music was acceptable. Crooning was similarly 'taboo' at the CBC, which asked the BBC to provide only military band music and 'light music with a "British" theme': material that would not offend Canadian tastes and could not be sourced locally.[93] Comedy and variety programmes had to be handled with similar care. Audiences in Australia, New Zealand, and Britain shared a sense of humour, and New Zealanders were especially well attuned to the UK regional accents that pervaded BBC comedy. Canadian audiences, however, had become used to North American quick-fire comedy routines, and found the 'restrained suggestiveness' of BBC comedy offensive, and UK regional accents virtually unintelligible.[94]

[89] Greene, 'Empire Programs for the C.B.C.', 20 September 1937, WAC, E1/522/4.

[90] *Sixth Annual Report and Balance Sheet of the ABC*, 31.

[91] Macgregor to W. A. Steel, 4 September 1936, WAC, E1/508. Shelley to NBS station managers, 11 January 1938, ANZ, AADL 564/31a 1/6/3 part 1.

[92] Murray to A. Plaunt, 5 December 1938, Plaunt fonds, 3/23.

[93] Clark, 'Notes on Visit to Ceylon, 30th May/6th June 1937', WAC, E4/21. Greene, 'Empire Programs for the C.B.C.', 20 September 1937 and R. T. Bowman to Macgregor, 24 February 1938, E1/522/4. Press cutting, 'Work of the Broadcasting Corporation', *Port Arthur Evening News-Chronicle*, 10 February 1937, LAC, C. D. Howe fonds, 204/6. For BBC attitudes toward crooning, see Paddy Scannell, *Radio, Television and Modern Life: A Phenomenological Approach* (Oxford, 1996), 63–5.

[94] R. T. Bowman to Cliffe, 24 February 1938, WAC, E1/522/4.

Voices were, of course, extremely important on radio, and empire broadcasting made it clear that people around the British world spoke in many different ways. The BBC had already faced a similar problem at home, as accents varied considerably within the UK itself according to divisions of region and class. In response, it had tried to create an accent of its own, BBC English, 'eliminating the extreme varieties' (comedy was one of the few areas in which regional accents were allowed freer rein).[95] Despite or perhaps because of this, BBC announcers and speakers almost invariably seemed to belong to a different place or social class than the majority of their listeners. The social exclusivity of BBC English was readily apparent to audiences in the dominions. The CBC argued during the Second World War that listeners would protest if 'definitely southern English' voices were included among its own staff of announcers, and that affected 'Oxford' English, with its 'unconscious overtones . . . of superiority', was never welcome, even from BBC announcers.[96]

In Canada Gladstone Murray spoke of the need for a distinctive announcing style, 'as different from American as from English announcing'.[97] The ABC, in contrast, tended to appoint announcers with 'English or near-English voices', to some extent masking differences between British and Australian programmes.[98] BBC English also made some headway at the NZBB and NBS, although Frost complained that many New Zealand radio voices were 'not only uncultured but uneducated': 'refined' speech continued to be stigmatized as 'the mark of a "Cissy"'.[99] Listeners may have found the voices of the BBC, ABC, NZBB, and NBS equally remote and unappealing, a too-obvious reminder of how public broadcasters sought to improve their audiences. Announcers on commercial stations 'put zip, and pep and humour into their work': the tones of NZBB announcers suggested 'that they are funereally conducting one to the morgue'.[100]

Should Empire Service announcers abandon their linguistic civilizing mission, and instead talk like, and try to win over, their listeners? Editor of the Sydney *Wireless Weekly* G. L. Blunden argued in an article for *BBC Empire Broadcasting* (which had replaced *World Radio* as the journal of the Empire Service) that although the BBC's 'impersonal civil-service ideal' and 'Big Ben point of view' might strike a chord with expatriates, it had little appeal for the Australian-born. They would only listen if the voice was friendly, and the programmes entertaining. The collective voice of the Empire Service needed to be

[95] A. Lloyd James, 'Spoken English: the Art of Announcing', *The Times* (London), 14 August 1934. Michael Bailey, 'Rethinking Public Service Broadcasting: the Limits to Publicness', in Richard Butsch (ed.), *Media and Public Spheres* (Basingstoke, 2007), 101–3.

[96] 'Commonwealth Radio Conference—Notes on Agenda', [*c.*1945], LAC, RG41, 357/19-12 part 2.

[97] Murray, 'Chatting with Listeners', 21 June 1937, WAC, E1/543.

[98] Inglis, *This is the ABC*, 22, 70.

[99] Frost to Graves, 17 June 1933, WAC, E4/50. Alan Mulgan, *The Making of a New Zealander* (Wellington, 1958), 193. In Scotland, there were similar complaints about 'emasculated' BBC voices: see Briggs, *History of Broadcasting in the UK*, iv, *Sound and Vision* (Oxford, 1995 [1979]), 326.

[100] *NZPD*, 238, C. L. Carr, 2 August 1934, 881.

> informal in tone...neither vain nor imposing; it might almost, in the case of the Australian transmissions at least, be the voice of an exiled Australian. Without some such humanising of the Empire programmes I can see no future for them except as a...service for homesick Englishmen exclusively.[101]

Similarly, Charles E. Wheeler argued that although New Zealanders generally preferred the 'restrained, non-hysterical style' of the BBC to that of European short-wave stations, BBC accents often sounded 'a trifle affected'. Why not include an occasional 'dinkum Aussie' voice?[102] Such criticisms chimed with the CBC's recommendation that the BBC should provide characteristically 'British' programmes, but adapted to a North American broadcasting style and framework: fast-paced, vivid, 'human', with more personality, and all timed to suit CBC schedules.[103] The Empire Service had to be imperially British, not UK British.

Those running the service were based in London, and were mostly Englishmen, with little experience of life overseas. They believed they had a duty to serve British expatriates in the tropical colonies as well as settler audiences in the dominions. They thus found it difficult to accept the validity of the suggestions coming from the dominions. Noel Ashbridge, for example, thought that the typical Empire Service announcer already sounded too friendly, like a 'motor salesman' or 'hotel receptionist'. What was needed, Ashbridge thought, was a male voice of authority: announcers should be 'considerably older men with a good deal of experience of the world and public speaking behind them, with strong voices, and a strict avoidance of modern accent'.[104] For Ashbridge, announcers needed to sound like they came from the upper echelons of Britain's class hierarchy if the Empire Service was to command respect. This might indeed have struck a chord with expatriate listeners in the dependent colonies. However, BBC officers who had been posted overseas, and absorbed something of their surroundings, were more open to the alternative suggestions coming from the dominions. In particular, Greene and Fielden repeatedly criticized the Empire Department's unwillingness to prioritize the needs of rebroadcasters. In their view, the Empire Service was spoiled by heavy doses of sentiment aimed at expatriates, which alienated whites in the dominions, and Indians as well. Like Blunden and Wheeler, they argued for an infusion of talent, voices, and experience from the overseas empire.[105]

This debate reflected unresolved problems stemming from the way that the Empire Service had been established: a single service, lacking government subsidies,

[101] G. L. Blunden, 'The Empire Service in Australia', *BBC Empire Broadcasting*, 14–20 November 1937.

[102] Charles E. Wheeler, 'New Zealand and the Empire Service', *BBC Empire Broadcasting*, 12–18 December 1937.

[103] 'Chapter 5—An Empire Service', [*c.*1939], LAC, W. E. G. Murray fonds, 2, 'Murray, W.E.G. Autobiography'.

[104] Ashbridge, 'The Present Position of the Empire Service', 1 October 1937, WAC, E4/10.

[105] Greene to Graves, 16 January 1936, WAC, E1/113/2. L. Fielden to Graves, 14 July 1937 and Greene to C. D. Carpendale, 15 July 1937, E4/10. Mansell, *Let Truth Be Told*, 36. Simon J. Potter, 'Who Listened when London Called? Reactions to the BBC Empire Service in Canada, Australia and New Zealand, 1932–1939', *Historical Journal of Film, Radio and Television*, 28/4 (October 2008), 475–87, esp. 482–3.

could not satisfy the disparate requirements of dominion and colonial expatriate listeners. As Ashbridge admitted, in trying to satisfy these different audiences, 'there is a possibility that we are falling between two stools'. BBC officers disagreed as to which audience to prioritize. Graves argued that the Empire Service's primary duty was to lonely white listeners in the tropical colonies: 'fairly simple people, who appreciate a simple form of broadcast programme'. However, Frost doubted whether a substantial audience of direct listeners was actually available. In an estimated empire market for short-wave receiving sets of 230,000, only 60,000 had been sold, due to technical difficulties experienced by users in powering, operating, and maintaining equipment in the colonies. Frost also argued that in the colonies the BBC was merely 'preaching to the converted'. Expatriates already had strong sentimental attachments to Britain, which needed little in the way of reinforcement. In the dominions, by contrast, the Empire Service could play a crucial role in combating both 'nationalist' sentiment and 'ignorance...about the mother country'.[106] However, Clark refused to abandon expatriate listeners in the colonies: 'Canada is not the Empire'.[107]

RECIPROCAL PROGRAMMES

During the late 1930s the BBC did take a greater amount of reciprocal programming from the dominions. Sport provided some key opportunities in this regard. During the 1936/7 tour of Australia by the English cricket team, for example, the BBC arranged radiotelephone coverage of the test matches from ABC commentators (including ABC General Manager Charles Moses) and its own man, Alan Kippax, whose 'tremendous Australian twang...ripped through the still morning air' in England.[108] The ABC similarly provided the BBC with a summary of highlights of the 1938 Sydney Empire Games, described in one ABC broadcast as

> the greatest Athletic carnival and Empire pageant yet organised or witnessed in the Southern Hemisphere...full proof of the existence of the spirit of comradeship and co-operation between members of the British Nation the world over.[109]

Attempts were also made to use sporting events to link the dominions up with each other. Radiotelephone feeds between Australia and New Zealand for sports coverage became relatively commonplace, and in 1937 it was even planned to transmit short-wave commentaries on a Springboks rugby tour of Australia and New Zealand back to South Africa. However, on this occasion reception proved abysmal,

[106] Graves, 'Report on Empire Service', 20 August 1937; Ashbridge, 'The Present Position of the Empire Service', 1 October 1937; and Frost, 'The BBC Empire Broadcasting Service: a Report on Certain Public Relations Aspects', 21 October 1937, all in WAC, E4/10.

[107] Clark to Greene, 20 December 1938, WAC, E1/207/1.

[108] G. V. Portus to Moses, 5 February 1937, NAA, Sydney, SP1558/2, 602, 'Professor Portus—Overseas Visit—Reports, 1935–38', 30513744. See also box 81, 'Australian Programmes for BBC—Test Matches 1936–37'.

[109] 'Broadcast by James S. Eve', 1938 and Cliffe to Moses, 14 February 1938, NAA, Sydney, SP1558/2, 814, 'Empire Games—1937–38', 12171157.

reflecting the underdeveloped nature of the radiotelephone infrastructure, which still focused on connecting each dominion to Britain, rather than with each other.[110]

Reciprocal elements were also built into the 1937 coronation broadcasts. The BBC and CBC collaborated on a joint programme, allowing listeners in both countries to hear actuality coverage of the proclamation of the accession of King George VI in Britain and then in Canada.[111] To accompany the coronation itself, the BBC arranged a series of talks by dominion ministers on *Responsibilities of Empire*. It also set up a transnational network programme in the style of the empire Christmas broadcasts, called *The Empire's Homage*. This featured messages from the dominion prime ministers and various colonial officials, from 'representative' citizens in the dominions and colonies, and from the new King.[112] Some of this material was produced in London by the BBC, the rest in the dominions for transmission to Britain by radiotelephone. Difficulties arose with the South African government, however, as Prime Minister J. B. M. Hertzog, and Finance Minister Nicolaas Havenga, refused to participate. Hertzog liked neither 'Empire' nor 'homage', and was reluctant to be involved in a broadcast that emphasized the unity of the Commonwealth under one King (South African legislation had recently established the divisibility of the Crown, further increasing the dominion's autonomy). While Hertzog was eventually persuaded to participate, Havenga remained aloof.[113] Meanwhile, the SABC contribution to *The Empire's Homage* also proved problematic. As with the ABC's earlier Christmas at Bondi recording, the SABC tried to pass dramatization off as reality. It soon leaked out that a purported actuality, supposedly including a segment from Kruger National Park, was in fact a recording featuring actors in a Cape Town studio and a lion in Pretoria Zoo.[114] Radio was prized for its ability to transmit authentic, live sounds. Fakery seemed to undermine its efficacy as a tool of imperial unification.

During the later 1930s the BBC's Christmas link-ups were disrupted by the abdication of King Edward VIII, and King George VI's reluctance to broadcast due to his stammer. However, anniversaries in the dominions, including the 150th anniversary of the arrival of the First Fleet in Australia (1938), and the centenary of the annexation of New Zealand (1939), continued to offer opportunities and excuses for the BBC to place programmes about the dominions in its domestic services.[115] On such occasions, the BBC could justify carrying reciprocal programmes,

[110] NBS to ABC, 7 May and 10 August 1937; Moses to Shelley, [11 August 1937]; Shelley to Moses, 16 August 1937; H. O. Collett to Shelley, 30 November 1937, all in ANZ, AADL 564/90a 1/3/10.

[111] Murray to Graves, 15 December 1936, LAC, RG41, 390/20-9 part 7. Clark to Murray, 15 December 1936, WAC, R47/678.

[112] Cliffe to Shelley, 19 February 1937, ANZ, AADL 564/17d 1/6/5. Hajkowski, *BBC and National Identity*, 90–1.

[113] W. H. Clark to Harding, 22 and 30 March 1937, UKNA, DO35/536/2.

[114] Cliffe to Murray, 12 June 1937, LAC, RG41, 240/11-37-2 part 3. M. E. Antrobus to Reith, 17 June 1937, UKNA, DO35/817/6.

[115] Macgregor to Shelley, 17 April 1939, ANZ, AADL 564/31a 1/6/3 part 1. *The Scotsman* (Edinburgh), 26 January 1938.

even if it believed them to be of an inferior standard, on the grounds of Britannic sentiment or unusual public interest. Similarly, on Empire Day, entire programmes devised and produced by dominion public broadcasting authorities continued to be commissioned for use on both the BBC home and empire services. The NBS produced the 1938 programme, along the lines that it thought most appropriate: while the BBC asked for something 'as representative as possible of the life and activities of New Zealand', the NBS decided instead to rely on 'novelty' value to overcome its 'comparatively limited resources of artistic talent'. Thus, after Maori and Pakeha speakers welcomed 'our kinsmen at home', listeners were treated to Maori greetings and music, and dramatized sketches about New Zealand history and natural history, as well as a more prosaic discussion of the wool industry. A comedy sketch about the thermal resort of Rotorua displayed the familiarity of New Zealand scriptwriters and producers with distinctive UK regional accents, as a Lancastrian and a Scotsman joined in butchering the name of the Whakarewarewa geyser. Savage provided a mildly socialist message for listeners 'at Home', and an announcer then presented the farewells of 'the younger members of the Great Empire Family', who had 'indeed been honoured to entertain our elders [in the] Motherland'.[116] NBS producers were perhaps unaware that the programme was intended for the Empire Service as well as listeners in the UK: or was this an unconscious reflection of the fact that, for most New Zealanders, 'the empire' really meant the link between New Zealand and Britain? At any rate, as with the CRBC's earlier Empire Day production, the NBS allowed old and new ideas about empire to mix and merge, in a fashion acceptable to a range of contemporary audiences.

During the early 1930s the BBC had arranged several series of talks about the colonies and dominions for home listeners, using speakers available in Britain. Some were light and anecdotal in tone, others more political, but few were critical of Britain's colonial record.[117] Most of the talks were given by Britons who had some link with the colonies: Africans and Asians were not granted much of a voice. However, a few speakers from the dominions were heard, and in 1937 more participated in a series called *World Affairs*.[118] Subsequently, the BBC commissioned a series of talks recorded in three of the dominions, with the help of the CBC, ABC, and NBS: *Canada Speaks*, *Australia Speaks*, and *New Zealand Speaks*. These included a mixture of background information and topical discussion, intended to help overcome the perceived ignorance of listeners in Britain and overseas about 'the various parts of the British world', and to make them 'more Empire minded—not in the old "Imperialistic" sense, but with a view to promoting mutual understanding'.[119] The conviction that UK listeners knew little about the empire was

[116] Cliffe to Shelley, 30 December 1937 and Shelley to Cliffe, 22 March 1938, AADL 564/31a 1/6/3 part 1. 'A New Zealand Panorama: Random Sound Shots from New Zealand Life', Radio New Zealand Sound Archive (RNZSA), D444/DCDR444/DCDR479/D479.

[117] Hajkowski, *BBC and National Identity*, 28–35.

[118] Clark to Hands, 27 October 1936 and Cliffe to Shelley, 28 June 1937, ANZ, AADL 564/92f 1/2/5. Cliffe to C. A. Siepmann, 3 September 1937, WAC, E4/10.

[119] Clark to Moses, 18 August 1938, WAC, R46/32.

reiterated by J. C. S. Macgregor, who had become Empire Service Director following Clark's promotion to the position of overall Director of Overseas Services: 'people over here are rather apt to lie content in a state of abysmal ignorance about the Dominions, and they will have to be carefully handled, if they are to be induced to listen to six talks!'[120] In 1939 efforts to overcome this perceived ignorance continued, through a series entitled *Dominion Commentary*, designed to complement Raymond Gram Swing's successful talks on US affairs.[121]

Macgregor asked the NBS to arrange talks about New Zealand that would 'interest the man-in-the-street', and avoid boosterism, foreign policy issues, and 'partisan references to internal politics'.[122] The BBC had in the past received complaints from the Indian and New Zealand governments about talks, and while it had become more confident in telling overseas listeners about UK and international affairs, it remained reluctant to report local controversies raging in the dominions. BBC coverage of the empire thus largely remained focused on the 'safely picturesque'.[123] South African affairs proved particularly problematic. When the BBC asked for a talk on the 1938 South African general elections, Caprara was reluctant to oblige, claiming that South African politics were so 'explosive' that the SABC chose to ignore them entirely in its domestic service, and did not wish to be implicated in any on-air discussion of them overseas. Eventually, a talk by a South African newspaperman was arranged from the radiotelephone station at Cape Town. It is unclear whether the SABC provided any assistance.[124]

Meanwhile the BBC began to branch out and seek other types of programme material from the dominions, including entertainment programmes from the CBC and musical scores and drama scripts from the ABC.[125] The BBC also sought to assemble a stock of recorded features from the dominions, for periodic use on the home and empire services: again, the emphasis was on providing ignorant listeners with basic information, rather than covering controversial topics.[126] Features were a type of programme that had been developed in Britain and the US during the 1920s and 1930s. Dramatized where necessary, but normally with plenty of recorded actuality material, features aimed to communicate ideas and realities with

[120] Macgregor to Moses, 24 November 1938, WAC, R46/32.

[121] Macgregor to Moses, 28 March 1939, NAA, Sydney, SP1558/2, box 81, '"Dominion Commentary" Programme for BBC, 1939–40'.

[122] Macgregor to Shelley, 22 March 1939, ANZ, AADL 564/18b 1/8/1 part 1.

[123] Siân Nicholas, '"Brushing Up Your Empire": Dominion and Colonial Propaganda on the BBC's Home Services, 1939–45', 209, in Carl Bridge and Kent Fedorowich (eds), *The British World: Diaspora, Culture and Identity* (London, 2003), also published as a special issue of *Journal of Imperial and Commonwealth History*, 31/2 (May 2003). For the correspondence with the New Zealand high commissioner see WAC, R44/295.

[124] Caprara to Macgregor, 7 April 1938 and Clark to Caprara, 22 April 1938, WAC, E1/76/1.

[125] Greene to Murray, 2 June 1938, LAC, RG41, 378/20-3-5/2. Clark to Moses, 28 November 1935 and Moses to Cliffe, 29 September 1937, NAA, Sydney, SP1558/2, box 81, 'ABC Progs for BBC part 1—1933–39'.

[126] Macgregor to Shelley, 15 September and 1 December 1938; Shelley to Macgregor, 22 December 1938; Macgregor to Shelley, 10 January 1939, all in ANZ, AADL 564/18b 1/8/1 part 1. T. Lucas to Moses, 9 September 1938 and Moses to Clark, 27 September 1938, NAA, Sydney, SP1558/2, box 81, 'ABC Progs for BBC part 1—1933–39'. Memorandum by Lucas, 24 February 1939, NAA, Melbourne, B2111, PRE37, 342359.

'emotional and dramatic impact', playing to the strengths of the medium.[127] However, as with earlier attempts to secure live actuality material from the dominions, the BBC found that, in some of the dominions, feature programmes had yet to become an established part of the broadcasting repertoire. Australian writers and producers had limited experience with the genre, and severe criticism of what they produced for the BBC came from within the ABC's own Overseas Department. It judged the programmes pedestrian, marred by poor scripting, and indicative of an Australian inability to speak or write even 'reasonably well'. The ABC also encountered serious difficulties in making recordings of a satisfactory technical standard.[128] Meanwhile, the SABC's only proposed contribution to the BBC scheme proved even more problematic. Caprara agreed to provide a recorded feature about the construction of the *Voortrekker* monument at Pretoria and the centenary of the Great Trek, subjects dear to the hearts of Afrikaner nationalists. While the BBC was enthusiastic, the SABC soon found itself embroiled in a 'minor racial feud' between English- and Afrikaans-speaking script writers, and eventually abandoned the entire project.[129]

By the late 1930s, BBC officers probably felt that their requests were reaching the limits of technical possibilities in the dominions, and of perceived UK audience interest. In July 1939 the BBC informed the CBC that

> Talks from Canada were considered acceptable, but on the whole public opinion over here held little interest as yet in Dominions' affairs. The suggestion of an Eskimo feature programme to be put on at Christmas was, however, welcomed.[130]

Was this attitude comical, condescending, or simply realistic? It certainly contrasted markedly with the BBC's approach to reciprocal programmes from the US: by 1939 it was 'not unusual to have as many as three programmes broadcast to Great Britain from America in a single day'.[131] The belief that British listeners were more interested in the US than in the dominions would survive the war, and thrive thereafter.

TRANSCRIPTIONS

The BBC's mixed record of achievement in terms of developing audiences for rebroadcasts in the dominions, and securing effective supplies of reciprocal material, was replicated in its attempts to supply dominion broadcasters with transcriptions (recorded programmes on disc). Although the BBC continued to insist that it was

[127] Laurence Gilliam (ed.), *B.B.C. Features* (London, 1950), 9–11.

[128] Lucas to Moses, 20 and 26 February 1940, and Barry to Moses, 24 July 1940, NAA, Sydney, SP1558/2, box 81, 'ABC Progs for BBC part 2—1940–44'.

[129] Caprara to Macgregor, 1 October 1938; Macgregor to Caprara, 8 November 1938; Gladys Dickson to Macgregor, 18 March 1939; Caprara to Macgregor, 4 April 1939, all in WAC, E1/76/1.

[130] Minutes, 'Meeting to Discuss C.B.C./B.B.C. Co-operation', 21 July 1939, WAC, E1/522/5.

[131] Greene, 'Report on Activities of BBC New York Office during the Past Year', 6 July 1939, WAC, E1/113/3. Valeria Camporesi, *Mass Culture and National Traditions: The BBC and American Broadcasting, 1922–1954* (Fucecchio, 2000), 99–115.

issuing transcriptions as a non-commercial service, subscribers still had to pay substantial sums for the discs, and thus demanded that the product meet their own particular requirements. The ABC paid the most for the BBC transcriptions and, unsurprisingly, had taken the lead in calling for a tailor-made service, consisting only of material that it could not produce itself, and featuring famous British speakers and performers. By the time a fourth series was offered in 1936, resentment at the BBC's unresponsive attitude had reached boiling point. Eric Sholl, ABC programme executive in New South Wales, argued that the BBC's claims to be operating a non-commercial transcriptions service were disingenuous: '[the] attitude of an omniscient benefactor is a curious one to be adopted by an institution which has goods to sell'. Sholl thought that the BBC had used 'Empire sentiment for commercial purposes', milking the ABC to pay for a service that was aimed mainly at the dependent colonies: unless the BBC met the requirements of dominion broadcasters, then subscriptions should be withdrawn. Moses duly advised Clark that the ABC would not subscribe to another series *in toto*, but would only pay for those programmes it thought suitable.[132]

Fortuitously, Clark was available in Australia during his world tour to discuss the matter face to face with Moses and the commissioners, and persuaded them to take the entire series. However, the ABC still pressed for programmes that would more closely match its own requirements, as did the NBS in a simultaneous (possibly coordinated) approach. Clark felt that although these requirements did not match those of subscribers in India and the dependent colonies, they would have to be met; for without funding from Australia, the BBC simply could not afford to make transcriptions.[133] As with the BBC's early plans for the short-wave Empire Service, the idea of getting the dominions to subsidize services for the colonies had proved a delusion.

In Canada the CRBC had refused to buy BBC transcriptions. To the BBC's satisfaction, the CBC initially reversed this decision.[134] However, while this brought more revenue to pay for the scheme, it introduced yet another set of particular local demands. As with Empire Service programmes, the CBC pressed for transcriptions to be arranged in fifteen- or thirty-minute blocks, to suit its inflexible scheduling requirements. This increased the BBC's production costs.[135] Moreover, after taking the fourth series of transcriptions, the CBC gave notice that it would not subscribe to another. Its officers argued that BBC transcriptions compared unfavourably with those provided by American companies, in terms of price and content. The BBC service cost around £1,200 for fifteen hours of programmes, while NBC charged a flat annual fee of £240 for access to a library of 2,500 selections. The CBC deemed a good portion of the BBC recordings unsuitable, because they included incomprehensible regional accents, relied on background knowledge that Canadian listeners would not normally possess, were not timed precisely

[132] E. Sholl to Moses, 11 August and 7 October 1936, NAA, Sydney, SP1558/2, box 80, 'BBC Recorded Programs 2 1935–40'. Moses to Clark, 5 February 1937, WAC, E5/5/4.

[133] Shelley to Clark, 22 December 1936, WAC, E5/54 and Clark to Cliffe, 22 July 1937, E5/5/4.

[134] Greene to Graves, 4 January 1937, WAC, E1/113/3. Clark to Greene, 8 December 1936, E5/9.

[135] Greene to Clark, 2 April 1937 and Cliffe to Greene, 26 April 1937, WAC, E5/9.

enough to suit CBC scheduling requirements, or fell short of North American standards of 'showmanship'. The differences between CBC demands and those of subscribers in Australia and New Zealand were clear. The ABC and NBS wanted plenty of light entertainment, but this was precisely the field in which UK and North American programming most obviously diverged. The CBC meanwhile wanted more music, but this was one of the areas that was surplus to ABC and NBS requirements.[136] Macgregor lamented that 'The amount of common ground between all our subscribers seems in fact to be distinctly limited.'[137]

Clark thought that the BBC would only generate the revenues it needed to expand the transcription service and compete with American suppliers if, like them, it sold its wares in the US.[138] However, the tastes of this vast audience could be expected to diverge significantly from those of listeners in the dominions and colonies. If the BBC tried to sell its transcriptions more widely overseas, would the needs of the ABC and NBS eventually be sacrificed to those of broadcasters in Canada and the US, as those of listeners in India and the dependent colonies had previously been sacrificed to Australian and New Zealand requirements? Despairing, Clark thought it might be best to shut down the service entirely. However, without BBC transcriptions, American and even German recorded programmes would occupy the field uncontested.[139] Perhaps as a result of the lingering uncertainty over how best to respond to conflicting feedback from different subscribers, and over the future prospects for transcriptions, the fourth series ultimately seems to have conformed rather unimaginatively to the pattern set by its predecessor. Programmes of widely varying types were provided (with an emphasis, as before, on music and tradition from the different parts of the UK) along with a series of talks: the actor-manager H. Granville Barker on *Shakespeare*; the popular author Philip Guedalla on *Writing Biography*; the eminent physician Lord Horder on *National Health*; Harold Abrahams, the former Olympic athlete, on *Breaking Athletic Records*; and Harold Nicolson MP on *The Coronation of King George VI*.[140] The coming of war eventually prevented the series from running or, rather, meandering along its full course.

ANTICIPATING WAR

The Empire Service had not been established explicitly to serve any projected wartime requirements, although the BBC's proposals of 1929 had noted that it was 'not impossible to conceive of a situation in which deliberate recourse to

[136] Report by Greene, *c.* June 1939, WAC, E4/11. Greene to Clark, 6 June 1939; 'A Selection of Notes of some Programmes of the Empire Transcription Service', n.d.; G. A. Taggart to Macgregor, 7 June 1939, all in WAC, E5/9.

[137] Macgregor to Taggart, 26 June 1939, WAC, E5/9.

[138] Clark to Greene, 11 May 1939, WAC, E5/9.

[139] Frost, 'The BBC Empire Broadcasting Service: a Report on Certain Public Relations Aspects', 21 October 1937, WAC, E4/10.

[140] W. R. Baker to Shelley, 7 and 11 July, 9 September, and 8 and 18 November 1938; 4 and 20 January, 25 April, 19 May, and 18 August 1939; 9 February 1940, all in WAC, E5/54.

propaganda value... might become desirable'.[141] Inevitably, as the European crisis intensified, contemporaries began to consider using short wave to disseminate information and propaganda to audiences outside the empire, in languages other than English. The need to compete with German and Italian short-wave efforts became urgent.

Some at the Dominions Office and within the BBC worried that the reputation of the Empire Service would suffer in the dominions if it were to become associated with propaganda. However, the Foreign Office disagreed, and Reith believed that the BBC could broadcast in foreign languages without compromising its independence or tarnishing its image. Thus in 1937 the BBC began short-wave services in Arabic and Spanish, accepting financial support from the state to do this, but rejecting Foreign Office control over what was broadcast. A BBC Monitoring Service was also set up, to analyse propaganda broadcasts from other countries, and share this information with the government. However, foreign language broadcasts for empire audiences were not inaugurated until the eve of war. Services in French for Canada, and Afrikaans for South Africa would, it was believed, be rejected by listeners as the instrument of imperial political interests, generating suspicion and hostility while only winning tiny additional audiences.[142]

In the years before the outbreak of war, the BBC also began to consider how to improve reception of signals from Daventry in the colonies and dominions. Engineers had, by the late 1930s, developed a better understanding of some of the peculiarities of short-wave broadcasting, and become aware of the possibilities offered by relay stations, i.e. intermediate installations equipped with short-wave receivers and transmitters. These could be used to boost the power of signals en route to their final destination, to redirect signals so as to avoid zones of serious disturbance in the ionosphere, and to switch frequencies so that a signal that crossed the day–night boundary could be transmitted with minimal loss of quality. Short-wave transmitters were already being constructed at Singapore to facilitate British broadcasting to audiences in the Far East: the BBC recommended the construction of dedicated relay stations in Singapore and the West Indies, high-power short-wave transmitters in the dominions, and short-wave receiving stations in all parts of the empire. Rediffusion or medium-wave facilities needed to be built in Jamaica, Uganda, Tanganyika, and Northern Rhodesia, which still lacked any broadcasting infrastructure. The BBC emphasized that while it would take at least two years to carry out this work, the benefits would be enormous, in war or in peace.[143] However, little progress was made in implementing the BBC's suggestions.

[141] 'The British Broadcasting Corporation—Empire Broadcasting', November 1929, UKNA, DO35/198/2.

[142] Mansell, *Let Truth Be Told*, 40–54. Callum A. MacDonald, 'Radio Bari: Italian Wireless Propaganda in the Middle East and British Countermeasures, 1934–38', *Middle Eastern Studies*, 13/2 (May, 1977), 195–207. UKNA, DO35/199/1. J. E. Stephenson to W. H. Clark, 13 January 1939, DO35/818/2. 'French Bulletins in the Empire Service', n.d., WAC, E4/51.

[143] Minutes, 'Committee on Broadcasting Services in the Colonies', 19 December 1938, WAC, E2/127/1. 'Communication by Broadcasting in the Empire', [*c.*18 May 1939] and 'Communication by Broadcasting in the Empire—Second Report', 8 June 1939, E2/360/1.

News broadcasts meanwhile became an increasingly important feature of the Empire Service, paralleling the expansion of BBC news broadcasting for home audiences. During the Munich crisis the Empire Service devoted 50 per cent more time to news, and bulletins were allowed to run for up to thirty-five minutes.[144] Nevertheless, the BBC continued to rely heavily on agency news, and to approach coverage of foreign affairs gingerly, fearful of provoking international controversy or misunderstanding, or the ire of the British government. Even within the BBC, many felt that coverage of the Munich crisis was woefully inadequate, particularly compared with the on-the-spot, eyewitness accounts provided by the foreign correspondents of the US networks.[145] Reuters did temporarily suspend the embargo on the rebroadcasting of its news overseas during the crisis, for which the NBS was particularly grateful: it was otherwise reliant on bulletins compiled locally by the New Zealand government.[146] The ABC also rebroadcast Empire Service bulletins, but began to supplement BBC and news agency material with telegraphic reports from its own newly appointed London correspondent.[147] Although the CBC carried Empire Service news and commentary (talks by Henry Wickham Steed, the veteran foreign correspondent and editor, were particularly well received), Felix Greene was appalled by the relative poverty of BBC coverage of the European crisis, and argued that the BBC risked being driven out of Canada and other empire news markets by US broadcasters.[148]

More generally, how would public broadcasting authorities in Britain and the dominions cooperate in time of war? Close and effective cooperation between the CBC and the BBC during the 1939 royal tour of Canada and the US provided an indication of what might be achieved.[149] Might such ad hoc, bilateral cooperative arrangements be replaced by more formal, multilateral structures? From as early as 1936 the idea of an empire broadcasting conference to discuss collaboration was mooted, although the Colonial Office was unhappy at the prospect of state-controlled colonial broadcasting services mixing with the more independently minded public authorities of the British world.[150] In 1936 Reith also suggested that each dominion broadcasting authority might appoint a senior officer to act as their London representative, liaise with the BBC, and assist with the development

[144] Mansell, *Let Truth Be Told*, 59.

[145] Siân Nicholas, '*War Report* (BBC 1944–5) and the Birth of the BBC War Correspondent' in Mark Connelly and David Welch (eds), *War and the Media: Reportage and Propaganda, 1900–2003* (London and New York, 2005), 140–2. Alice Goldfarb Marquis, 'Written on the Wind: The Impact of Radio during the 1930s', *Journal of Contemporary History*, 19/3 (July, 1984), 385–415.

[146] Macgregor to Shelley, 23 September 1938 and Shelley to Macgregor, 13 October 1938, ANZ, AADL 564/31a 1/6/3 part 1.

[147] NAA, Sydney, SP613/1, 15/13/1 part 1, 3190211, 'News—London News Service'.

[148] R. T. Bowman to Macgregor, 24 February 1938, WAC, E1/522/4. Report by Greene, 8 June 1938, E1/113/3. Report by Greene, [*c.* June 1939], E4/11.

[149] Simon J. Potter, 'The BBC, the CBC, and the 1939 Royal Tour of Canada', *Cultural and Social History*, 3/4 (October 2006), 424–44; Mary Vipond, 'The Mass Media in Canadian History: The Empire Day Broadcast of 1939', *Journal of the Canadian Historical Association*, new series, 14 (2003), 1–21; Vipond, 'The Royal Tour of 1939 as a Media Event', *Canadian Journal of Communications*, 35 (2010), 149–72.

[150] Minutes of 24 February 1936, 11 May 1936, and 22 January 1937, UKNA, DO35/199/1.

of the Empire Service.[151] Bushnell and Charles Jennings of the CBC made a similar suggestion in 1938, but no action was taken before the outbreak of the war.[152]

Meanwhile there was some uncertainty about the role that the CBC, in particular, might play in any wartime collaborative effort. Alan Plaunt, still a CBC governor, had grown wary of Canadian entanglement with European affairs and possible involvement in a future war.[153] Plaunt wanted his views to be reflected in CBC programmes: radio debates should be organized to consider the merits of 'purely Canadian' defence as opposed to integrated imperial planning, and of freer trade with the US rather than continued commercial links with Britain.[154] Murray, still by all appearances a committed empire man, fundamentally disagreed with Plaunt's views. As conflict within the CBC intensified, the federal government itself sought information from the BBC about how it covered divergent perspectives on international affairs.[155] Murray was meanwhile drinking heavily again, and his hold of administrative detail began to slip. Plaunt launched an exhaustive investigation into CBC management and recruitment practices. In July 1939 Murray travelled to London, against Plaunt's wishes, ostensibly to attend a conference of 'Empire and American broadcasting officials'. No such conference was actually planned. Murray seemed to be fleeing Plaunt's investigation, and possibly trying to line up a new job for himself in the UK. On arrival in Britain, Murray publicly declared his intention to meet with the BBC and Lord Perth, who had been charged with preparing British official 'information' services in anticipation of war. Leonard Brockington, the CBC Chairman, had travelled to Britain with Murray, and had to reassure a furious Plaunt, by radiotelephone, that he would not commit the CBC to participate in any British propaganda scheme. He later reported that he had turned down an invitation to join 'a proposed Empire consultative committee' but, to Plaunt's dismay, also claimed that Murray had met with the British Cabinet.[156]

As far as can be ascertained, Brockington and Murray's discussions with the BBC centred on rebroadcasting. Murray repeated his earlier recommendations about the types of programmes Canadian listeners wanted, and the Empire Department agreed to fall in with them.[157] Discussions of plans for wartime were 'deferred', but it was agreed that Spry would be replaced as the CBC's London commentator, possibly because of his left-wing political views. Brockington also

[151] Report by Bearup, 17 December 1936, NAA, Sydney, SP1558/2, box 76, 'ABC Overseas Representative—policy London Office opening of (Bearup), 1945–46'.

[152] Bushnell, 'Report on Visit to BBC, 1938', Bushnell papers, 14, 'CBC—Bushnell appointed as North American correspondent of BBC'. C. Jennings to Murray, 6 July 1939, LAC, RG41, 378/20-3-5/2.

[153] See for example Plaunt to King, 25 August 1939, Plaunt fonds, 8/18.

[154] Plaunt to 'Frank', 15 November 1938, Plaunt fonds, 8/9.

[155] V. Massey to Tallents, 18 May 1938, WAC, E1/556. O. D. Skelton to Massey, 17 January 1939 and Ogilvie to Massey, 24 February 1939, University of Toronto Archives and Records Management, Vincent Massey papers, 166/9.

[156] Unattributed memo re: Murray, n.d., Plaunt fonds, 15/8. Plaunt, 'Trans-Atlantic Telephone Calls with Mr Brockington', 29 July 1939, 15/6.

[157] Minutes, 'Interchange of Programmes between C.B.C. and B.B.C.', 25 July 1939, WAC, E1/493/1.

suggested that, to cement the links established by the earlier exchange of producers between the CBC and the BBC, a continuous stream of Canadian personnel should be seconded to the BBC. These men could act in an 'advisory capacity', and help ensure that BBC programmes were properly 'adapted for Canadian minds'.[158]

CONCLUSIONS

By 1939 the future for broadcasting in the British Empire must have seemed reasonably clear to contemporaries. In Britain a monopoly public broadcasting authority would occupy a central coordinating position, but also operate as an 'Empire Exchange', bringing material from the dominions back to Britain, and transmitting it on to other parts of the empire via short wave.[159] In the dominions, autonomous public broadcasting authorities, controlled by the state remotely in Canada, Australia, and South Africa, and directly in New Zealand, would collaborate with the BBC. Private broadcasting would continue in Canada and Australia, but would play only a secondary role as a channel of imperial communication. Meanwhile, in the dependent colonies, broadcasting would either be run by private monopolies under the scrutiny of the colonial state, or would more likely follow the Indian model of direct state control. Where available, resources would be allocated to improving medium-wave and rediffusion services, and establishing short-wave transmitters and receivers in order to facilitate imperial communication. Possibly, more formal mechanisms for discussion and cooperation would be established, linking up the empire's assorted broadcasting authorities in some sort of overarching consultative body.

In this context, the improving relationship between public broadcasting authorities in Britain and the dominions seemed reassuring. Mobility had created personal connections and mutual understanding of the problems faced by public broadcasting authorities around the British world; some shared and others peculiar to local circumstance. The Empire Service seemed to be working more effectively, with better programmes and improved transmitters, although it still seemed unable satisfactorily to serve both colonial expatriate and dominion listeners with the scant resources available. At least programme makers and engineers now knew more about their work. Tendencies towards centralization remained, and tensions and miscommunication continued, but recognition that the world was sliding towards another global war sharpened minds and encouraged increased collaboration. During and after the war many of the cooperative initiatives that had been proposed during the 1930s would be implemented: dominion secondments to the BBC; conferences of broadcasting officials; short-wave broadcasting from the dominions; and dominion broadcasting representation in London. However, the idea

[158] Minutes, 'Meeting to Discuss C.B.C./B.B.C. Co-operation', 21 July 1939, WAC, E1/522/5.

[159] Clark, 'Speech at Inauguration of New Transmitter and Studio, Colombo', 6 June 1937, WAC, E4/21.

of a more formal, centralized structure to secure improved collaboration in the sphere of empire broadcasting would largely be abandoned. Less predictably, after the war the predominance of public broadcasting would be seriously undermined.

4

War, 1939–45

Between 1939 and 1945 mass communication by radio played a significant role in the war effort of the British world, by virtue of its ubiquity, its timeliness, and its direct and intimate appeal to listeners. During the 1930s Sir Stephen Tallents had emphasized how 'national projection' might be accomplished through the adoption of public relations methods by the state.[1] During the war such principles were applied more comprehensively to the task of disseminating information to the public. As contemporaries well understood, information and national projection shaded easily into propaganda, in Allied as well as in Axis countries. Wickham Steed described propaganda as the 'fifth arm' of the war effort, complementing the army, navy, air force, and the tools of economic warfare.[2] During the early years of the war, and particularly as its embryonic listener research operations grew, the British Broadcasting Corporation (BBC) also recognized the strategic importance of entertainment, to boost the morale of civilians and service personnel. At home, the BBC moved to provide more variety and light music programmes than ever before.[3] It is hard to know how listeners responded to the resultant blending of information, propaganda, and entertainment. The available evidence suggests that neither a healthy scepticism, nor periodic resistance to various propaganda messages led listeners in Britain entirely to reject the idea of a 'people's war' in which all made equal sacrifices regardless of divisions of class or gender, or of a 'people's empire' dedicated to peaceful progress towards mutual welfare and liberty.[4]

Radio offered a means to communicate with audiences abroad as well as at home: maintaining contact with troops stationed overseas; waging 'political warfare' against enemy combatants and civilians; and exchanging news and programmes with distant allies. During the war the patterns for broadcasting collaboration established during the 1930s continued to shape, and indeed to restrict, how radio operated to link up the component parts of the British world. However, some changes were noticeable. The superior, centralizing BBC attitudes, which had

[1] Stephen Tallents, *The Projection of England* (London, 1932).

[2] [Henry] Wickham Steed, *The Fifth Arm* (London, 1940). Some grouped economic and psychological warfare together as the 'fourth arm'. See Charles Cruickshank, *The Fourth Arm: Psychological Warfare, 1938–1945* (London, 1977).

[3] Siân Nicholas, *The Echo of War: Home Front Propaganda and the Wartime BBC, 1939–45* (Manchester, 1996), 40–65.

[4] Sonya O. Rose, *Which People's War? National Identity and Citizenship in Wartime Britain, 1939–1945* (Oxford, 2003). Wendy Webster, *Englishness and Empire, 1939–1965* (Oxford, 2005), 19–54. Similar points have been made about wartime cinema: see Anthony Aldgate and Jeffrey Richards, *Britain Can Take It: The British Cinema in the Second World War* (Edinburgh, 1994, 2nd edn.).

marred cooperation with dominion public broadcasters in the past, endured; although BBC officers did begin to see the problems that they caused, and tried to develop more reciprocal and decentralized structures of collaboration. As in the 1930s the BBC also still preferred to work with dominion public broadcasting authorities, rather than with private commercial stations and networks. Yet during the war it did seek more actively to cooperate with private broadcasters: this was necessary if BBC programmes were to be heard by more than the minority of Canadian and Australian listeners who regularly tuned to the public stations. In general, the BBC continued its attempts to extend the domestic principles of public broadcasting to its operations overseas, rejecting commercialism, and emphasizing its autonomy from the state. Nevertheless, it was obliged to broadcast more 'popular' types of programmes, and to accept an increasing degree of government interference and funding.

Less change was apparent when it came to broadcasting to and in the dependent colonies. BBC officers remained frustrated by the relatively low priority assigned to radio by the Colonial Office, but were themselves also partly responsible for the continued neglect of the imperial potential of radio in Africa, Asia, and the West Indies. Even in India, only around 80,000 licences had been issued to households by 1939, covering an estimated 250,000 listeners. All India Radio (AIR) suspended its village listening schemes for the duration of the war.[5] Broadcasting was deemed to be well established only in Palestine, Ceylon, and Hong Kong (where it was under government control), and in British Guiana, Kenya, and Fiji (where it was run by private companies: we might add Egypt, a crucial wartime territory of informal empire, to this list). Rediffusion services ('wired wireless') operated in some urban centres, mainly carrying programmes from BBC short-wave services. Elsewhere, public and private stations were deemed 'experimental' at best.[6] Some attempts were made to use radio to stimulate African and Asian support for the empire. Transcriptions were particularly significant in this regard, and short-wave services in Asian languages, and in English for East and West Africa, were also expanded.[7] Nevertheless, throughout the war years empire broadcasting remained largely something for the white British world. BBC officers were aware of this sin

[5] Sanjoy Bhattacharya, *Propaganda and Information in Eastern India, 1939–45: A Necessary Weapon of War* (Richmond, Surrey, 2001), 69.

[6] F. W. Ogilvie to Malcolm MacDonald, 26 February 1940; note by J. Megson, 23 August 1940; and J. B. Clark to C. G. Graves, 3 March 1942, all in BBC Written Archives Centre (WAC), E2/127/2. During the war, the SABC also operated a rediffusion service for urban African listeners in Zulu, Xhosa, and Sotho. See Graham Hayman and Ruth Tomaselli, 'Ideology and Technology in the Growth of South African Broadcasting' in Ruth Tomaselli, Keyan Tomaselli, and Johan Muller (eds), *Broadcasting in South Africa* (Bellville, 1989), 39.

[7] Gerard Mansell, *Let Truth Be Told: 50 Years of BBC External Broadcasting* (London, 1982), 122–3. Asa Briggs, *The History of Broadcasting in the United Kingdom*, iii, *The War of Words* (Oxford, 1995 [1970]), 466. Fay Gadsden, 'Wartime Propaganda in Kenya: the Kenya Information Office, 1939–1945', *International Journal of African Historical Studies*, 19/3 (1986), 401–20. Rosaleen Smyth, 'Britain's African Colonies and British Propaganda during the Second World War', *Journal of Imperial and Commonwealth History* (*JICH*), 14/1 (October 1985), 65–82. Kate Morris, *British Techniques of Public Relations and Propaganda for Mobilizing East and Central Africa during World War Two* (Lampeter, 2000), esp. 111–62.

of omission, but continued to focus their external broadcasting efforts on audiences in Europe, the dominions, and crucially, the English-speaking world's other core area: the US.

These broadcasting patterns reflected broader wartime contexts and requirements. The British Empire mounted an impressive war effort, loyally supported in some places, but elsewhere undermined by dissent. Australia and New Zealand followed Britain to war immediately. In South Africa General J. C. Smuts won parliamentary support with a narrow majority for a declaration of war, and replaced Hertzog as prime minister. Nevertheless, many Afrikaner nationalists continued to oppose participation in the war, and some even hoped for a Nazi victory, believing that Germany would look favourably upon their claims for local supremacy. In Canada Mackenzie King delayed the declaration of war for a week to allow parliamentary discussion and approval: a move calculated to underline Canadian autonomy and secure domestic unity. Nevertheless, in 1942 a referendum on conscription heightened divisions between English and French Canadians. Dependent colonies in Africa and Asia automatically went to war alongside Britain, but in India the viceroy's decision to declare war without consulting Indian politicians led the Indian National Congress to withdraw from participation in government in protest. Later, the Japanese attack on Pearl Harbor and American entry into the war brought the prospect of an eventual Allied victory. However, interim defeats at the hands of the Japanese, and the loss by February 1942 of Hong Kong, Malaya, and Singapore, subsequently revealed the weakness of the old conception of a single plan of imperial defence. Without Singapore, Australia and New Zealand seemed badly exposed to a Japanese invasion and, symbolically, in 1942 Australia adopted the Statute of Westminster, confirming its full legislative autonomy. Australian forces were redeployed to the Pacific theatre and increasingly subsumed into the American war machine.[8] Canada had already agreed to a Permanent Joint Board on Defence with the US, and war also brought closer North American economic integration. In India Gandhi dismissed the offer of post-war dominion status (effectively, independence) as 'a post-dated cheque on a crashing bank', and demanded unconditional and immediate British withdrawal, launching the abortive 'Quit India' campaign. This all-out programme of civil disobedience quickly degenerated into violence and repression. Yet despite all these setbacks, the dominions, India, and the colonies made a massive contribution to the Allied war effort, in terms of troops, labour, equipment and munitions, raw materials, and food. Conscription and compulsion were accompanied by a substantial element of voluntary collaboration.[9]

In Britain and the dominions participation in an imperial war effort was for many underpinned by a shared sense of British identity. Broadcasters may have consciously encouraged this, but it was also a pervasive background presence, largely taken for granted. When the Canadian journalist Matthew Halton described

[8] David Day, *The Great Betrayal: Britain, Australia & the Onset of the Pacific War, 1939–42* (North Ryde, NSW, 1988).

[9] Keith Jeffery, 'The Second World War', in Judith M. Brown and Wm. Roger Louis (eds), *Oxford History of the British Empire*, iv, *The Twentieth Century* (Oxford, 1999). Ashley Jackson, *The British Empire and the Second World War* (London, 2006).

for Canadian Broadcasting Corporation (CBC) listeners the aerial bombardment of London in 1940, he spoke of 'the flowering of the British spirit to face the darkest but grandest hour of our race'.[10] Five years later a Canadian airman, interviewed for a BBC victory programme, explained how '[w]e Canadians have had to work with men from all the British world. We must go on like this—understanding and working together... or else we may have to do this job again.'[11]

BROADCASTING AND THE WARTIME STATE

During the 1930s the creation of more collaborative, less centralized structures for empire broadcasting had been proposed on several occasions. War provided the incentive to act. Pioneering work was undertaken by individual broadcasting officers who moved around the British world, establishing new flows of information and programmes, spreading mutual knowledge, and creating a more reciprocal set of connections. Perhaps paradoxically, this less centralized set of relationships was created as public broadcasting authorities in Britain and the dominions came under more direct state control, as governments expanded their general remit over wartime life.

Well before the outbreak of war measures for imposing and coordinating state censorship had been drawn up in Britain and communicated around the empire. As early as October 1935 planning began for the establishment in Britain of a Ministry of Information (MoI) to take responsibility for information and propaganda policy in the event of war. After the outbreak of hostilities other more shadowy propaganda bodies were established in Britain. Overseas the Australian Department of Information (DoI), the Canadian Department of Public Information, the New Zealand Publicity Department, the South African Bureau of Information and Controller of Censorship, the government of India's Department of Information and Broadcasting, and the publicity departments of the Indian provincial governments, all played a key role in shaping mass media coverage of the war. These agencies remained in close touch with each other, forming a set of transnational connections that mirrored and influenced the structures of collaboration established by public broadcasting authorities.[12]

At the BBC, rapid wartime expansion was accompanied by closer relations with various branches of government. Between September 1939 and March 1943 (when

[10] Matthew Halton, CBC talk 'Britain the Citadel', 14 July 1940, Library and Archives Canada (LAC), Matthew Halton fonds, 17.

[11] 'Victory Programme', 8 August 1945, National Library of Australia (NLA), Chester and Edith Wilmot papers, MS8436, 3/21.

[12] 'CID paper no. 1446-B—Regulations for Censorship 1938', 15 July 1938, Hocken Library, John Thomas Paul papers, MS-0982/472. Philip M. Taylor, '"If War Should Come": Preparing the Fifth Arm for Total War, 1935–1939', *Journal of Contemporary History*, 16/1 (January 1981), 27–51. Cruickshank, *Fourth Arm*. Ian McLaine, *Ministry of Morale: Home Front Morale and the Ministry of Information in World War II* (London, 1979). Edward Louis Vickery, 'Telling Australia's Story to the World: the Department of Information, 1939–1950' (D.Phil. thesis, Australian National University, 2003).

a peak was reached) staff numbers rose from 4,889 to 11,663. Programme hours trebled, and total available transmitter power virtually quintupled. While the principle of remote state control was retained, the powers of the BBC's governors were reduced. Several BBC staff members were seconded to the MoI, and in 1941 MoI foreign and home advisers were placed inside the corporation. BBC responses to the state's increased interest in broadcasting were mixed. Smoother relations with officialdom were often welcomed, but any loss of autonomy was generally resisted. The wartime BBC directors-general found it difficult to strike a workable balance, and few lasted long. Reith had left the BBC in 1938. His successor, the former academic F. W. Ogilvie, resigned in 1942, to be replaced by a businessman, R. W. Foot. Graves acted as joint director-general until he retired in 1943. In March 1944 Foot was succeeded by William Haley, a former director of Reuters and joint managing director of the *Manchester Guardian* and *Evening News*.[13]

Similar turbulence characterized the upper echelons of public broadcasting in the dominions, as governments increased their powers of supervision and control. The Australian Broadcasting Commission (ABC) retained its statutory independence, but worked closely with the new DoI. Charles Moses volunteered for active military service, and William Cleary appointed T. W. Bearup acting general manager. Following the British imperial defeat at Singapore, pressure was placed on the ABC by John Curtin's federal Labor government to adhere to an 'Australia first' policy, and to emphasize the primacy of the Pacific War. Moses had escaped Singapore as it fell, and while recuperating in Australia used his influence with Robert Menzies (now leader of the opposition) to modify a new broadcasting act, so as to strengthen the general manager's authority relative to the ABC commissioners. Following another tour of duty with the Australian army, this time in New Guinea, Moses was recalled to the general manager's chair by the prime minister. Curtin directed Moses to broadcast more entertainment and fewer talks, and to adopt a more 'aggressive national policy'. Cleary, sick of government pressure, and compromised by Moses's behind-the-scenes negotiations with politicians, resigned in February 1945. Richard Boyer, one of the commissioners, agreed to become chairman, but insisted that Curtin first publicly reaffirm the ABC's independence.[14]

The CBC likewise remained under remote state control while entering into a close *de facto* relationship with the government. Bypassing the governors, Gladstone Murray increasingly dealt direct with the Liberal government, and loaded CBC schedules with speeches by ministers and programmes prepared in government departments. In the name of national unity, critical voices were suppressed and opposition political parties largely kept off air. CBC Chairman Leonard Brockington resigned, as did the Head of the Talks Department.[15] Alan Plaunt meanwhile continued his pre-war campaign against Murray, fearing the general manager's

[13] Briggs, *History of Broadcasting in the UK*, iii, 18, 22–39, 92, 149–51, 499–502.

[14] K. S. Inglis, *This is the ABC: The Australian Broadcasting Commission, 1932–1983* (Carlton, Vic., 1983), 78–127, quote at 111.

[15] Michael Nolan, *Foundations: Alan Plaunt and the Early Days of CBC Radio* (Toronto, 1986), 166–71.

links with the Canadian government, and the consequences of his British connections. Just before the war, Murray had travelled to London, apparently to discuss the CBC's wartime role with senior British officials. Subsequently, in New York Murray met with an old friend and fellow-Canadian, William Stephenson, chief liaison officer for British intelligence in North America. Contemporaries suspected the two of planning how the CBC might help prepare US public opinion for entry into the war. In one of the more bizarre contributions made by public broadcasting to the war effort, Murray apparently allowed Stephenson to use the CBC as cover for 'Camp X' (a training base for the British secret services) and 'Station M' (a document-manufacturing facility for covert operations and 'black' propaganda). It seems that Eric Maschwitz, former Head of BBC Variety, was among those posted to Station M.[16]

Murray meanwhile portrayed Plaunt as a frustrated 'anti-British' intellectual who wished to use radio to subvert the Canadian war effort.[17] Plaunt, already seriously ill with cancer, resigned in August 1940; he died the following year. However, Plaunt's friends carried on his vendetta against Murray. Following the emergence of damaging evidence of financial impropriety, and amidst heavy bouts of drinking, Murray was forced out. Echoes of his earlier fall from grace at the BBC were clear. However, neither of Murray's two immediate successors did much to restore the CBC's independence. Dr James Thomson, President of the University of Saskatchewan, lasted for only a year as general manager, while Dr Augustin Frigon (a former member of the Aird Commission) interpreted his position as that of a public servant, and scrupulously carried out instructions handed down by ministers.[18]

At the South African Broadcasting Corporation (SABC), concerns about the pro-Nazi sympathies of some Afrikaner and German-born officers meanwhile led to infighting between English- and Afrikaans-speaking staff. Ultimately, several officers were dismissed, and at least two were interned. These purges helped ensure the SABC's support for the Smuts government's pro-imperial political agenda. While in theory at least the corporation retained much of its autonomy, personnel were nominally integrated into the army signal corps for the duration of the war.[19]

[16] Eric Maschwitz, *No Chip on my Shoulder* (London, 1957), 144–5. H. Montgomery Hyde, *The Quiet Canadian: The Secret Service Story of Sir William Stephenson (Intrepid)* (London, 1962), 135. Peter Stursberg, *Mister Broadcasting: The Ernie Bushnell Story* (Toronto, 1971), 103. *British Security Coordination: The Secret History of British Intelligence in the Americas, 1940–45*, introduction by Nigel West (London, 1998), 104–9. Bill Macdonald, *The True Intrepid: Sir William Stephenson and the Unknown Agents* (Vancouver, 2001), 77. David Hunt, 'Stephenson, Sir William Samuel (1897–1989)', rev., *Oxford Dictionary of National Biography* (Oxford, 2004; online edn, January 2010) <http://www.oxforddnb.com/view/article/40177> accessed 6 July 2010.

[17] Memorandum, [*c.*1940?], LAC, Ernest L. Bushnell Papers, 1/10.

[18] Knowlton Nash, *The Microphone Wars: A History of Triumph and Betrayal at the CBC* (Toronto, 1994), 162–208. B. Claxton to L. Brockington, 7 November 1942, LAC, Brooke Claxton fonds, 23.

[19] Hayman and Tomaselli, 'Ideology and Technology', 37–9. T. W. Bearup, 'Report on Broadcasting in the Union of South Africa', 5 January 1946, ABC Document Archives, Charles Moses correspondence. Central Head Office files, 2/3, 2/3.1, and 2/3.2, United Party Archives, University of South Africa Library.

Elsewhere pre-war structures of direct state control remained in place. In New Zealand the National Broadcasting Service (NBS) and National Commercial Broadcasting Service (NCBS) were amalgamated in 1943 for technical and administrative purposes, and James Shelley was given executive authority over all New Zealand broadcasting.[20] Meanwhile, AIR remained under the control of the government of India's Information and Broadcasting Department.[21] In the Middle East both the Palestine Broadcasting Service (PBS) and the more autonomous Egyptian State Broadcasting Service (ESBS) carried information and propaganda broadcasts for local listeners, facilitated the onward transmission of material from the North African front to Britain and the dominions, and provided time on their transmitters for programmes for imperial forces stationed in the region. PBS transmitters were also used to broadcast Free French propaganda to Vichy-controlled Syria and Lebanon. In Palestine and India officials attempted to clamp down on listening to enemy short-wave stations.[22]

LONDON TRANSCRIPTIONS AND THE BBC OVERSEAS SERVICE

In dominions, colonies, mandates, and territories of informal empire, and in Britain itself, state control was thus applied to varying degrees, to ensure that broadcasting remained a reliable weapon in the imperial war effort. The new Director of Empire Services (later Assistant Controller of Overseas Services), R. A. Rendall, had already gained experience of working closely with government officials as the first manager of the PBS, and also knew what it was like to work outside Britain, in one of the BBC's target areas. Back at the BBC, Rendall presided over the expansion and reorganization of overseas services. This involved greater state funding, supervision, and, to some extent, intervention.

One of the first areas to be affected was transcriptions. A subsidy was provided to allow the issuing of discs free of charge, but strings were attached: the BBC lost full control of the service, and instead produced programmes on behalf of the MoI and in consultation with the Joint Broadcasting Committee (an obscure body created by MI6's espionage and propaganda wing). The Dominions Office, India Office, Burma Office, and Colonial Office all played advisory roles. Programmes previously broadcast on short wave were issued as recordings, and supplemented with specially commissioned material. The new service was named the London Transcription Service (LTS) 'in order to avoid stressing its official origin'.[23] In May

[20] Patrick Day, *A History of Broadcasting in New Zealand*, i, *The Radio Years* (Auckland, 1994), 249–82.

[21] Bhattacharya, *Propaganda and Information*, 124.

[22] Andrea L. Stanton, 'A Little Radio is a Dangerous Thing: State Broadcasting in Mandate Palestine, 1936–1949' (Ph.D. thesis, Columbia University, 2007), 160–9.

[23] R.A. Rendall to J. Shelley, 25 January 1940, Archives New Zealand (ANZ), AADL 564/134d 1/8/2 part 1. Minutes, 'Meeting to Discuss Ministry of Information Transcription Scheme', 15 November 1939, WAC, E5/47. On the Joint Broadcasting Committee see Philip M. Taylor, *British Propaganda in the Twentieth Century: Selling Democracy* (Edinburgh, 1999), 122 and 136.

1941 direct state control was relaxed and overall control of the LTS passed to the BBC. Nevertheless, state subsidization continued.[24]

Rendall hoped that the LTS would allow dominion listeners to perceive the reality and urgency of the war effort, and thus help to 'build up and to foster a public opinion which is informed, loyal and resistant to the attacks of hostile propaganda'. South Africa was in particular need of such an inoculation. LTS programmes aimed at Afrikaners stressed the moral case against Nazism, and the similarities between British and Afrikaner values: 'rural, agricultural, religious, democratic'. However, as Rendall admitted, the LTS was primarily aimed at India and the dependent colonies; this was a departure from the pre-war tendency to adapt transcriptions to the needs of dominion broadcasters and white listeners.[25] Many LTS programmes now explicitly targeted Africans and Asians, stressing social and economic development under benevolent British colonial rule. A 'simple' series of talks called *How England is Governed* sought to show, for example, 'that the paying of taxes and other inconveniences are not limited to Colonial peoples'.[26] LTS recorded programmes in English were supplemented with scripts for translation into other languages.

War also necessitated rapid expansion of the BBC's short-wave services in order to compete with formidable Axis efforts.[27] By the end of 1940 the BBC was broadcasting in thirty-four languages (mainly European) and was in desperate need of funds. From May 1941 rather than continue to give the BBC a fixed share of listener licence-fee receipts, the Treasury instead approved a substantial grant-in-aid, initially set at £6.7 million per annum, and rising in 1942 to £8.4 million. The state subsidies previously provided to open up other channels of imperial communication were now finally forthcoming for radio.

By November 1943 the BBC had the use of forty-three short-wave transmitters at various UK sites, as well as a number of powerful medium-wave transmitters to reach European listeners. Short wave and Daventry were no longer synonymous, and funding for both short wave and transcriptions had been moved onto an entirely new basis. Expansion was accompanied by further administrative change. The Empire Department had already been incorporated into the new Overseas Department: in November 1939 the name 'Empire Service' was officially dropped in favour of 'Overseas Service'. The following October broadcasts to the Continent were separated out into a new European Service, which was placed under closer state supervision.[28]

The change from Empire Service to Overseas Service proved more than cosmetic. During the 1930s the BBC Empire Department had been unsure how to

[24] Briggs, *History of Broadcasting in the UK*, iii, 314.

[25] [Rendall], 'Empire Transcription Scheme: Notes on Propaganda Policy', *c.* April 1940 and Rendall to J. C. S. Macgregor, 10 April 1940, United Kingdom National Archives (UKNA), INF1/165.

[26] [Rendall?] to N. J. B. Sabine, 5 April 1940, UKNA, INF1/166.

[27] For a useful contemporary survey of Axis and Allied activities see Albert A. Shea and Eric Estorick, *Canada and the Short-Wave War* (Toronto, 1942).

[28] Extract from Control Board minutes, 1 November 1939, WAC, E4/11. Mansell, *Let Truth Be Told*, 83, 102–4, 120–3.

balance the divergent needs of listeners in the colonies and dominions. During the war the gap narrowed: all audiences wanted as much up-to-date material from Britain as possible. Direct listeners in the colonies no longer demanded regular, fixed schedules, and bespoke programmes for dominion rebroadcasters could be inserted more easily.[29] Moreover, both the MoI and the BBC now recognized the overwhelming importance of reaching a large audience in the dominions and, for the first time (officially at least), in the US. There were few direct short-wave listeners in these places, so medium-wave rebroadcasting would be crucial, and the BBC now prioritized the requirements of local stations and networks.[30] After a decade of indecision, the role of the BBC's short-wave services had at last become clearly defined.

In 1940 two further changes were implemented. First, greater specialization was introduced to meet the requirements of particular audiences. Overseas Service schedules were split into four distinct strands: North American; African (primarily for South Africa, with more modest offerings for listeners in East and West Africa); Eastern (including India, and broadcasting in a range of Asian languages); and Pacific (including Australia and New Zealand). From November 1942 the Overseas Service was also responsible for a new service aimed at British and empire troops fighting in North Africa and the Middle East: the General Forces Programme. This was progressively expanded to reach troops stationed in other parts of the world, and also to serve North and South America: it was more widely known as the General Overseas Service (GOS). The GOS came to fulfil some of the functions of the pre-war Empire Service, broadcasting to widely separated groups of expatriates and other colonial and overseas listeners, as well as soldiers. It allowed the four regional overseas services to focus even more specifically on their particular target audiences. For a while, towards the end of the war, the GOS was also made available to UK listeners; this was the only time when the boundaries between internal and external audiences, home and empire, were comprehensively removed.[31]

Second, officers with first-hand experience of broadcasting in the dominions were now recruited to tailor BBC short-wave services to the requirements of target audiences. Ernie Bushnell had visited the BBC before the war, and again in spring 1940 when he helped establish a CBC Overseas Unit in London. His forthright advice on short-wave transmissions to North America (previously ignored by the BBC) was now heeded: indeed, he was soon seconded to the BBC's new North American Service (NAS) to act as its programme organizer in London. Bushnell introduced new schedules, methods of presentation, and types of programmes, to make the NAS more appealing to listeners and rebroadcasters in Canada and the US. He also encouraged the use of North American voices: an indirect response to

[29] 'What the BBC is Doing—Oversea Listeners' Wartime Programmes', *London Calling*, 4 October 1939.

[30] Nicholas John Cull, *Selling War: The British Propaganda Campaign against American 'Neutrality' in World War II* (New York and Oxford, 1995), 85. Susan Brewer, *To Win the Peace: British Propaganda in the United States during World War II* (Ithaca, NY and London, 1997), 49.

[31] Briggs, *History of Broadcasting in the UK*, iii, 446–7, 533–40.

criticisms of the metropolitan British accents that had previously dominated the Empire Service. The Scots-inflected Canadian voice of J. B. 'Hamish' McGeachy (London correspondent of the *Winnipeg Free Press*) was now heard reading daily news commentaries.[32]

After Bushnell returned to Canada, the BBC continued attempts to tailor the NAS to local idiom, pronunciation, and listening habits, and even to target particular cities and regions in the US and Canada with bespoke programmes, in order to maximize rebroadcasting. The new Director of the NAS Maurice Gorham (an Irishman) worked closely with the BBC's North American representatives and (particularly following America's entry into the war) the US networks.[33] Meanwhile, following the NAS's lead, schedules for all BBC external services were divided up into precisely timed fifteen-minute segments, to facilitate rebroadcasting and to allow more flexible use of transmitters.[34] Thus, largely due to the importance of reaching US audiences, earlier uncertainty as to whether BBC short-wave services should prioritize direct listeners or rebroadcasters was resolved in favour of the latter, at least for the duration of the war.

In 1940 Lionel Fielden returned from India and rejoined the BBC for six months to help improve the Eastern Service: his protégé Z. A. Bokhari came with him and was eventually appointed Indian programme organizer. Robert McCall, the ABC's Scottish-born manager for Victoria, was similarly brought in to help improve the BBC's Pacific Service. McCall worked hard to maximize rebroadcasting by adapting BBC programmes to the requirements of Australian rebroadcasters, particularly those of the ABC. On his return to Australia he was succeeded in London by an ABC Talks Producer, George Ivan Smith. Around fifty men and women 'representing every part of the Empire' were working at the Overseas Service by 1942, with others also employed in domestic branches of the BBC. The presence of broadcasters from the dominions was less noticeable than that of the 'foreign' staff simultaneously taken on by the European Service, because many were regarded as 'British', and indeed some, like McCall, were British-born.[35]

In London BBC officers also worked closely with members of the CBC's Overseas Unit, established by Bob Bowman (as commentator) and Arthur Holmes (as recording engineer).[36] By August 1940 Bowman and Holmes had been joined by five other CBC officers, and subsequently more were sent to provide programmes in French. Until the liberation of Europe the unit was bottled up in Britain. Staff

[32] Simon J. Potter, 'The Colonization of the BBC: Diasporic Britons at BBC External Services, *c.* 1932–1970' in Marie Gillespie and Alban Webb (eds), *Diasporas and Diplomacy: Cosmopolitan Contact Zones at the BBC World Service* (forthcoming). Script for 'The World of McGeachy' radio interview, 1963, LAC, James Burns McGeachy fonds, 2/2-4 and 2-5.

[33] Eldon Moore to N. C. Tritton, 24 December 1943, WAC, E1/428. Briggs, *History of Broadcasting in the UK*, iii, 369–71. Brewer, *Win the Peace*, 73–5. Nicholas J. Cull, 'Radio Propaganda and the Art of Understatement: British Broadcasting and American Neutrality, 1939–1941', *Historical Journal of Film, Radio and Television* (*HJFRT*), 13/4 (1993), 403–31.

[34] Mansell, *Let Truth Be Told*, 119.

[35] Potter, 'Colonization of the BBC'. T.O. Beachcroft, *Calling all Nations* (London, [*c.*1942]), 19.

[36] A. E. Powley, *Broadcast from the Front: Canadian Radio Overseas in the Second World War* (Toronto, 1975), 1–3.

were sent over from Canada on rotation for tours of duty of approximately fifteen months, allowing as many as possible to 'gain fresh broadcasting experience and... sample life in Britain in wartime'.[37] Bowman was replaced as unit head by Rooney Pelletier, who later moved to the BBC to become its North American programme organizer.[38] Pelletier argued that it was necessary to have a CBC presence in London, rather than relying on the BBC, because 'members of the Unit are Canadians with Canadian viewpoint, background, and knowledge of Canadian broadcasting'.[39] Nevertheless, while the unit remained autonomous, it relied on the BBC to provide accommodation, clerical support, and equipment, to the value of almost £23,000 by the end of the war. Moreover, while the CBC paid its unit's salaries, the BBC covered all costs associated with programmes included in BBC domestic and overseas services, including staff living allowances and expenses.[40] As most of the unit's reports for Canada were sent via the NAS, these costs were substantial.

The ABC, NBS, and SABC also sent reporting units overseas in the early years of the war, to provide first-hand coverage of the activities of Australian, New Zealand, and South African troops in North Africa and the Mediterranean. Reporting the Desert War was particularly difficult; the front was extended, and far removed from onwards transmission facilities in Cairo.[41] To help overcome these obstacles, an Empire Broadcasting Coordination Committee was formed at the front in November 1940. It was agreed that while each broadcasting unit should retain its autonomy, they would pool equipment, technical expertise, and war reports.[42]

The BBC meanwhile despatched more of its own staff abroad, and established new offices overseas. Towards the end of 1940 a BBC Intelligence Division was formed, and European, Latin American, Near Eastern, and Empire intelligence services were all established under its aegis to gather information about the impact of BBC programmes. Existing sources of information (newspaper cuttings, correspondence, official reports, and personal contacts) were supplemented by material gathered by 'intelligence representatives' and eventually, it was planned, by systematic listener research.[43] In 1942 N. Corbett Tritton, former private secretary to the Australian prime minister, was appointed BBC representative in Australia, with a brief covering New Zealand.[44] The BBC New York Office was also expanded, under the direction of Lindsay Wellington from September 1942, and new offices were established in Cairo in February 1943, and in Delhi a year later.[45]

[37] H. R. Pelletier, 'Notes on Staff Training Lecture', 18 July 1941, LAC, RG41, 488/2-3-3-2.

[38] Powley, *Broadcast from the Front*, 36. Mansell, *Let Truth Be Told*, 189.

[39] Pelletier, 'Notes on Staff Training Lecture', 18 July 1941, LAC, RG41, 488/2-3-3-2.

[40] 'CBC—Expenditure and Facilities provided by the BBC', n.d., WAC, E1/524. J. Yorke, 'Mr R.T. Bowman', 21 June 1940, E1/483.

[41] R. McCall to T. W. Bearup, 8 February 1941, National Archives of Australia (NAA), Sydney, SP312/1, box 1, 666886, 'AIF-ABC Field Unit—BBC'.

[42] Summary of letters/cables received by the ABC Field Unit, 3 July 1941, Wilmot papers, 1/48. N. R. Palmer to Smith, [*c.*3 October 1940] and Palmer to Shelley, 20 November 1940, both in ANZ, AADL 564/37c 2/4/43.

[43] [Stephen Fry], 'Recommendations for an Empire Intelligence Service', 27 November 1940, WAC, E1/312. Briggs, *History of Broadcasting in the UK*, iii, 386.

[44] UKNA, DO35/600/6.

[45] Briggs, *History of Broadcasting in the UK*, iii, 371. Mansell, *Let Truth Be Told*, 203, 208.

Several senior Overseas Service officers visited Canada during the early stages of the war, including Graves, Gorham, and Rendall.[46] Arriving in early 1943 in the wake of the conscription referendum, Stephen Fry, head of the new Empire Intelligence Service and formerly programme director at the PBS, judged the CBC a poor corrective to Canadian disunity: 'The CBC is a reflection of Canada. It has no unity of purpose. It has no policy. It lacks direction.' Fry and Wellington agreed that the NAS should step into the breach, encouraging Canadians to focus on their common interests in the wider world, rather than on their domestic differences.

> [O]ne of Canada's great handicaps is that the people are a nation of quarrelsome introverts with interests that conflict so long as they are considered within Canada. Turn Canadians into extroverts, persuade them to see themselves as part of a world set-up, and playing a pretty responsible part in that set-up, and you've gone a long way.

To promote programme exchange Seymour J. de Lotbinière was appointed BBC Canadian representative in September 1943, and accommodated in CBC offices in Toronto.[47] De Lotbinière had ten years' BBC experience, and Canadian roots. His father, a French Canadian, had served in the British army.[48] However, he was also thoroughly English, perhaps too English; Wellington thought that 'a rather more vulgarly convivial' man might have been better suited to the job, and de Lotbinière was soon seeking a programme assistant with good social mixing skills, but without a 'fancy English accent'.[49] After some delay, Gilbert Harding (later a well-known BBC television personality) was appointed. A hard-drinking straight-talker, Harding made some good friends in Canada: tellingly, when he returned to Britain, Murray and Charles Jennings gave him a case of whisky as a farewell gift.[50] However, when moving in less bibulous circles, his 'social mixing skills' were not always so good. Harding recalled arriving in Canada 'cross and tired, a state from which I rarely seemed to recover during the rest of my stay'. The passage of time only reinforced his conviction that '[c]ulturally, Canada stinks'. He had little regard for Canadians who affected sophistication: when told by a fellow guest at a dinner party in Toronto that his temporary home on Jarvis Street was 'not a very fashionable address', he replied, 'Neither is Toronto, madam.'[51]

Harding notwithstanding, all these new structures facilitated collaboration among the British world's public broadcasting authorities. They also encouraged a greater sense of equality, reducing the friction caused by the BBC's centralizing agenda for empire broadcasting. Crucially, the new structures that were established were relatively informal, emerging largely on an ad-hoc basis, and capable of adaptation to suit the interests of different organizations, even if London remained the

[46] Rendall, 'Relations with Canada and the C.B.C.', 17 September 1942, WAC, E1/503.

[47] Fry to R. E. L. Wellington, 23 March 1943, WAC, E1/584/1. Rendall, 'Report on a Visit to the United States, Canada, Mexico and Jamaica', 27 July 1943, E1/207/3.

[48] Rendall to J. S. A. Salt, 20 September 1943, WAC, E1/519.

[49] Wellington to Clark, 5 October 1943, WAC, E1/207/3. S. J. de Lotbinière, 'Report on Transcontinental Trip', 11 January 1944, E1/509/1.

[50] McGeachy fonds, 1/1–19.

[51] Gilbert Harding, *Along my Line* (London, 1953), 146–53.

empire's radio hub. However, enthusiasm for more formal collaborative structures, centralized under UK leadership, did not die out altogether at the BBC.

AN EMPIRE BROADCASTING NETWORK?

Prior to the war BBC officers had pressed British, dominion, and colonial governments to improve the empire's broadcasting infrastructure. While the dominions were relatively well served by private and public stations, and AIR had achieved some success in developing services in India, broadcasting in most of the dependent colonies remained rudimentary, and outside of Britain there were few short-wave transmitters capable of long-range broadcasting. Following the outbreak of war the BBC thus pushed even harder for the more systematic development of radio facilities. It also suggested that it should act as a coordinating agency for short-wave broadcasting in the empire. This was in many ways the last gasp of the old centralizing conception of empire broadcasting.

In Australia a war-time short-wave service had been established under the broadcasting division of the DoI, with the ABC providing many of the programmes. However, after an ambitious start, the DoI service was cut back to focus on the Pacific, North America, and South-East Asia, until more powerful short-wave transmitters could be installed.[52] Meanwhile, the CBC and BBC continued their joint campaign to have a short-wave station built in Canada. This now seemed a wartime necessity: a Canadian station jointly funded by Britain could replace UK facilities if they were knocked out by bombing or invasion.[53] However, worried that Canada might be seen as a conduit for British propaganda to the US, King vetoed the proposals.[54] More rapid progress was made with the construction of a high-power short-wave station at Singapore. This allowed the UK MoI to direct broadcasts in Asian languages to the Far East, including China and Japan, with the local assistance of the British Malaya Broadcasting Corporation. It was also hoped that the Singapore transmitters could be used by the BBC as a relay station for the Pacific Service, improving reception quality in Australia and New Zealand.[55] By mid-1941 AIR had announced its own plans to construct powerful short-wave transmitters.[56]

[52] 'Radio Australia—Summary of History, Activity, Aims, Results, etc.', University of Melbourne Archives, Tom Hoey papers, 1/12. *The Ninth Annual Report and Balance Sheet of the Australian Broadcasting Commission Year Ended 30th June 1941*, 14. The ABC ran the Australian short-wave service for a short period (July 1942 to April 1944) but responsibility was then transferred back to the DoI.

[53] Ogilvie to Sir Kenneth Lee, 24 June 1940, WAC, E2/160. 'Memorandum on British Shortwave Station Overseas (Canada)', 9 September 1940 and 'Canadian Transmitter', 11 September 1940, both in E1/576/1.

[54] Despatch from UK high commissioner, 15 October 1940 and Ogilvie to N. Ashbridge and Tallents, 27 December 1940, both in WAC, E1/576/1.

[55] High Commissioner's Office, London [to DoI], 29 August 1940, NAA, Melbourne, MP272/2, 31/10/4, 345345. McCall to Bearup, 17 June 1941, NAA, Sydney, SP1558/2, box 81, 'BBC Programmes for ABC—Shortwave Pickups—Part 2—1939–42'.

[56] L. W. Hayes, 'Some Technical Considerations in the Development of Inter-Imperial Broadcasting', 9 July 1941, WAC, E2/360/1. [Macgregor], 'Broadcasting Facilities at the Disposal of His Majesty's Government', 20 May 1942, WAC, E2/160.

Could the development of these short-wave stations be accelerated and coordinated? Could medium-wave operations in the dependent colonies be expanded to facilitate rebroadcasting? Frustrated by the perceived torpor of the MoI and the Colonial Office, and by rivalries between them, in August 1941 a BBC 'Expansion Committee' attempted to force the pace of change.[57] It recommended that medium- and short-wave transmitters in Britain, the dominions, and the colonies be augmented and linked into an 'Empire Broadcasting Network', relaying and rebroadcasting BBC transmissions. Such a network, it was argued, could also be used to feed reciprocal material back to Britain for rebroadcast to UK and overseas listeners. The report incorporated a suggestion from the MoI that a 'British Empire Broadcasting Commission' be established to take 'full executive control' of the network, probably headquartered in London, but possibly also holding meetings in the dominions 'in order to emphasize that the whole project is intended to be of a reciprocal and cooperative character'. Ultimately, such a commission might develop a cooperative imperial news service, and establish branches in each of the empire's main centres. Each broadcasting authority could second a senior representative to each branch to facilitate liaison.[58]

Sir Stephen Tallents, Controller of BBC Overseas Services, doubted whether the degree of centralization and formal collaboration demanded by the scheme was either workable or desirable. Forwarding a simplified plan to the MoI, he dispensed with the idea of a central commission.[59] The MoI's response was nevertheless vague and non-committal. The BBC's Control Board resolved to push the matter, pressing in particular for the construction of short-wave relay stations in the West Indies and either Ceylon or Kenya.[60]

The fall of Singapore, and the consequent loss of the empire's only major short-wave station outside the UK, made action even more urgent, particularly given the threats posed to AIR's planned transmitters in India by internal unrest and the prospect of Japanese invasion.[61] Yet little was done to coordinate the empire's short-wave activities, and progress with constructing stations in the dominions remained slow. The first of the new Australian short-wave transmitters was not operational until May 1944.[62] Similarly, while the Canadian government had formally authorized construction of two transmitters and a receiving station at Sackville, New Brunswick, in September 1942, it was not until February 1945 that the CBC's International Service was officially inaugurated. This broadcast in English, German,

[57] Clark to Rendall and Tallents, 21 December 1940, WAC, E2/161. On relations between the Colonial Office and the MoI see Smyth, 'African Colonies and British Propaganda'.

[58] Briggs, *History of Broadcasting in the UK*, iii, 448–50. 'The British Broadcasting Corporation—Empire Broadcasting Network', 13 August 1941, WAC, E2/360/1. For the MoI's suggestion see Wellington to Clark, 11 January 1941, E2/161.

[59] Tallents to Ogilvie, 18 August 1941 and 'G.61/41—An Empire Broadcasting Network', 21 October 1941 [originally 8 September 1941], WAC, E2/360/2. Control Board Minute of 3 September 1941, E2/160. Tallents' circumspect attitude probably reflected his earlier experiences and frustrations at the Empire Marketing Board.

[60] Radcliffe to Ogilvie, 23 October 1941, WAC, E2/360/2. R. T. B. Wynn to Ashbridge, 20 November 1941, E2/360/1.

[61] [Ashbridge?], 'Notes on Short-Wave Transmitter', *c.* April 1942, WAC, E2/160.

[62] 'Radio Australia—Summary of History, Activity, Aims, Results, etc.', Hoey papers, 1/12.

French, Czech, and Dutch, and was operated by the CBC on behalf of and with direct funding from the Canadian government.[63] Southeast Asia Command (SEAC) meanwhile established a short-wave service from Ceylon, and at the end of the war short-wave operations were re-established at Singapore, under the aegis of the British Far Eastern Broadcasting Service, to be operated by the BBC.[64] The Empire Broadcasting Network remained a pipe dream.

BBC DOMESTIC SERVICES AND THE IMPERIAL WAR EFFORT

The sounds of war, as broadcast by public authorities around the British world, were varied. They reflected the different arenas in which troops from Britain and the dominions were deployed, the places that listeners were particularly interested in, and the specific agendas of different governments and broadcasting authorities. Nevertheless, public broadcasters were still able to work together to render an imperial war effort audible and convince audiences around the empire that they were part of a single great collaborative venture. Producers sought to capitalize on the special qualities of broadcasting: the 'immediacy and reality' of radio was now employed in 'bringing the war to life'.[65] The threat posed by the enemy had to be made real to those living far from the seat of battle, and the promise of support from overseas rendered palpable to those in immediate danger.

Prior to the war BBC home schedules were only seasoned with a modest sprinkling of programmes about the empire. Several influences had conspired to limit imperial content: the perceived low general standard of contributions from the dominions and colonies; difficulties in transmitting topical material to Britain; the reluctance of the BBC and its overseas collaborators to provoke controversy about imperial affairs; and, perhaps most important, the British public's perceived lack of enthusiasm for programmes about the empire. However, the desire to boost morale in Britain by highlighting imperial contributions to the war effort provided a powerful stimulus for change. Similarly, following the entry of the US into the war, the need to balance the prevalence of American material encouraged producers to turn to the empire for inspiration. However, technical difficulties in getting material from the dominions to Britain continued to plague empire-related series intended for home listeners, such as *Dominion Commentary*, and poor audience figures (both for *Dominion Commentary* and a replacement series of talks on the colonies and dominions produced in Britain, *Palm and Pine*) indicated the limited progress made in overcoming perceived public ignorance about and apathy towards the empire. As the Colonial Office lamented, 'when quite a

[63] Wellington to Graves, 20 September 1942, WAC, E1/576/1. 'The International Service of the Canadian Broadcasting Corporation' [*c.* January 1950], Hoey papers, 1/12.

[64] E. D. Robertson, 'British Broadcasting for Asia', *Asian Affairs*, 2/1 (February 1971), 34–45.

[65] Quoted in Brian P. D. Hannon, 'Creating the Correspondent: how the BBC reached the frontline in the Second World War', *HJFRT*, 28/2 (June 2008), 175–94, quote at 176.

large number of home listeners hear the magic word "Empire" their automatic reaction is to switch off'.[66]

The series *Red on the Map* offered serious discussion of imperial topics. However, producers also attempted a lighter treatment of empire themes, in the songs, sketches, and anecdotes of the series *Travellers' Tales*, and in the interview programme *Brush up your Empire*. Springing empire-related material on unwary listeners in stand-alone 'loosebox' programmes, rather than in predictable regular series, was another way around perceived audience apathy. During the war the BBC also tried to trick listeners by 'infiltrating' imperial material into other programmes: introducing dominions speakers into talks on non-empire topics, and producing special empire-themed editions of successful programmes such as *The Brains Trust*. However, none of these efforts convinced BBC officers that they had stimulated much enthusiasm for the empire among UK audiences. As with the development of medium- and short-wave facilities around the empire, the BBC argued that progress had been retarded by unhelpful official attitudes. Specifically, the Colonial Office's reluctance to countenance on-air discussion of controversial topics made it hard to produce interesting programmes about the empire.[67]

As during the 1930s, listeners thus seemed uninterested in the empire: a problem compounded by the BBC's continuing failure to tackle controversial imperial issues. Another major difficulty was the limited availability of fresh actuality material from the dominions and colonies: a consequence of the inadequacy of the empire's short-wave infrastructure, and wartime pressure on resources and communication links. The suspension of international sporting events robbed the BBC of one of the staples of short-wave traffic to and from the dominions. Actuality material for inclusion in features and documentaries was also in short supply. In October 1939 Laurence Gilliam hastily assembled a feature to communicate to British and overseas listeners the loyal response of the colonies and dominions to the declaration of war, *The Empire's Answer*. However, this suffered badly from the lack of actuality material, and ended up as an uninspiring series of 'colourful platitudes'.[68]

To cover empire themes, the BBC was thus often forced to fall back on talks. The pool of potential speakers was generally restricted to those who happened, by chance or design, to be available in Britain at the time. It proved almost impossible to secure a steady supply of compelling talks and commentaries, presented by capable speakers from the dominions and colonies, and matching material from other parts of the world in terms of intrinsic appeal. As the BBC's Director of Talks

[66] G. R. Barnes to Sir Richard Maconachie, 6 August 1941 and Sabine to L. Gilliam, 13 September 1941, both in WAC, R51/91/1.

[67] Thomas Hajkowski, 'The BBC, the Empire, and the Second World War, 1939–1945', *HJFRT*, 22/2 (June 2002). Hajkowski, '*Red on the Map*: Empire and Americanization at the BBC, 1942–50' in Joel H. Wiener and Mark Hampton (eds), *Anglo-American Media Interactions, 1850–2000* (Basingstoke, 2007). Siân Nicholas, '"Brushing Up Your Empire": Dominion and Colonial Propaganda on the BBC's Home Services, 1939–45', in Carl Bridge and Kent Fedorowich (eds), *The British World: Diaspora, Culture and Identity* (London, 2003), also published as a special issue of *JICH*, 31/2 (May 2003). For the special issue of *The Brains Trust* see Rendall to V. Massey, 28 April 1943, University of Toronto Archives and Records Management, Vincent Massey papers, 247/6.

[68] Nicholas, '"Brushing Up Your Empire"'. For the script of the programme see *Listener*, 12 October 1939, 695–702.

admitted, 'the Dominions are neither so important nor so interesting to the British public as is the United States of America and the commentators are not so good as the American commentators'.[69]

When available, actualities could evoke a strong audience response, as demonstrated by a CBC feature *Canada Carries On*. This was relayed to Britain by radiotelephone, and rebroadcast on the BBC's domestic and overseas services on 19 June 1940. The programme was derived from a series about the Canadian war effort normally aimed at Canadian listeners. Opening the special edition for the BBC, the strains of *O Canada* were followed by an announcement:

> This is Canada calling... Canada calling the Motherland!... a message that comes from the heart of every one of our eleven million people as we cry out to you of the Motherland... 'Canada Carries On!'

The sound of marching feet built into the sound of a marching army, and a mixture of dramatized and actuality segments followed, portraying Canada as a diverse country united in a common enterprise of military service overseas, and civilian production on farms and in factories at home: 'we're but one people in this struggle'. The theme was developed in ways that now seem crassly stereotyped:

> JACQUES (French Canadian) What that you say?
> ALEX (Scotch) The word just came over the air that war has been declared.
> JACQUES That man Hitler... she got to be stop! Here... take my axe! The great North Woods she will see Jacques no more for a little while... me, I have the date with a gun in the Van Doos!
> ALEX Hold on you crazy French Canadian! Alex MacDonald has a date too... a date with me own kilted kin of the fightenest outfit that ever won the right to wear a bit of red in their bonnets... the Black Watch!
> [...]
> GIRL What is it darling... you've scarcely spoken a word all evening... what's troubling you dear?
> BOY Look, Helen... would you mind terribly if we postponed our wedding?
> GIRL But I don't understand...
> BOY Poland has fallen... beaten, bombed, smashed... and there's no telling where Hitler will strike next!
> GIRL (UNDERSTANDING) And you can't sit here... looking at plans for houses... talking about furniture and drapes while...
> BOY While the very things that make our planning and dreaming possible are threatened. I've got to get into it, Helen!
> GIRL Oh Jack, Jack... I'm so proud of you![70]

Nevertheless, to many contemporaries *Canada Carries On* seemed an inspiring piece of propaganda: 'a first class programme, and one that comes to this country most opportunely... Every listener in this country ought to hear it.'[71] The BBC

[69] Barnes to Maconachie and B. E. Nicolls, 22 August 1941 and Barnes to Pelletier, 2 November 1942, both in WAC, R51/91/1.

[70] Script for *Canada Carries On*, broadcast 19 June 1940, LAC, RG41, 252/11-39-12.

[71] De Lotbinière to Murray, 19 June 1940, LAC, RG41, 745/18-16-1-14. This file also contains a digest of letters of appreciation from UK listeners.

asked both the ABC and NBS to produce similar features. However, the outcome again illustrated the difficulty of obtaining suitable material from overseas. Production of both features was long delayed, and recordings only reached Britain in May 1941. Gilliam thought neither was of a high standard. While *Australia Carries On* was nevertheless broadcast in full on both the domestic and overseas services, *New Zealand Front*, which adopted production techniques 'so outmoded that the impression, if broadcast now, would almost amount to burlesque', was carried in only an edited version.[72]

On Christmas Day the BBC continued to attempt to link home and empire audiences with special round-the-empire programmes, now focusing on the imperial war effort. Great care was taken to ensure that at least some actuality material was available from overseas. For the Christmas 1943 programme, for example, the NBS recorded messages from New Zealand troops in the Solomon Islands, and airmailed copies on disc to both the BBC's New York Office and to the ABC, to ensure the best chance of onwards short-wave transmission to Britain.[73] One of the BBC's other successes in linking domestic and empire audiences were the 'message' programmes arranged between evacuee children in the dominions and their families back home in the UK. Again, however, it often proved difficult to get this actuality material to Britain: after eleven failed attempts to transmit one edition of *Hello Parents* from Australia to Britain by short wave, the ABC was obliged to despatch a recording by surface mail, which was eventually broadcast a year behind schedule.[74]

UK listeners also heard about the imperial war effort via the BBC's Forces Programme, which had replaced the Regional Service following the outbreak of war. Aimed principally at troops, the Forces Programme focused on light entertainment and soon attracted large numbers of civilian listeners.[75] Among the target audience were dominion troops stationed in the UK; the BBC asked dominion public broadcasting authorities to send material by radiotelephone or airmail, initially mainly sports reports, and later news, talks, interviews, and entertainment programmes.[76] The CBC Overseas Unit in London also produced up to five programmes for the Forces Programme per week, including variety shows featuring performances by Canadian troops.[77] The BBC made a similar programme using Australian troops, and from April 1940 also broadcast a special series for Indian soldiers.[78] All these programmes would also have been heard by British troops and civilians, bringing the empire home.

[72] Bearup to C. Conner, 28 March 1941 and Gilliam to G. D. Adams, 9 May 1941, both in WAC, E1/321/2. George Ivan Smith, script for *Australia Carries On*, NAA, Sydney, series SP1558/2, box 81, 'Australia Carries On—Prog for BBC'.

[73] Shelley to Moses, 8 December 1943, NAA, Sydney, ST1790/1, box 33, 12203270, 25/1/25.

[74] Nicholas, '"Brushing Up Your Empire"', 212, 220.

[75] Briggs, *History of Broadcasting in the UK*, iii, 114–28.

[76] Rendall to Shelley, 15 February 1940, ANZ, AADL 564/18a 1/8/1 part 2. Conner to Shelley, 31 January 1941, AADL 564/135a 1/8/6. Bushnell, 'The Role of Radio in Wartime Canada', 5 May 1941, LAC, E. Austin Weir Papers, 1/8.

[77] 'Notes on Canadian Broadcasting Corporation', 22 August 1946, WAC, E1/493/2. Rendall, 'Relations with Canada and the C.B.C.', 17 September 1942, E1/503.

[78] K. Barry to Bearup, 20 April 1942, NAA, Sydney, SP341/2, NN, 1902553, 'BBC Features'. Briggs, *History of Broadcasting in the UK*, iii, 124.

BBC PROGRAMMES IN THE DOMINIONS

Getting suitable material from the dominions to Britain was difficult. Communication among the dominions was problematic for similar reasons. Bushnell and McCall did arrange CBC–ABC programme exchanges while they were both in London, and McCall also used the BBC Overseas Service to broadcast *Empire Exchange*, a series of dialogues between UK-based speakers from different dominions.[79] Yet in general, a limited short-wave infrastructure made programme exchange among the dominions a rarity. Outward flows from London were easier to arrange, and for the Overseas Service, the BBC could source plenty of actuality material in the UK, along with talks by prominent British personalities. Thus, while the echo of each dominion's war effort remained somewhat faint in Britain and in the other dominions, the UK's effort was clearly audible overseas. This was something new in an imperial war effort, and a further stimulant to a sense of Britannic community. As one broadcasting officer in New Zealand remarked: 'In the first [world] war there was no weekly talk by Wickham Steed to enlighten and comfort. There were no Churchill accents to stiffen up the sinews of faith and hope and fortitude. There was no voice of the King.'[80]

Wartime increases in rebroadcasting built on the upward trends of the later 1930s. By late 1942 stations in each of the dominions were rebroadcasting a substantial amount of material from the BBC Overseas Service (see Table 4.1). Unsurprisingly, news bulletins were carried in greater number. The BBC still relied heavily on Reuters reports in compiling its bulletins, but the agency finally agreed to relax copyright restrictions, allowing its news to be rebroadcast overseas.

Table 4.1. Rebroadcasting of BBC Overseas Service, October 1942

	No. of news bulletins	*Radio Newsreel*	No. of talks/other programmes	Total hours
	(per day)	(per week)	(per week)	(per month)
NZ NBS/NCBS	7	–	5 to 10	72
ABC	3	6	10	51
SABC	4 (3 English, 1 Afrikaans)	2	7	50
CBC	3 (2 English, 1 French)	7	9	40
AIR	1	–	2 (plus many others on irregular basis)	13

NB: Rediffusion services in Barbados, Gold Coast, Malta, Nigeria, and Sierra Leone were reported as depending almost entirely on BBC Overseas Service programmes. In the US there was a substantial, if irregular, amount of rebroadcasting of news, talks, and other programmes.

Source: 'Empire Service: Rebroadcasting', 7 December 1942, WAC, E2/360/1.

[79] McCall to Bearup, 24 June 1941, NAA, Sydney, SP1558/2, box 81, 'BBC Programmes for ABC—Shortwave Pickups—Part 2—1939–42'. 'Empire Exchange: a Maori tells a Canadian of his Race', 8 December 1941, RNZSA, D5750 and D5751.

[80] Alan Mulgan, *The Making of a New Zealander* (Wellington, 1958), 112.

For the ABC, rebroadcasting effectively meant the BBC. In September 1939, for example, of 167 rebroadcast items, two were from Germany, one was from the US, and the remainder (including 129 news bulletins) came from Britain.[81] Indeed, the BBC enjoyed an additional privilege: while Australian censorship regulations were applied to all rebroadcast programmes from allied, neutral, and enemy countries, programmes from Britain and the other dominions were exempt.[82] The ABC carried three hours of BBC talks and twelve hours of BBC news weekly by early 1940, and also rebroadcast all available British ministerial speeches.[83] The ABC's news department wanted to compile more of its own bulletins, and rebroadcast less from the BBC, but was over-ruled by Bearup and the commissioners, who thought the British bulletins popular with listeners and, more importantly, cheap.[84] As Bearup advised Ogilvie: 'Relays from London have now become so frequent and so satisfactory that one shudders to think what the psychological effect would be if for some reason they suddenly ceased.'[85] The ABC's freedom to rebroadcast the BBC was temporarily restricted in February 1942, as a result of Australian government dissatisfaction with the BBC's perceived failure to give sufficient coverage of the Pacific war. ABC state managers were instructed to check BBC talks and features before rebroadcasting, to ensure that their content did not 'unduly conflict with the high policy of the Federal Government'. However, staff shortages meant that this rule was soon allowed to lapse.[86]

In New Zealand, as in Australia, BBC news, talks, and ministerial speeches were rebroadcast extensively. BBC bulletins offered the NBS its only alternative source of news to the official reports provided by the New Zealand prime minister's department. Shelley, the director of the NBS, reassured the government that BBC bulletins would not subvert official information policies, for they would 'undoubtedly represent' what the British MoI thought it 'in the best interests of the community' to know.[87] BBC talks and news commentaries similarly allowed the NBS to cover controversial issues without taking full responsibility for what was said. Returning home after working at the BBC during the war, Ormond Wilson reported that in New Zealand 'the voice of the B.B.C. is about as near a thing to the voice of God as is to be found outside of Rome and the Kremlin'. NBS stations were willing to carry all the material the BBC could provide: indeed, they often

[81] Ogilvie to Massey, 16 November 1939, Massey papers, 247/6.

[82] 'Standing Orders for Publicity Censorship, 3 March 1940', Paul papers, MS-0982/663.

[83] Moses to Macgregor, 29 April 1940, NAA, Sydney, SP1558/2, box 80, 'BBC Recorded Programs 2 1935–40'. *The Ninth Annual Report and Balance Sheet of the Australian Broadcasting Commission Year Ended 30th June 1941*, 9.

[84] F. Dixon to McCall, 17 May 1945; Barry to McCall, 22 May 1945; and McCall to Dixon, 26 July 1945, all in NAA, Sydney, SP314/1, NN, 707882, 'News Bulletins—Presentations and Times'. Press cutting, Dixon, 'Inside the ABC', *Century*, 29 June 1962, NLA, Sir Richard Boyer papers, MS3181, box 1.

[85] Bearup to Ogilvie, 19 October 1940, WAC E1/362/1.

[86] Bearup to ABC state managers, 10 February 1942, NAA, Melbourne, B2111, TKS30, 339877. Barry to McCall, 22 April 1942, NAA, Sydney, SP1558/2, box 80, 'BBC London Transcriptions 3 1941–45'.

[87] Shelley to Paul, 13 September 1939, AADL 564/30b 1/6/3 part 2. *Annual Report of the National Broadcasting Services for the twelve months ended 31st March 1941* (Wellington, 1941), 2.

used material from transmissions aimed at other parts of the world, especially when reception conditions for the Pacific Service were poor. The only real problem, Wilson believed, was the NBS's reluctance to offer any criticism of what was sent, and its willingness to allow the Pacific Service to be shaped primarily by the requirements of the ABC.[88] As in the 1930s, the NBS remained an enthusiastic and uncomplaining collaborator.

In South Africa the opening of a new short-wave receiving station in June 1939 allowed a substantial amount of wartime rebroadcasting. By early 1943 the BBC's African Service was producing special programmes for the Union such as *Songtime in the Laager* and *Calling South Africa.* It was also providing twice-daily broadcasts in Afrikaans. However, the German short-wave service (which included talks and features in both English and Afrikaans) still found a ready audience.[89] In India, AIR lost many of its Indian listeners, as nationalist and communal unrest escalated. This limited the audience for rebroadcasts of BBC news and programmes. Indians tuned to Axis short-wave broadcasts for practical as well as ideological reasons: Axis stations provided a better signal in many parts of India than did any of AIR's nine medium-wave transmitters.[90]

The challenges facing the BBC in Canada were less pronounced, but daunting nonetheless. In the week following the British government's declaration of war, when Canada remained neutral, Plaunt criticized Murray for rebroadcasting BBC news, arguing that the CBC should carry international news from neutral America rather than belligerent Britain. Later, Murray's critics would claim that, for war-related material, the CBC had relied almost entirely on rebroadcast BBC talks and news commentaries, and had failed to produce distinctively Canadian programmes.[91] In fact, at the beginning of 1941 the CBC was producing 73 per cent of its own programmes: 26 per cent came from the US networks, and only 1 per cent from the BBC. Subsequently, improvements in the NAS encouraged increased rebroadcasting of BBC programmes. By March 1942 US programmes occupied only 16.5 per cent of network hours, and BBC programmes 6.5 per cent. BBC programmes may have had a disproportionate political impact, given that they were mostly news and talks, whereas US programmes carried by the CBC mainly consisted of light music.[92] However, there is still a danger of exaggerating the BBC's presence on the Canadian airwaves: many of the NAS programmes rebroadcast in Canada had actually been produced by the CBC's own Overseas Unit. Later, de

[88] Rendall to G. I. Smith, 12 December 1944, WAC, E1/1100.

[89] G. Heyworth to J. Rodgers, 26 August 1941, UKNA, INF1/538. 'Overseas Planning Committee—Plan of Propaganda to the Union of South Africa—Channels', 16 April 1943, INF1/563. Christoph Marx, '"Dear Listeners in South Africa": German Propaganda Broadcasts to South Africa, 1940–1941', *South African Historical Journal*, 27 (November 1992), 148–72.

[90] Partha Sarathi Gupta, *Radio and the Raj, 1921–47* (Calcutta, 1995), 33–8. Bhattacharya, *Propaganda and Information*, 66.

[91] Stursberg, *Mister Broadcasting*, 102. 'Lack of Co-ordination in War Programmes', n.d., Library of the University of British Columbia Special Collections Division, Alan Plaunt fonds, 17/1. These charges were made public in V. R. Hill, 'Broadcasting in Canada—Part 1', *Canadian Forum*, February 1941.

[92] CBC Station Relations Department, 'Programme Statistical Report for January, 1940' and 'Programme Statistical Report for March, 1942', both in WAC, E1/562.

Lotbinière did persuade the CBC to carry BBC programmes in a regular *Listen to London* period, four days a week; but the experiment was short-lived. Poor sound quality meant that only highly topical programmes, or material closely attuned to Canadian requirements, could generate a significant audience.[93]

Broadcasting to French Canadians was a new venture for the BBC. The NAS carried several regular series of programmes in French, including *Jean Baptiste s'en va-t-en guerre* and *Sur le qui vive* (both consisting of CBC's Overseas Unit interviews with French Canadian servicemen in Britain), and *La France combattante* (a series of features about the Free French forces in England, originally produced for the BBC European Service).[94] Stephen Fry argued that, as with many of its programmes aimed at UK audiences and English-speakers overseas, the BBC should stress the most basic and unexceptionable of shared values in its broadcasts to French Canadians, and infiltrate war-related material (with as much 'human interest' as possible) into entertainment programmes, in an attempt to reflect and influence the thinking of 'ordinary people'.[95]

However, resistance to the BBC's Quebec initiatives came not just from listeners, but from within the CBC's French network itself. The French network scheduled its own news report directly before BBC bulletins, seemingly in an attempt to reduce the news value of the rebroadcasts. Moreover, the French Canadian commentator in the CBC's Overseas Unit, Marcel Ouimet, refused to have his reports (which were transmitted to Canada by the BBC's NAS) included in a BBC programme which, he claimed, would find no listeners in Quebec due to the 'mentality [of the] French Canadian radio audience'. Ouimet insisted that his reports instead be presented in the earlier CBC bulletin. Eventually, the BBC news division cancelled its bulletin for Quebec.[96] In Canada, as elsewhere, broadcasting had its limitations as a means of encouraging a united imperial war effort.

In terms of the Overseas Service more generally, the BBC's 1944 *Overseas Presentation Handbook* provided a guide to what was, and what was not, now deemed acceptable. Earlier, some critics had urged the Empire Service to adopt a less formal tone. Now, 'Oxford' accents and modes of speech were firmly suppressed. Announcers were told to speak in a

> common idiom which does not infer the background peculiar to prosperous Englishmen... Do not add to overseas suspicion of 'privilege' in Great Britain. To suggest that there is anything remarkable in an Earl and a grocer finding themselves fellow Home Guardsmen cuts no ice overseas...

Class hierarchies in the UK were to be downplayed and, in a multiracial empire united most clearly by abhorrence of the racialist policies of its enemies, words relating to divisions of race were to be chosen with particular care:

[93] De Lotbinière to M. Gorham, 31 December 1943, WAC, E1/536. De Lotbinière, 'Report on the First Six Months' Working of the Canadian Office', 3 July 1944 and 'Third Report on the Working of the BBC's Canadian Office', 2 January 1945, E1/509/1.

[94] 'Notes on Canadian Broadcasting Corporation', 22 August 1946, WAC, E1/493/2.

[95] Fry to Wellington, 23 March 1943, WAC, E1/584/1.

[96] Marcel Ouimet to Omar Renaud, 22 April 1944; B. Moore to E. E. Bowerman, 29 September 1944; and Moore, 'French Canadian Service', 6 October 1944 all in WAC, E1/535/3.

> Avoid 'darky' jokes and references. Never use 'native' unless it is unavoidable. Prefer 'Indian' or 'African' or 'Jamaican' or 'Hindu'. Do not talk of 'Wops', 'Chinks', 'Dagoes', or 'Niggers'...
>
> Be careful in the use of 'coloured', 'black', or 'white' races. Never use 'yellow' as a racial adjective.

Religion was similarly problematic. Announcers were told not to conceal the fact that Britain was a Christian country, but to avoid offending the empire's non-Christian majority.[97]

All this was part of a broader British wartime attempt 'to find a new language of imperial authority, untainted by the notion of white racial superiority'. Planners at the Colonial Office and elsewhere stressed the need to emphasize equality and future economic development, as a means to ensure African and Asian support for the imperial war effort.[98] Nevertheless, smoothing out the complex politics of identity and race to suit such simplistic formulae was not a straightforward task. René S. Caprara of the SABC urged the BBC to refer to 'Afrikaans-speaking South Africans' rather than 'Dutch South Africans', and not to lump South African troops in with 'imperial' troops in its reports. The BBC had little problem with either suggestion. With some misgivings, it even agreed to Caprara's request that it bow to South African 'colour prejudice', and refrain from drawing attention to the fact that white and black troops were fighting side by side in East Africa.[99] The increasing regional specialization of the Overseas Service made it easier to juggle the prejudices of different target audiences.

The BBC's attempts to tailor services to local needs, and to include dominion as well as metropolitan British accents in its broadcasts, were perhaps most apparent in its ground-breaking *Radio Newsreel* programmes. These mixed 'reporting, news commentary, interviewing, and recorded actuality', in four daily editions, each devised to suit different overseas audiences.[100] Many of the contributions to the NAS edition came from the CBC Overseas Unit, and Bushnell brought the CBC's Stanley Maxted over to London as producer. Robert McCall's Pacific edition of *Radio Newsreel* meanwhile carried stories of special interest to Australian and New Zealand listeners, including items about Australian and New Zealand troops overseas.[101] As with the NAS version, the Pacific edition was used as a channel for transmitting reports from the dominion broadcasting units back to their parent authorities. With the assistance of the ESBS, despatches from the ABC and NBS units were recorded and transmitted from Cairo to Britain via radiotelephone, or

[97] *BBC Overseas Presentation Handbook* (2nd edn, August 1944), 23–5.

[98] Suke Wolton, *Lord Hailey, the Colonial Office and the Politics of Race and Empire in the Second World War: The Loss of White Prestige* (Basingstoke, 2000), 119. See also J. M. Lee and Martin Petter, *The Colonial Office, War and Development Policy: Organization and Planning of a Metropolitan Initiative, 1939–1945* (London, 1982).

[99] R. S. Caprara to Conner, 22 February 1941 and J. G. Williams to M. Barkway et al., 7 April 1941, WAC, E2/163.

[100] Maurice Gorham, *Sound and Fury: Twenty-one Years in the BBC* (London, 1948), 104.

[101] McCall to Bearup, 29 November 1940, WAC, E1/315/4. McCall to Bearup, 28 February 1941, NAA, Sydney, SP1558/2, box 34, 'Prime Minister's (Menzies') Tour Overseas 1941'.

sent by airmail, for use by McCall.[102] While few good actualities were forthcoming (the SABC unit had more success in this regard), the ABC's Chester Wilmot sent as much material as he could, and *Radio Newsreel* was able to include his reports on the fighting in Greece, Crete, and at Tobruk.[103] The techniques pioneered by *Radio Newsreel* would later be picked up by the BBC's *War Report* programme, covering the Allied invasion of Europe for home listeners. Maxted and Wilmot would both join the BBC's War Reporting Unit.[104]

The BBC's willingness to carry material from the dominion overseas units generated goodwill, and also increased rebroadcasting in the dominions. In effect, the dominion units were helping the BBC to adapt its overseas services to the needs of local audiences. Without having to appoint correspondents of its own, the BBC was getting Australian, Canadian, and New Zealand contributions to its programmes. The BBC was also able to include these reports in its domestic services and in transmissions aimed at other parts of the empire, to bring home to listeners the fact that this was a genuinely imperial war effort. Material from the ABC unit included in the Pacific edition of *Radio Newsreel* was, for example, also heard in Britain, South America, and Africa, while reports from the SABC unit were included in the Pacific, NAS, and African editions of the programme.[105]

The BBC's growing understanding of the local requirements of overseas audiences and rebroadcasters was facilitated by its connections with the empire's wider machinery of diplomacy, information, and propaganda. Political and economic briefings prepared by the UK high commissioners for the Dominions Office were made available to the BBC to help devise programme policies, and close liaison was maintained with the MoI and other branches of the British and dominion governments.[106] Many officers were also able to develop first-hand knowledge of the empire as they travelled overseas, supplementing information and analysis provided by wartime visitors to Britain from the dominions and colonies. Later, BBC overseas representatives also offered useful advice. Nevertheless, overseas audience responses remained unpredictable, and listeners did not always appreciate attempts to render the voice of the BBC less metropolitan. When de Lotbinière visited the northern Ontario town of Timmins, he asked a man in a miners' club, listening to a rebroadcast of the BBC news over the canteen's loudspeakers, whether the announcer's voice was offensive to Canadian ears. The miner turned out to be a Londoner: '"Offensive? That's not offensive, that's Michael Brooke; 'e was at

[102] BBC to ABC, 19 July 1940, NAA, Sydney, SP312/1, box 1, 666886, 'Mobile Unit—Relations with BBC 1940–41'. Palmer to Shelley, 16 December 1940, ANZ, AADL 564/37c 2/4/43. McCall to Bearup, 8 February and 2 May 1941, NAA, Sydney, SP1558/2, box 82, 'Robert McCall—Secondment to BBC—1940–42'.

[103] McCall to Bearup, 30 May and 17 June 1941, NAA, Sydney, SP1558/2, box 81, 'BBC Programmes for ABC—Shortwave Pickups—Part 2—1939–42'.

[104] Siân Nicholas, '*War Report* (BBC 1944–5) and the Birth of the BBC War Correspondent' in Mark Connelly and David Welch (eds), *War and the Media: Reportage and Propaganda, 1900–2003* (London and New York, 2005).

[105] McCall to L. Cecil, 18 July 1941, NAA, Sydney, SP1558/2, box 82, 'Robert McCall—Secondment to BBC—1940–42'. Williams to G. Dickson, 24 September 1941, WAC, E2/163.

[106] Rendall to de Lotbinière, 8 August 1944, WAC, E1/509/1.

Oxford and you can 'ear every word 'e says".'[107] Meanwhile, the NBS passed on complaints from listeners about the unintelligibility of Hamish McGeachy's 'Scotch Canadian' accent: he seemed to be speaking 'from inside a drum with his mouth full of Jews Harps'.[108] Wilson thought that the wartime BBC Overseas Service carried too many dominion voices for the liking of New Zealand listeners.[109]

While rebroadcasting of BBC short-wave programmes increased in the dominions, LTS recordings met with a mixed response, even though most records were supplied free of charge. Transcriptions continued to be welcomed in New Zealand and, more surprisingly, in South Africa. They were less well received in Canada, where the only series carried regularly by the CBC was *Frontline Family*, in fifteen-minute episodes broadcast five times a week. Bushnell had in fact helped devise this programme during his time in London: BBC writers, producers, and actors were making what was in essence a North-American-style soap opera.[110] The ABC meanwhile worried that accepting LTS recordings would reduce its ability to reject other forms of official propaganda, and argued that short wave, anyway, provided more up-to-date material than transcriptions ever could.[111] It was only when the BBC began to include entertainment recordings in the LTS, at heavily subsidized prices (see below), that the ABC agreed to take a regular supply of discs.

THE FRIENDLY INVASION OF THE AIRWAVES

Around the British world, the entry of the US into the war at the end of 1941 had significant but varied effects on radio. American programmes were already commonplace in Canada, but in Britain, Australia, and New Zealand, the arrival of US troops meant the provision of American material in unprecedented quantities. The Special Services Division (SSD) of the US War Department supplied recordings of entertainment programmes for American troops which, broadcast by local stations on medium wave, were also accessible to civilians. Stations in South Africa (where American troops were much rarer) and India (where officials worried about the effects of American broadcasts on a restive civilian population) carried less of this material. The US Office of War Information (OWI) meanwhile distributed more 'serious' news, commentaries, talks, and features relating to the US war effort

[107] 'Report on the First Six Months' Working of the Canadian Office', 3 July 1944, E1/509/1.

[108] Shelley to Tritton, 9 April 1943, ANZ, AADL 564/134a 1/8/7.

[109] Rendall to Smith, 12 December 1944, WAC, E1/1100.

[110] *Annual Report of the National Broadcasting Services for the twelve months ended 31 March 1942* (Wellington, 1942), 3. P. A. Clutterbuck to J. E. Stephenson, 7 December 1939, UKNA, INF1/166. 'Overseas Planning Committee—Plan of Propaganda for Canada—Channels', 19 September 1944, INF1/566. Howard B. Chase, script of CBC talk, 25 March 1945, WAC, E1/490. Michele Hilmes, 'Front Line Family: "Women's Culture" Comes to the BBC', *Media, Culture and Society*, 29/1 (January 2007), 5–29.

[111] Barry to Moses, 5 January 1940 and Moses to Macgregor, 29 April 1940, NAA, Sydney, SP1558/2, box 80, 'BBC Recorded Programs 2 1935–40'. Moses to J. L. Treloar, 8 January 1940, NAA, Canberra, SP112/1, 31/4/1, 267702.

(on disc and by short wave) as well as some classical music recordings, all targeted at civilians in Allied countries.[112]

As with American entry into the war more generally, the increased US presence on the airwaves was often initially greeted with enthusiasm around the British world. Relations between US and UK information and propaganda authorities became close and largely cordial.[113] In programmes for listeners at home and overseas, the BBC was keen to emphasize the role played by the US in what was now an Allied rather than purely imperial war effort. This was depicted as the 'United Nations' working together, but also, echoing Churchillian rhetoric and a well-established strand in British thinking, as a sign of the long-promised return of English-speaking unity. The boundary between the British world and the English-speaking world blurred:

> Broadcasting day in and day out from this land that stands in the spearhead of our battle-line, we think of you daily, hourly, in Canada, Australia, New Zealand, South Africa, in the Colonies and remote outposts, and in the United States of America; and we are made aware, as it is given to few of our countrymen to be aware, of the solidity and the resolution of the world-wide Commonwealth of English-speaking peoples.[114]

Such slippages (an English-speaking 'Commonwealth' that included the US?) were perhaps conscious. US broadcasters sometimes picked up on similar themes: one American programme for Australia described US aid to Britain as the 'friendly gifts of Anglo Saxon comrades to brothers and sisters who are fighting the family battle'.[115]

However, behind the rhetoric, contemporaries recognized that a wide gulf still separated the two main branches of the English-speaking peoples. Undertaking wartime work for the NAS, BBC Features Producer Geoffrey Bridson argued that US audiences had to be treated fundamentally differently from their empire counterparts. Colonial listeners, Bridson believed, saw Britain as 'a leader in a common policy or way of life'. Dominion audiences regarded her as 'an equal and a partner in common aims of national development'. By contrast, the BBC approached US listeners as

[112] Bhattacharya, *Propaganda and Information*, 67–8. 'Overseas Planning Committee—Plan of Propaganda to the Union of South Africa—Channels', 16 April 1943, UKNA, INF1/563.

[113] Holly Cowan Shulman, *The Voice of America: Propaganda and Democracy, 1941–45* (Madison, 1990), 42–9. On the OWI see also Allan M. Winkler, *The Politics of Propaganda: The Office of War Information, 1942–1945* (New Haven and London, 1978); Clayton D. Laurie, *The Propaganda Warriors: America's Crusade against Nazi Germany* (Lawrence, Kansas, 1996); and Gerd Horten, *Radio Goes To War: The Cultural Politics of Propaganda during World War II* (Berkeley and Los Angeles, California, and London, 2002).

[114] Press cutting, Michael Barkway, 'Accuracy and Speed, but Above All Accuracy', *London Calling*, n.d., Wilmot papers, 3/31.

[115] F. T. Birchall, script for 'Broadcast to Australia', 27 March 1942, NAA, Sydney, SP1558/2, box 95, 'American Programmes for Australia'. For a discussion of the complex relationship between American and British identities at this time, see David Goodman, 'Loving and Hating Britain: Rereading the Isolationist Debate in the USA', in Kate Darian-Smith, Patricia Grimshaw, and Stuart Macintyre (eds), *Britishness Abroad: Transnational Movements and Imperial Cultures* (Carlton, Vic., 2007).

> a commercial traveller with something which his livelihood depends upon his selling to a suspicious buyer who distrusts him, doesn't feel that he wants his goods, and is probably engaged on different salesmanship in his own right... Great Britain's radio relationship with the United States calls for the highest form of radio sales talk, and has to cope with the deepest form of sales resistance.[116]

US propaganda authorities were meanwhile engaged in their own 'sales talk' in Britain and the dominions. The OWI established a London 'outpost' in September 1941. News and information about the US and its war effort were fed to the British mass media. In January 1943 a dedicated British division of the OWI was established in Washington, DC.[117] The BBC carried significant amounts of OWI material, an increasing number of programmes supplied by the American commercial networks, and more programmes of its own about the US. From July 1943 the American Forces Network also broadcast to the main US army camps in Britain from UK medium-wave transmitters: around 10 per cent of the UK civilian population could tune in, and experience broadcasting as almost pure entertainment.[118] In August 1942 an OWI outpost was also set up in Sydney, and offices were subsequently established in Brisbane and Wellington.[119] The OWI beamed news, features, and entertainment programmes at Australia and New Zealand by short wave, such as the series *RAAF Voices*, carrying messages from Australian airmen training in the US.[120] Recorded OWI and SSD programmes meanwhile benefited from the willingness of US celebrities to broadcast free of charge to overseas audiences. As part of a series called *America Answers Australia*, Mickey Mouse and his creator Walt Disney appeared at the microphone at the request of a Sydney schoolboy.[121] Similarly, in the series *America Talks to Australia and New Zealand*, Spencer Tracy treated Australian and New Zealand listeners to an 'eloquent restatement of American ideals through the words of Thomas Jefferson'. Other speakers in the same series included Yehudi Menuhin, Bing Crosby, Greer Garson, Paul Robeson, and Cole Porter.[122]

[116] D. G. Bridson, 'A Consideration of Short-wave Broadcasting to America', 10 April 1942, WAC, E1/128.

[117] 'Outposts, Office of War Information', 12 April 1945, National Archives and Records Administration (NARA), RG208, Records of the Historian, area files 1943–45, box 3, 'Outposts—General'. 'Report on the British Division of OWI, 1943', name files 1944, box 1, 'Agar, Herbert'. 'Chapter VII—Outposts', f. 21, draft historical reports, 1941–48, box 3.

[118] Hajkowski, '*Red on the Map*', 179. Valeria Camporesi, *Mass Culture and National Traditions: The BBC and American Broadcasting, 1922–1954* (Fucecchio, 2000), 31, 155–70.

[119] N. T. Johnson to H. V. Evatt, 6 August 1942, NAA, Canberra, SP112/1, 429/3/20, 269411. Ruth Walter, 'Wellington, New Zealand, 15th July–31st August 1944', 31 August 1944, NARA, RG208, outpost records, 1942–46, box 5. Outposts were also set up in South Africa and India, and for a short time in Canada.

[120] 'Chapter IV—Developing Overseas Operations', f. 34, NARA, RG208, draft historical reports, 1941–48, box 3. T. Lucas, 'Monthly Reports of the Overseas Department—September 1942', 8 October 1942, NAA, Sydney, SP1558/2, box 83, 'Monthly Reports of Overseas Dept 1938–43'.

[121] 'America Answers Australia no. 27—broadcast date September 24th', NARA, RG208, scripts for Australian broadcasts, 1943–44, box 3501, 'America answers Australia'.

[122] Script, 'America Talks to Australia and New Zealand', 7 June 1943, NARA, RG208, scripts for Australian Broadcasts, 1943–44, box 3501, 'America talks to Australia'. 'Personalities used to date on "America Talks to Australia and New Zealand"', box 3501, 'America answers Australia'.

Both public-service and commercial broadcasters carried plenty of American material, and in Australia censorship regulations were relaxed to facilitate this.[123] Reoccupying the ABC general manager's chair, Moses wrote to the head of the OWI Sydney outpost that:

> I have just returned from New Guinea, where my regiment fought side by side with American troops against our common enemy. I cannot tell you how pleased I was on my return to take up my duties as General Manager of the Australian Broadcasting Commission to find it engaged in a similar common effort...After the war the bonds tied by blood and common thought can never be broken.[124]

The ABC already retained an agent in New York, and Moses was keen to establish a full ABC office in the US. Curtin, however, was unhappy about the large amount of American material being carried by the ABC, and vetoed Moses' plan, stipulating that the ABC should first establish a London office.[125]

Indeed, to some, the large amount of American material placed on the Australian and New Zealand airwaves during the war seemed a significant threat to embryonic national identities and sentimental and cultural connections with Britain. *New Zealand Truth* complained about a wartime rebroadcast of an American programme (*You Can't Do Business with Hitler*) which implied that the US had won the First World War single-handedly.

> That might be good history for the American children and people—travesty though it is—but it is quite another matter when the home of broadcast propaganda decides to regale a section of the British Commonwealth of Nations with stories that are demonstrably false.[126]

The same newspaper later claimed that the sheer volume of American programmes being rebroadcast threatened to swamp New Zealand listeners, and sap the country's 'British' culture:

> British ideas of entertainment are being willy-nilly submerged by the present policy... It should be the duty of the broadcasting authorities to help New Zealanders preserve their identity by giving them the entertainment of their choice over the air.[127]

In private, some station directors also called for more programmes from Britain, and complained about programmes that implied the US was fighting the war

[123] D. Wilson to R. E. Knox, 16 April 1943, ANZ, AADL 563/14a 4/39 part 2. J. H. Hall, *The History of Broadcasting in New Zealand, 1920–1954* (Wellington, 1980), 139. McCall to H. P. Brown, 12 August 1942, NAA, Sydney, SP613/1, 8/1/33, 3188857.

[124] Moses to M. Stiver, 29 March 1943, NAA, Sydney, SP1558/2, box 95, 'American Programmes for Australia'.

[125] 'Proposal to establish ABC rep in America' [19 October 1943], NAA, Sydney, SP314/1, box 4, 'Mr McCall—Federal Superintendant (Old Files)'.

[126] Cutting, *New Zealand Truth*, 31 March 1943, ANZ, AADL 563/14a 4/39 part 2. For similar complaints in Britain about the Hollywood war film *Operation Burma*, see I. C. Jarvie, 'Fanning the Flames: Anti-American Reaction to *Operation Burma* (1945)', *HJFRT*, 1/2 (1981), 117–37.

[127] Cutting, 'US Programmes Get on the Nerves', *New Zealand Truth*, 15 September 1943, ANZ, AADL 563/14b 4/39 part 3.

single-handedly.[128] To make matters worse, American broadcasters and propagandists seemed to display little interest in taking reciprocal material from Australia or New Zealand.[129]

The UK MoI was also alarmed by American radio penetration of Australia and New Zealand. While the BBC might be well represented on ABC stations, most Australians listened to private stations, which carried plenty of American transcriptions and virtually nothing from the BBC.[130] Senior officers at the BBC similarly feared that the war would establish a long-term demand for US material in Australia and New Zealand, possibly at the expense of BBC programmes. Rendall of the Overseas Service urged that the amount of 'straight' propaganda in BBC transcriptions should thus be reduced, in favour of more entertainment, in order to compete with US offerings.[131] As American OWI and SSD transcriptions were gradually withdrawn from 1943, the BBC increased the flow of LTS material to Australian stations, and added recorded entertainment programmes taken from its domestic services. The ABC agreed to carry four hours of LTS programmes a week, and the Federation of Australian Commercial Broadcasters distributed two hours of programmes to its members weekly, to be used in unsponsored slots.[132] After the war the BBC would build on this new-found success with transcriptions.

THE FIRST COMMONWEALTH BROADCASTING CONFERENCE

The idea of holding an empire broadcasting conference had been mooted as early as 1936. It was revived in September 1943 by Rendall, as a means to relaunch the stalled campaign for a coordinated Empire Broadcasting Network. A conference would, Rendall argued, allow the BBC to exert a guiding influence over the wider pattern of broadcasting in the empire, capitalizing on high levels of wartime rebroadcasting and general goodwill, without seeming to violate the autonomy of dominion broadcasting authorities. Attendance would be restricted to public broadcasters, bolstering their status in the face of private rivals, not only in the dominions, but also potentially in the UK, where the BBC feared the introduction of commercial competition. Government officials would also be excluded from the conference, thus helping to re-establish and preserve the autonomy of the BBC and dominion broadcasting authorities over external services (the fact that broadcasters in New Zealand and India were public servants was overlooked). Colonial broadcasting authorities were not invited: the BBC would discuss plans

[128] See for example W. Elliot to NCBS Head Office, *c.*18 October 1943, ANZ, AADL 563/14b 4/39 part 3.

[129] H. C. Brown to C. Scrimgeour, 14 April 1943, ANZ, AADL 563/14a 4/39 part 2.

[130] D. Anderson to Smith, 26 August 1943, WAC, E1/342.

[131] Rendall, 'Report on a Visit to the United States, Canada, Mexico and Jamaica', 27 July 1943, WAC, E1/207/3.

[132] Tritton to Smith, 15 October 1943, WAC, E1/341/4. Tritton to Smith, 3 December 1943 and Smith to Clark, 11 April 1944, E17/9/1. Tritton to Clark, 1 May 1944, E1/428.

for broadcasting in the dependent colonies separately, with the Colonial Office.[133]

In May 1944 the BBC thus invited senior representatives of the ABC, CBC, NBS, SABC, and AIR to attend a broadcasting conference in London 'or other convenient Empire centre'. The BBC highlighted three particular areas of common concern: news, programme exchange, and 'the linking up of short-wave services between various parts of the Commonwealth and Empire' (a nod in the direction of the Empire Broadcasting Network idea).[134] Dominion broadcasters also suggested the more systematic exchange of information on a wide range of programming, engineering, and administrative matters: the CBC recommended the formation of a joint technical committee to pursue specialist matters in detail, particularly regarding short-wave broadcasting.[135]

On 15 February 1945 the first Commonwealth Broadcasting Conference opened in London. Sessions were held over a three-week period, and delegates were able to visit many of the BBC's production, transmission, and monitoring facilities, including Alexandra Palace for a demonstration of television. They were guided through blitzed London and various provincial centres in England and Scotland, and watched Tommy Handley record an episode of *ITMA* (the BBC's flagship comedy and one of the LTS entertainment programmes). They also visited the Western Front, and got to know one another under conditions of wartime camaraderie. A V-2 rocket exploded in the skies above Broadcasting House on the first day of the conference, but according to the chairman of the CBC, the delegates 'just took it as a salute from Jerry and carried on'.[136] Henceforth, contact among senior broadcasting officers around the Commonwealth became more intimate: they no longer regarded each other merely as 'names at the end of cables and letters'.[137] This smoothed the path of broadcasting collaboration. Indeed, the conference provided a foretaste of the sort of international mobility that would soon become commonplace. During the war many broadcasting officers had braved the hazards of shipping lanes harassed by enemy submarines, but in 1945 Shelley availed himself of the chance to travel between Britain and New Zealand by air.[138]

The conference had no executive powers, and individual broadcasting authorities would decide for themselves whether to implement its recommendations.[139] The idea of establishing an Empire Broadcasting Network was quickly abandoned. Indeed, while the SABC, NBS, and AIR favoured the creation of a central

[133] Rendall to Clark, 7 September 1943, WAC, E2/160. On the prospect of the introduction of commercial broadcasting in the UK see Briggs, *History of Broadcasting in the UK*, iv, *Sound and Vision* (Oxford, 1995 [1979]), 29–37 and Camporesi, *Mass Culture*, 31–9.

[134] Clark to A. Frigon, 22 May 1944, LAC, RG41, 357/19-12/1. Michael Stephens, 'Some Notes on the Evolution of the Commonwealth Broadcasting Conference', CBA archive, box 1, CBA/1.

[135] Frigon to Rendall, 4 October 1944, LAC, RG41, 357/19-12/1. 'Commonwealth Conference: Agenda Items', 14 November 1944, CBA archive, box 1, C/1/1.

[136] Chase, script of CBC talk, 25 March 1945, WAC, E1/490.

[137] A. Powell to W. J. Cleary, 16 March 1945, NLA, William James Cleary papers, MS5632, scrapbook 'ABC (Resignation, 19 February 1945)'.

[138] *New Zealand Listener*, 1 June 1945, 7.

[139] Minutes, Commonwealth Broadcasting Conference, 15 February 1945, CBA archive, box 1, C/1/1.

Commonwealth secretariat in London (to 'assist in routine work such as copyright coverage, purchase of music and scripts, engagement of artists, etc.') even this suggestion for a modest cooperative organization was rejected by the CBC and ABC delegates, who insisted that their organizations would have separate representation in London. It was agreed that future cooperation would be best carried out on an informal basis, in what Haley described as a 'partnership of goodwill'.[140]

Other divergences of opinion were also recorded at the conference, although few seemed serious. Delegates discussed issues such as programme policy, controversial broadcasting (political and religious), television, listener research, studio equipment, and educational and agricultural broadcasting. It was agreed that existing cooperative arrangements (such as the reciprocal provision of studio facilities and office accommodation, and the exchange of information on artists' fees and copyright and performing rights payments) would continue after the war. The conference also regularized administrative arrangements for secondments, training, and staff exchanges, and a gentlemen's agreement was struck, according to which the organizations would refrain from poaching each other's staff.[141] Delegates were able to proclaim the importance of maintaining 'a true standard of objectivity' in political broadcasting, with a unanimity that would disappear in later years.[142]

The war had made the dissemination of news one of the key functions of broadcasting, around the British world, and also in the US. Everywhere, BBC news had been rebroadcast extensively, and while in India and New Zealand the development of domestic news broadcasting had been hampered by direct state control, in Canada and Australia public broadcasters had expanded their news departments considerably. Senior news officers at the CBC and ABC expected that, in the post-war period, these newly developed news-gathering structures would interact, and that there would be 'a general exchange of news between the national networks', initially between Britain and the dominions, and eventually perhaps also with the US networks.[143] However, as the conference made clear, the BBC was averse to any measure that would dilute its editorial control of overseas news. Instead of taking dominion news from the dominion broadcasting authorities, it planned to supplement agency news by expanding its wartime news-gathering structures and appointing foreign correspondents of its own. Rather than a free exchange of news, it was thus agreed that Commonwealth broadcasting authorities should provide reciprocal access to recording and transmission facilities for 'own correspondents', and offer one another advance notice of events of special interest. Actual news reports would only be exchanged on request.[144]

[140] Minutes, Commonwealth Broadcasting Conference, 26 February 1945, LAC, RG41, 357/19-12-1/1. 'Commonwealth Radio Conference—Notes on Agenda', LAC, RG41, 357/19-12/2.

[141] See the minutes and papers relating to the conference contained in LAC, RG41, 357/19-12-1/1 and 19-12/2, and CBA archive, box 1, C/1/1.

[142] 'Commonwealth Broadcasting Conference—Part I—Final Summary of Conclusions', 9 March 1945, LAC, RG41, 357/19-12/2.

[143] D. C. McArthur to Frigon, 27 February 1943; Dixon to McArthur, 5 July 1944; McArthur to Dixon, 27 September 1944, all in LAC, RG41, 174/11-17-4-4. Bushnell to BBC, [*c.* January 1944], WAC, E1/493/2.

[144] 'CC6—Commonwealth Broadcasting Conference—Part I—Notes by BBC on Item 1—News Exchanges: Future BBC plans', LAC, RG41, 357/19-12-1/1.

The final report of the conference's technical subcommittee meanwhile emphasized the pressing need for the construction of short-wave transmitting stations in New Zealand and South Africa, and of two short-wave relay stations: one at Georgetown, British Guiana (possibly running in conjunction with a medium-wave regional transmitter for the British West Indies); and another either at Delhi or in Ceylon.[145] The Commonwealth Broadcasting Conference was clearly a chance to exert further pressure on British officials to act, and shortly afterwards the BBC sent an engineer to British Guiana to survey possible sites.[146]

The CBC chairman claimed that the 1945 conference would result in 'a pooling of the entire broadcasting resources of the Commonwealth'.[147] This was overly optimistic. As Bushnell remarked, the results of the conference were not 'earth shaking': rather, they largely confirmed existing wartime collaborative practices that had developed on an ad-hoc basis. At the conference, the dominion delegates tended to divide into two groups. Bushnell thought that the New Zealand, South African, and Indian broadcasting authorities were highly dependent upon the BBC for programme material, and that their delegates therefore tended to treat the BBC as 'the "Great White Father" of broadcasting'. By contrast, the Australians and Canadians were cooperative, but keen to promote their own agendas.[148]

Moses was certainly determined to preserve the ABC's autonomy, and to promote greater reciprocity. The ABC press release issued on his return from the conference was headlined 'Greater Publicity for Australia through Radio', and emphasized that: 'The Dominions would, in future, contribute more to the B.B.C. home programme than in the past.' Moses immediately ordered the despatch to the BBC and other Commonwealth public broadcasting authorities of recordings of a thirteen-episode ABC documentary series on the New Guinea campaign (*Theirs be the Glory*) and of samples of the ABC's innovative *Kindergarten of the Air* programmes for young children.[149] However, the BBC was not convinced. Haley warned that, while the BBC planned to increase the amount of reciprocal material from the Commonwealth included in its schedules, available space was limited.[150] While programme exchanges could promote 'understanding and goodwill among the peoples of the Commonwealth', they had to be justified in terms of 'their intrinsic appeal to the listening public of the receiving country'. Exchanges could not be undertaken for the sake of Commonwealth relations, at the expense of

[145] Minutes, 'Commonwealth Broadcasting Conference—Technical Sub-Committee', 20 February 1945, CBA archive, box 1, C/1/1. 'Commonwealth Broadcasting Conference—Report of the Technical Sub-Committee', 1 March 1945, LAC, RG41, 357/19-12/2.

[146] A. E. Barrett, 'Broadcasting Survey of the British West Indian Colonies in connection with the establishment of a Commonwealth Relay Station and Caribbean Areas Station', February 1946, WAC, E2/360/1.

[147] Chase, script of CBC talk, 25 March 1945, WAC, E1/490.

[148] Bushnell, 'Preliminary report on British Commonwealth Broadcasting Conference', 3 April 1945, LAC, RG41, 357/19-12/2.

[149] Press release, 'Greater Publicity for Australia through Radio—A.B.C. General Manager back from Empire Broadcasting Conference', 27 May 1945, ABC Document Archives, Charles Moses correspondence.

[150] Minutes, Commonwealth Broadcasting Conference, 20 February 1945, LAC, RG41, 357/19-12-1/1.

programme standards.[151] Many of the BBC's earlier reservations about scheduling material from the dominions had survived the war intact. There were clearly limits to the BBC's appetite for Commonwealth cooperation, particularly where this involved a threatened loss of control over programme standards or supplies of overseas news.

CONCLUSIONS

Wartime attempts to include voices from the British world in BBC programmes reached a peak with the Allied invasion of Europe. No longer reliant entirely upon news agencies, the BBC established its own War Reporting Unit to provide bulletins and actuality material. For the landings in France and the Low Countries, the BBC assembled a team of reporters from the UK, the US, and the dominions, equipped with new portable recording devices. The key BBC commentators included Howard Marshall as director, Frank Gillard, and Richard Dimbleby, all English, and Robert Barr, a Scot. Robert Dunnett, formerly an editorial assistant at the *Scotsman*, was BBC correspondent with the US forces. The Canadian members of the unit were Stanley Maxted (who covered Arnhem) and Stewart MacPherson, a former sports writer and commentator who had joined the BBC's outside broadcasts department in 1941. French-language coverage from a North American perspective was provided by a New Yorker, Pierre Lefevre. From Australia came Chester Wilmot, previously with the ABC, and Colin Wills, a former Queensland newspaperman and regular BBC voice before and during the war. Richard North, a British journalist who had worked in Australia for six years, also joined the unit. *Radio Newsreel* was thus able to carry actualities from the front, accompanied by a broad sample of the accents of the British world. For the NAS edition of *Radio Newsreel*, even the linking continuity scripts were read by a Canadian on loan from the CBC, Byng Whittaker.[152] During the Allied invasions the BBC was also given almost unlimited access to material produced by the CBC Overseas Unit in Sicily, Italy, and Western Europe. CBC correspondents like Matthew Halton became well-known voices on the BBC's domestic and overseas services; their reports were accompanied by plenty of actuality material.[153]

On D-Day stations in New Zealand rebroadcast every BBC news bulletin they could. Later, at the NBS, Alan Mulgan thought the BBC's VE Day programme its 'very finest effort'. 'We had sat up all night "bringing in the world" to our system, and then about six in the morning came this profoundly moving salute to all who

[151] 'Commonwealth Broadcasting Conference—Part I—Final Summary of Conclusions', 9 March 1945, LAC, RG41, 357/19-12/2.

[152] Cutting, Macdonald Hastings, 'How the BBC Covers the Invasion', *Picture Post*, 17 June 1944, Wilmot Papers, 3/3a. Powley, *Broadcast from the Front*, 130.

[153] 'CBC Overseas Unit', 11 January 44, LAC, RG41, 488/2-3-3-2. Memorandum by Moore, 14 March 1944, WAC, E1/503. A. E. Powley, 'General Memo on Receipt and Retransmission of Field Broadcasts', 19 July 1944, LAC, Andrew Gillespie Cowan fonds, 16, 'CBC War Correspondent, correspondence 1943–44'.

had wrought the victory.'[154] From Australia, Tritton reported that, thanks to its invasion coverage, 'the prestige of the Corporation has never been higher', and the BBC had become a household name. In February 1944 Tritton saw a board posted outside a church in Melbourne: 'Gossip is the Devil's BBC—Are You an Announcer?'[155]

While it is unlikely that the overall contribution made by broadcasting decisively transformed the empire's war effort, it is clear that the war played a major role in reshaping collaborative links among the British world's public broadcasting authorities. Earlier, patchy and under-resourced structures of collaboration were rejuvenated, becoming more formidable and vigorous, and also to some extent less centralized. They seemed more inclusive, flexible, and satisfying to broadcasters in the dominions.

Yet contemporaries were also aware of some of the failures of wartime empire broadcasting. As Siân Nicholas has argued, BBC officers seldom felt that they had overcome the perceived apathy of British audiences towards matters imperial. Indeed, the nature of the war compounded this tendency: colonial and dominion contributions seemed less spectacular, more familiar and taken for granted, than those of the US and USSR. UK listeners also retained much of their traditional insularity, preferring to hear above all about 'their plucky selves'.[156] UK audiences heard something of the doings of their fellow-Britons in Canada, Australia, New Zealand, and South Africa, but the lack of actuality material from the empire was a problem. Meanwhile, the 'Quit India' campaign offered a far from edifying spectacle, and the war effort in other parts of Asia, and in Africa, was covered in an uneven fashion, reflecting established lines of inclusion and exclusion in the sphere of empire broadcasting. Africans, Asians, and other non-white listeners were segregated and neglected. Contemporaries linked the British world not primarily with the colonial empire, but rather with the US, Britain's most significant cultural rival yet also her most important diplomatic and military ally. Attempts to change BBC priorities in the post-war years, to strengthen links with African and Asian broadcasters, and to restrict American influence, would meet with only limited success.

[154] Memorandum by Smith, 7 July 1944, WAC, E1/1097/1. Mulgan, *Making of a New Zealander*, 191.

[155] Tritton to Smith, 3 February and 14 June 1944, both in WAC, E1/428.

[156] Nicholas, '"Brushing Up Your Empire"', 225.

5

Continuities, 1945–59

Even before peace had returned, the director-general of the British Broadcasting Corporation (BBC) set about restoring the pre-war status quo. William Haley told officers to curb American influences on British programmes, and augment inward flows of programme material from Europe and the empire.[1] During the late 1940s scarce resources were focused on the familiar world of radio: in a time of great austerity, television was still regarded as a luxury. Freedom from government intervention was restored, the BBC's broadcasting monopoly defended, and the Reithian mission of cultural uplift pursued once more. The BBC also reasserted its Britannic and imperial roles. The promise of continuity, of broadcasting as a stabilizing influence after a period of upheaval at home and abroad, perhaps seemed comforting in much the same way as it had to Reith and his colleagues in the 1920s. The BBC would be one of the institutions that, in the early 1950s, 'seemed to provide a stable and secure framework within which the decent life could be lived'.[2]

Public broadcasters in the dominions, as in Britain, assumed that radio would continue to be used to encourage the unity of the British world and of its component parts, and the perpetuation of British rule in the dependent colonies. Australian Broadcasting Commission (ABC) Chairman Richard Boyer envisaged the British Empire/Commonwealth as a modern, progressive, unifying influence in an increasingly interdependent post-war world. For Boyer, the Commonwealth was a model for an 'inter-racial' global community, despite its internal divisions.[3] Public broadcasting was a product of the 'characteristic genius for compromise [of] the British people', and could help strengthen and spread this sense of unity.[4] The theme of international interdependence was also taken up by James Shelley, the New Zealand director of broadcasting. Shelley emphasized that, in a world characterized by improved long-range civil and military communications, Commonwealth cooperation would be vital, and national independence unrealistic.[5]

[1] Thomas Hajkowski, '*Red on the Map*: Empire and Americanization at the BBC, 1942–50' in Joel H. Wiener and Mark Hampton (eds), *Anglo-American Media Interactions, 1850–2000* (Basingstoke, 2007), 183–7. W. J. Haley to R. E. L. Wellington, 16 August 1944, BBC Written Archives Centre (WAC), E1/207/3.

[2] Brian Harrison, *Seeking a Role: The United Kingdom, 1951–1970* (Oxford, 2009), 69.

[3] R. Boyer, 'The British Empire and the World of Tomorrow—Talk, Royal Empire Society', 8 September 1948, National Library of Australia (NLA), Sir Richard Boyer papers, MS3181, 3/31.

[4] Boyer, [script for broadcast talk?], 'Do We Need National Broadcasting?', 21 August 1946, Boyer papers, 1/2.

[5] Minutes, Commonwealth Broadcasting Conference, 22 February 1945, Library and Archives Canada (LAC), RG41, 357/19-12-1/1.

Indeed, through their day-to-day work, broadcasters were made particularly aware of improvements in international airmail, telephone, and recording and transmission technologies, which made it easier (in theory, at least) to discuss, coordinate, and execute joint broadcasting initiatives. Faster and cheaper international travel (especially by air) meant that administrators, producers, and performers alike could tour the Commonwealth in increasing numbers, and make shorter and more frequent visits. In 1944 Stephen Fry anticipated that BBC 'flying production teams' would soon bring direct coverage of world affairs to British and overseas listeners.[6] Everyday experiences of improving practical cooperation seemed to indicate that closer imperial unity was an achievable goal for the future, not a wistful glance towards the lost worlds of the past.

Indeed, in considering how broadcasters envisaged the goals towards which they were working, it is worth emphasizing that the terminal decline of the British world and the British Empire became apparent to contemporaries only gradually, even after 1945. No one could deny that the post-war empire was different from what had gone before. India, Burma, and Palestine were stripped away in an early violent phase of 'decolonization', marked in particular by the bloody partition of India and creation of Pakistan. With the pace set by India and Burma, peaceful negotiations led to the granting of independence to Ceylon in 1948. British informal influence, meanwhile, collapsed in China and the Middle East, culminating in the latter case in the disastrous invasion of Suez in 1956. Yet to contemporary observers, particularly before the Suez crisis, the damage inflicted on Britain's world-system did not necessarily seem catastrophic. India, Pakistan, and Ceylon joined the Commonwealth. Following the Japanese surrender, British authority was successfully reimposed in Malaya, Singapore, and Hong Kong. Communist insurgents in Malaya were neutralized through a massive policing and military operation, underlining to US policymakers the potential Cold War benefits of Britain's imperial presence. A 'second colonial invasion' meanwhile took place in tropical Africa, where Britain sought to exploit raw materials as the basis for its own post-war economic recovery, and for the economic advancement of Africans too. In Kenya Mau Mau violence posed only a limited threat to a colonial state able and willing to deploy lethal force, and elsewhere in Africa it seemed possible either to contain nationalist movements, or divert them into channels compatible with British interests. Moreover, while the victory of the Afrikaner-dominated National Party in the 1948 South African elections presented clear problems, the member states of the British world seemed likely to continue to cooperate with one another for the foreseeable future.

If this post-war revival of the British Empire ultimately proved largely illusory, then the illusion at least cushioned the process of decline and facilitated gradual adjustment.[7] British politicians and civil servants remained remarkably 'imperially minded' throughout the decolonization period, and a wider cultural acknowledgment

[6] S. Fry to J. W. MacAlpine, 19 August 1944, WAC, E1/207/3.

[7] Wm. Roger Louis and Ronald Robinson, 'The Imperialism of Decolonisation', *Journal of Imperial and Commonwealth History*, 22/3 (September 1994), 462–511.

of retreat was even further delayed. Perversely, the cultural impact of empire on Britain perhaps became most noticeable after imperial decline had already set in.[8] In Canada, Australia, and New Zealand, Britannic identities similarly remained entrenched throughout the 1950s.[9]

Public broadcasters in the dominions tended to share a broader belief that it was still possible to reconcile national and Britannic identities, without either seeming 'too English' at one extreme, or 'playing down our Empire' at the other.[10] The BBC's representative in Australia was aware that the country was 'marching daily towards greater independence and nationhood', with economic development and closer links with the US encouraging 'a nationalistic frame of mind', particularly among the younger generation. However, he believed that there remained many Australians 'to whom the traditional links with Britain are fundamental'.[11] Similarly, the BBC's Canadian representative noted the passing of many Britannic emblems, and the 'increasing awareness among Canadians of their own identity', but claimed that this did not mean that Canada would inevitably break with Britain. The BBC could help Canadians in their new quest for a national culture by setting first-class standards: otherwise Canada would continue to suffer from 'a plethora of the second-rate in all fields'.[12] The BBC clearly retained something of its imperial civilizing mission. Visiting Britain in 1949, Keith Barry, ABC controller of programmes, still thought he could detect a patronizing BBC tendency 'to regard Australia as a place where missionary effort is required in most things cultural'.[13]

The late 1940s and the 1950s marked the high point of Britannic broadcasting cooperation. A second Commonwealth Broadcasting Conference was held in Britain in 1952, and a third in Australia in 1956. Increasing numbers of broadcasting officers moved around the British world. Transcriptions and short-wave transmissions continued to link audiences together across intervening oceans. The broadcasting of news, sport, and great royal ceremonial occasions provided impressive evidence of underlying practical and sentimental connections. However, at the same time, powerful forces were mobilizing to undermine Commonwealth broadcasting collaboration. The threatened post-war flood of US radio exports never came, as the resources of American private broadcasting were instead transferred into television. This gave the BBC something of a breathing space, and allowed it to maintain its global

[8] Stephen Howe, 'Internal Decolonization? British Politics since Thatcher as Post-Colonial Trauma', *Twentieth-Century British History*, 14/3 (2003), 286–304. John M. MacKenzie, 'The Popular Culture of Empire in Britain', *in* Judith M. Brown and Wm. Roger Louis (eds), *Oxford History of the British Empire*, iv, *The Twentieth Century* (Oxford, 1999).

[9] Phillip Buckner, 'The Long Goodbye: English Canadians and the British World' and David Lowe, 'Australia's Cold War: Britishness and English-Speaking Worlds Challenged Anew', both in Buckner and R. Douglas Francis (eds), *Rediscovering the British World* (Calgary, 2005). Stuart Ward, *Australia and the British Embrace: The Demise of the Imperial Ideal* (Carlton South, Vic., 2001).

[10] E. J. Turner to [A. D. Dunton], 14 March 1951 and Dunton to Turner, 21 March 1951, LAC, RG41, 906/PG10-9/3.

[11] P. Jubb to G. Looker, 1 June 1950, WAC, E1/332/1. Jubb to R. McCall, 4 April 1952, E1/332/2.

[12] T. Sloan, 'Canadian Representative's Report', 1 May 1952, WAC, E1/509/5.

[13] K. Barry to C. Moses, 14 January 1949, National Archives of Australia (NAA), Sydney, SP724/1, 8/1/25, 1360456.

supremacy in radio, into the 1960s and beyond. In the emergent mass medium of television, however, the BBC faced more serious challenges.

RE-ESTABLISHING PUBLIC BROADCASTING

After the war direct state control of broadcasting continued in New Zealand, and in newly independent India and Pakistan (with partition, three AIR stations were transferred to a new authority: Radio Pakistan).[14] Meanwhile, in terms of their domestic broadcasting operations at least, the BBC, ABC, Canadian Broadcasting Corporation (CBC), and South African Broadcasting Corporation (SABC) were all essentially restored to a position of autonomy, albeit with some caveats. In Australia, where many politicians seemed 'profoundly suspicious of "culture"' and thus of the ABC, two new 'official' seats were added to the commission: one to be filled from the Postmaster-General's (PMG's) Department and the other from the Treasury.[15] The ABC was no longer to receive a guaranteed share of the listener licence fee, but would rather rely on an annual grant from parliament, reducing its financial autonomy. An Australian Broadcasting Control Board was also established with regulatory powers over both public and private stations.[16] The future independence of the SABC seemed even more uncertain. The responsibilities assigned to the SABC expanded after the war, notably with the introduction in 1950 of commercial broadcasting under SABC auspices: Springbok Radio. However, under National Party influence, the SABC's board of control was increasingly dominated by members of the *Broederbond* (an Afrikaner nationalist secret society) and seemed set to become a tool of republicanism and apartheid. Both René S. Caprara and his successor as director-general, Gideon Roos (a Cape Afrikaner), were liberals, and relatively enthusiastic about the Commonwealth link. Nevertheless, from 1950 the SABC ceased rebroadcasting BBC news, a sign of things to come.[17]

At the BBC, the Reithian ideal of uplift was reasserted, with Haley offering only a modestly revised version of older arguments about broadcasting and cultural standards. Instead of viewing the audience as a myriad collection of individual listeners, all of whom could be taught to appreciate the finest cultural products at the same time and in similar ways, Haley presented 'the community as a broadly based cultural pyramid slowly aspiring upwards'. The BBC would target different strata of the pyramid separately: individual listeners would work their way up through the hierarchy over time, and cultural standards would improve overall.[18] Haley's BBC gave listeners a choice of three domestic services, reflecting this

[14] H. R. Luthra, *Indian Broadcasting* (New Delhi, 1986), 164–72.

[15] N. Hutchison to J. B. Clark, 29 April 1947, WAC, E1/356/2.

[16] K. S. Inglis, *This is the ABC: The Australian Broadcasting Commission, 1932–1983* (Carlton, Vic., 1983), 131–3.

[17] Graham Hayman and Ruth Tomaselli, 'Ideology and Technology in the Growth of South African Broadcasting', 46–54, in Ruth Tomaselli, Keyan Tomaselli, and Johan Muller (eds), *Broadcasting in South Africa* (Bellville, 1989).

[18] Sir William Haley, *The Responsibilities of Broadcasting: The Lewis Fry Memorial Lectures delivered in the University of Bristol, 11, 12 May 1948* (BBC, 1948), 11.

philosophy of stratification, and also in an attempt to fend off the introduction of outside competition. The Home Service would offer a broad middle group a mixture of different types of programmes, 'paying attention to culture at a level at which the ordinary listener can appreciate it', and also some regional variations. The Light Programme would cater to those at the base of the pyramid, 'those who look to broadcasting purely for relaxation and amusement'. But it would also maintain high 'standards of integrity and taste', and infiltrate into the overall pattern of light music and entertainment some news, documentary, and current affairs material (notably in the new *Woman's Hour* session), and 'light classical music'. For the 'serious listener' at the top of Haley's pyramid, from September 1946 the Third Programme broadcast, 'without regard to length or difficulty, the masterpieces of music, art, and letters which lend themselves to transmission in sound'. Listeners were encouraged to move between the networks, and some programmes were broadcast on more than one network.[19] Despite its essential conservatism, Haley's new approach did diverge somewhat from the Reithian ideal of mixing programme output on a single service so as to expose listeners to varied types of material. Some at the BBC worried that the new approach acted to immobilize the tastes of individual listeners, rather than improve them, and to create a minority Third Programme ghetto for the 'best listening'.[20]

To some extent, Haley's ideas about the pyramid of taste were exported overseas. After the war New Zealand's National Broadcasting Service and National Commercial Broadcasting Service were amalgamated to form the New Zealand Broadcasting Service (NZBS). In 1950 the NZBS followed the BBC's lead, dividing its stations into 'light', 'middlebrow', and 'third' services.[21] From 1946 the ABC similarly placed 'serious' material on its National Network, and 'light' material on its Interstate Programme. The ABC also wished to established a third network of stations, in order to carry parliamentary broadcasts, talks, and other 'higher quality' offerings (including BBC transcriptions of Third Programme material). However, the government repeatedly refused to grant the ABC authorization for this.[22] Meanwhile, the CBC had introduced a second network (the Dominion Network) at the end of the war, to carry light, commercial material. In 1947, on the old Trans-Canada Network, it introduced *Wednesday Night*: a weekly slot for 'serious' programming intended as something akin to the Third Programme. Championed amongst others by Rex Lambert (CBC supervisor of school broadcasts and former editor of the BBC's *Listener*) the new initiative chimed with broader demands from cultural critics for a 'new cultural order' in post-war Canada. Citizens would be

[19] *BBC/1—Confidential—Broadcasting Committee 1949—General Survey of the Broadcasting Service* (BBC, May 1949), 29, Manchester Archives and Library Services, Lord Simon of Wythenshawe papers, M11/6/5c.

[20] D. G. Bridson, *Prospero and Ariel: The Rise and Fall of Radio, a Personal Recollection* (London, 1971), 177–9.

[21] Patrick Day, *A History of Broadcasting in New Zealand*, i, *The Radio Years* (Auckland, 1994), 308–11.

[22] Inglis, *This is the ABC*, 141–6. Moses to ABC commissioners, 23 January 1950, NAA, Sydney, SP724/1 8/1/48, 3159905. *Twenty-First Annual Report and Financial Statements of the Australian Broadcasting Commission Year Ended 30th June 1953* (Sydney, 1953), 7.

encouraged to discharge their democratic responsibilities, and to enjoy 'folk' and 'high' culture as much as US-produced mass culture.[23] Initially, the BBC's Canadian representative judged the mix of *Wednesday Night* programmes to be more like 'a rather light [BBC] Home Service', particularly as no one programme was of over an hour in duration.[24] This dismissive response perhaps reflected the fact that *Wednesday Night* was also designed to boost 'Canadian content', and at first made only very limited use of BBC transcriptions.[25] However, by the mid-1950s the BBC had become more positive about the standard of music and drama broadcast in the slot: it was not, perhaps, a coincidence that the amount of BBC material included had increased.[26]

Meanwhile, broadcasting in Britain's tropical colonies followed a different path: Haley argued that there was still 'vital pioneer work' to be done.[27] Until the mid-1950s, at least, colonial broadcasting policies continued to reflect a wider British assumption that self-government was something for the medium to long term, only to be achieved after a lengthy period of economic 'development', education, and political tutelage under continued British rule.[28] During the war some progress had been made in establishing rediffusion and medium-wave services, but white expatriates still constituted the bulk of the audience, particularly in tropical Africa. As far as the Colonial Office was concerned, the pre-war report of the Plymouth Committee thus remained a key document. Broadcasting was to be a tool of colonial policy, a means of promoting economic development and suppressing political subversion. £1 million was set aside in 1949 under the Colonial Development and Welfare Act to support broadcasting initiatives, to be supplemented by local funds. Most of this money was earmarked for the African colonies, where broadcasting was least developed. A further £250,000 was made available in 1952. More than forty projects were sponsored, providing studios, transmitting stations, and listening facilities in twenty-seven countries.[29]

The BBC played a key supporting role in these developments. Between 1945 and 1953 its engineers undertook technical surveys of broadcasting requirements and possibilities around the empire, paying particular attention to the West Indies, Burma, Singapore, and tropical Africa. An *English by Radio* language teaching

[23] Peter Stursberg, *Mister Broadcasting: The Ernie Bushnell Story* (Toronto, 1971), 133. L. B. Kuffert, *A Great Duty: Canadian Responses to Modern Life and Mass Culture, 1939–1967* (Montreal and Kingston, 2003), 67.

[24] M. Barkway to MacAlpine, 27 November 1947, WAC, E1/586/1.

[25] J. Polwarth to A. E. McDonald, 11 June 1948, WAC, E1/586/2.

[26] P. Gallagher to H. C. Greene, 9 July 1956; B. Thornton to O. J. Whitley, 1 February 1957; M. Frost to D. Stephenson et al., 13 June 1957, all in WAC, E1/1786.

[27] Sir William Haley, *Broadcasting as an International Force: The University of Nottingham, Montague Burton International Relations Lecture, 1950–1* (Nottingham, *c.*1951), 19.

[28] Michael Havinden and David Meredith, *Colonialism and Development: Britain and its Tropical Colonies, 1850–1960* (London, 1993). Ronald Hyam, *Britain's Declining Empire: The Road to Decolonisation, 1918–1968* (Cambridge, 2006), 139–46.

[29] Minutes, Commonwealth Broadcasting Conference, 1 March 1945, CBA archive, box 1, C/1/1. 'Minutes, reports and summaries of meetings of Commonwealth Broadcasting Conference 1952, BBC, November 1952', extended summaries of discussion, 25 June 1952, LAC, RG41, 367/19-31-1/2. Eliot Watrous, 'Broadcasting in the Colonies', *London Calling*, 10 December 1953.

project had already been introduced during the war, and a BBC Colonial Schools Transcription Unit was now established.[30] At one point, the BBC even suggested that it should assume direct control of broadcasting in the African colonies: the BBC's empire-building tendencies were still apparent. However, while the prospect of the BBC parlaying its domestic monopoly into a colonial monopoly was welcomed by some colonial officials, it was rejected by the British governor of Nigeria. Instead, the BBC would second staff to fledgling colonial state-run broadcasting authorities, and offer access to training facilities in Britain.[31]

In another continuation of pre-war and wartime patterns, BBC domestic and overseas services were meanwhile mobilized to publicize 'the modern British conception of Empire and Commonwealth' among UK and world audiences, and particularly to draw attention to British plans for the economic development of the tropical colonies.[32] J. Grenfell Williams, Head of the BBC's Colonial Service, urged that in order to achieve this, more information about the empire had to be 'infiltrated' into BBC news and current affairs programmes, including coverage of 'the bright side to the picture' of colonial rule.[33] The Colonial Office similarly urged the BBC that, in view of the financial burden that would have to be imposed on the UK taxpayer to foster colonial development, 'the British public should be induced to take a greater interest in colonial affairs'.[34] Gradually, the BBC began to work towards this goal, starting in 1949 with a series (*The Colonial Dilemma*) introduced by the historian and scholar of colonial administration Margery Perham.[35]

The BBC also helped deploy radio to reinforce the authority of the colonial state and, where necessary, to complement violent counter-insurgency campaigns. Hugh Carleton Greene (a cousin of Felix Greene, and brother of the novelist Graham Greene) had joined the BBC's German Service during the war, and subsequently became Head of the East European Service. His experience of conducting propaganda campaigns, particularly against Communists, qualified him for a year's secondment to Malaya to help oversee the 'hearts and minds' campaign. Meanwhile, in Cyprus the facilities of the colonial radio service were clearly perceived as a prop of British rule, and were twice blown up by insurgents in 1955.[36]

In the tropical colonies radio was also linked with processes of peaceful constitutional change. Before the Second World War remote state control of broadcasting, like British parliamentary institutions more generally, had not been deemed

[30] Briggs, *The History of Broadcasting in the United Kingdom*, iv, *Sound and Vision* (Oxford, 1995 [1979]), 483–90.

[31] Charles Armour, 'The BBC and the Development of Broadcasting in British Colonial Africa, 1946–1956', *African Affairs*, 332/83 (July 1984), 359–402, esp. 374–5 and 378–81.

[32] J. A. Camacho, 'Empire and Commonwealth Series', 15 January 1948, WAC, R51/91/5. Thomas Hajkowski, *The BBC and National Identity in Britain, 1922–53* (Manchester, 2010), 69–71.

[33] J. G. Williams, 'Programmes on the Commonwealth and Empire', 12 July 1948, WAC, R51/91/5.

[34] Arthur Creech Jones to Haley, 14 January 1949, ibid.

[35] Christopher Holme to C. Conner, 12 January 1950, ibid.

[36] Greene, 'Report on Emergency Information Services, September 1950–September 1951', 14 September 1951, Bodleian Library, Sir Hugh Carleton Greene papers, box dep.c.903. 'Opening Addresses to the Commonwealth Broadcasting Conference', 27 May 1963, Commonwealth Broadcasting Association (CBA) archive, box 1, C/1/5.

appropriate to the colonies. However, during the 1950s as preparations began to be made for the transition to self-government, these assumptions changed and the BBC helped deliver a matching litter of public broadcasting corporations. In 1953 Williams led a detailed investigation of broadcasting in Gold Coast, and recommended the establishment of an interim broadcasting council under state auspices, which would ultimately hand power to a BBC-style independent broadcasting corporation. It was hoped that this would set a precedent for other colonies, and the following year Williams produced a similar report for the Kenyan colonial government. In Nigeria an autonomous broadcasting corporation was actually set up before the transfer of power: in 1957 the Nigerian Broadcasting Service became the Nigerian Broadcasting Corporation.[37] In the Central African Federation of Rhodesia and Nyasaland a commission of inquiry headed by Greene (who had since become Controller of the BBC Overseas Service) recommended the establishment of a public corporation under remote state control, and the Federal Broadcasting Corporation of Rhodesia and Nyasaland duly began operations in 1958. The following year autonomous broadcasting corporations were inaugurated in both Cyprus and Jamaica.[38]

These corporations were established hastily, as the pace of more general political change began to accelerate in the colonies in a way that few had previously predicted.[39] British officials hoped that, despite decades of paternalistic and authoritarian colonial rule, democratic values could be grafted quickly onto local societies. Public broadcasting, with its emphasis on preparing listeners for active citizenship, might help achieve this. British policymakers also feared that many colonies, in which widely differing groups had been brought together for strategic or administrative convenience, would disintegrate into warring statelets following independence. Again, public broadcasting offered a possible solution, through its nation-building functions. The new public broadcasting authorities would also, it was hoped, resist the commercialization of broadcasting, limit American influence, and sustain cultural and political links with Britain. In characteristically forthright and unreconstructed terms, Haley's successor as director-general, Ian Jacob, wrote of the risk that private stations in Africa and Asia would become 'very third rate, & 90 per cent American. The growth of organizations eager to establish such stations at no cost to the country concerned reminds me of the days when we sold rum to the natives!'[40] In contrast, fostering public broadcasting in the colonies perhaps seemed a modern variation on the theme of nineteenth-century 'legitimate commerce'. Nevertheless, in some colonies, such as Hong Kong, privately owned commercial broadcasting continued to spread.[41]

[37] Watrous, 'Broadcasting in the Colonies'. Briggs, *History of Broadcasting in the UK*, iv, 488. Ian K. Mackay, *Broadcasting in Nigeria* (Ibadan, 1964).

[38] 'Report on the Commission of Inquiry into the Organization of Broadcasting within the Federation', 31 May 1955, Greene papers, box dep.c.903.

[39] Graham Mytton, *Mass Communication in Africa* (London, 1983), 64–5.

[40] I. Jacob to Boyer, 10 December 1959, Boyer papers, 1/12.

[41] David Clayton, 'The Consumption of Radio Broadcast Technologies in Hong Kong, *c.*1930–1960', *Economic History Review*, 57/4 (2004), 691–726.

Like the BBC, public broadcasters in Canada, Australia, and New Zealand tried to impress upon their domestic audiences new ideas about colonial rule and colonial development. During 1952/3, for example, the ABC *Nation's Forum of the Air* session treated listeners to discussions about 'The British Commonwealth To-day', the Colombo Plan, British policy in Kenya, 'Planning within the British Commonwealth', and the diverse forms of government to be found around the Commonwealth.[42] For the CBC, Matthew Halton (also a regular BBC contributor) provided frequent reports from London on what he called Britain's 'creative abdication' of imperial power. For Halton, writing before the Suez crisis, Canadians might regret the end of empire ('Who among us, with *our* upbringing, is not sad to see the red receding slowly down the map?') but at least British decolonization seemed to be occurring on a peaceful, consensual basis, and not 'a series of last-ditch bloody stands ending in catastrophe, obloquy and hate'.[43]

The CBC, ABC, and NZBS also joined the BBC in promoting the development of broadcasting in the colonies, offering training, advice, and programmes.[44] The CBC's Chairman A. Davidson Dunton travelled to Jamaica to advise on radio licensing policy, and in 1959 two CBC officers visited Ghana to assist with the introduction of television.[45] For these broadcasting authorities, such efforts were partly aimed at developing a new leadership role for themselves in a Commonwealth context. Charles Jennings argued that, by cultivating links with India, Pakistan, and Ceylon, Canada could form a much-needed bridge between 'West' and 'East': 'as a leader of such a movement, the CBC would gain for itself a very great deal of prestige'.[46] The ABC and NZBS meanwhile focused their efforts on those countries deemed to be of particular importance to Australia and New Zealand. From 1950, under the auspices of the Colombo Plan and UNESCO, the ABC offered training to officers from Asian broadcasting organizations, and from 1956 also assisted the establishment of schools broadcasting outside the Commonwealth, in the Philippines.[47] Intervention in the Pacific region was more direct, and by 1952 the ABC had established a station at Port Moresby, Papua New Guinea, and the NZBS had erected a transmitter and was providing programme and technical assistance in Western Samoa. Subsequently, the NZBS also provided advice to the new Fiji Broadcasting Commission.[48] Like the BBC's work in the colonies,

[42] *Twenty-First Annual Report and Financial Statements of the Australian Broadcasting Commission Year Ended 30th June 1953* (Sydney, 1953), 54.

[43] M. Halton, script for 'Capital Report', 22 February [1953], LAC, Matthew Halton fonds, 4, 'Capital Reports, News Roundup, Report on Britain Series, 1948–50'. Emphasis in original.

[44] 'Commonwealth Broadcasting Conference—Part I—Final Summary of Conclusions', 9 March 1945, LAC, RG41, 357/19-12/2.

[45] G. A. Brown to Dunton, 26 October 1956, LAC, RG41, 989/12. R. D. Cahoon and S. R. Kennedy, 'Recommendations on the establishment of Television Service in Ghana', December 1959, RG41, 391/20-13.

[46] C. Jennings to E. Bushnell, 31 January [1952], LAC, RG41, 367/19-31/1.

[47] *28th Annual Report and Financial Statement of the Australian Broadcasting Commission Year Ended 30th June 1960* (Sydney, 1960), 7–8. NAA, Sydney, SP724/1, 14/1/12 part 1, 3160898.

[48] 'Minutes, Reports and Summaries of Meetings of Commonwealth Broadcasting Conference 1952, BBC, November 1952', extended summaries of discussion, 25 June 1952, LAC, RG41, 367/19-31-1/2. R. Stead to I. E. Thomas, 29 March 1954, WAC, E1/332/3.

all this raised an important question: were public broadcasters perpetuating 'neo-colonialism', or preparing their neighbours for independence?

MOBILITY

Links between public broadcasting authorities were strengthened in this period by the increased mobility of personnel, continuing trends established during the war, when many broadcasting officers from the dominions had experienced life and work in Britain. Some acquired a taste for expatriation, and the contacts to facilitate a permanent move. Rooney Pelletier and Stanley Maxted both stayed on in Britain after 1945; so did the New Zealander Tahu Hole, who was promoted from Manager of Overseas Talks to become BBC News Editor. Robert McCall returned after a short interlude back in Australia, and worked as Controller of the BBC Overseas Service and assistant director of television before becoming regional controller for Northern Ireland. Many others who lacked wartime experience in Britain also came to the BBC from the dominions after 1945. Some stayed only temporarily; others became permanent migrants. Senior executives tended to come for short periods, seeking information on issues of pressing importance back home: for example, in 1946 the SABC Director-General Caprara travelled the English-speaking world gathering information prior to the introduction of commercial radio in South Africa. Junior officers often visited Britain for longer periods, sometimes working while on unpaid leave, and seeking to gain new expertise that would improve their promotion prospects at home.[49]

A significant number of Australian and New Zealand radio performers also migrated to London, which after the war re-established itself as an imperial cultural capital. One disappointed hopeful lamented that Australians were regarded 'as a species of bandits and bounders' in London.[50] However, others had notable success, including Joy Nichols and Dick Bentley, stars of *Take it From Here*, one of the BBC's most successful comedy shows of the early post-war years.[51] Bentley, along with script-writers Frank Muir and Denis Norden, visited Australia on two occasions to make a special series for the ABC: *Gently Bentley*.[52] Canadians had easy access to alternative opportunities in the US, but some did rise to prominence

[49] Simon J. Potter, 'Strengthening the Bonds of the Commonwealth: the Imperial Relations Trust and Australian, New Zealand, and Canadian Broadcasting Personnel in Britain, 1946–52', *Media History*, 11/3 (December 2005) and 'The Colonization of the BBC: Diasporic Britons at BBC External Services, *c.* 1932–1956' in Marie Gillespie and Alban Webb (eds), *Diasporas and Diplomacy: Cosmopolitan Contact Zones at the BBC World Service* (forthcoming). *South African Broadcasting Corporation—Annual Report, 1946*, 6–7.

[50] E. Roland to L. Rees, 26 February 1953, Mitchell Library (ML), MSS5454, Leslie Rees papers, 1.

[51] Obituary for Dick Bentley, *The Guardian* (London), 29 August 1995.

[52] *Twentieth Annual Report of the Australian Broadcasting Commission year ended 30th June 1952* (Sydney, 1952), 10. *Twenty-third Annual Report of the Australian Broadcasting Commission Year Ended 30th June 1955* (Sydney, 1955), 16.

in Britain: most notably Bernard Braden and his wife Barbara Kelly, who in 1949 left Canada to launch UK careers as radio and television personalities.[53]

Presiding over the 'colony of A.B.C. folk' in the austere surroundings of a blitzed post-war London was T. W. Bearup, head of the ABC's new London office.[54] Bearup had acted as ABC general manager during the early stages of the war, but was sidelined upon Charles Moses' return. The London posting gave Bearup a role that was suitably senior and yet conveniently far from Sydney. It was suggested that Bearup might also represent the SABC. However, stopping off in South Africa en route, he was struck by the antipathy between English- and Afrikaans-speakers, and the strength of anti-British sentiment among the latter. The requirements of the SABC and ABC seemed too different to permit of joint representation. Subsequently, neither the SABC nor the NZBS established a London office, due to lack of enthusiasm in the former case, and lack of resources in the latter.[55]

In London Bearup was instructed to liaise with the BBC and other broadcasting organizations, and to persuade them to take more Australian material 'so that the Australian scene can be projected in and from Britain and the Commonwealth in increasing measure'. Bearup was tasked with selecting material from Britain for use by the ABC, channelling information about broadcasting in Britain back to Australia, and supervising the commission's London news operation.[56] While the BBC offered office accommodation, Moses initially preferred to set Bearup up in his own offices, to preserve the ABC's autonomy.[57] By the mid-1950s, however, both the ABC and the CBC had situated their London offices in the Langham building, opposite the BBC's Broadcasting House. The BBC leased the Langham and covered many of the basic costs in the building. It also offered the CBC and ABC access to radio facilities on a relatively free and informal basis. Similar privileges were granted to BBC representatives in Canada and Australia in return.[58]

The CBC London office was initially headed by Andrew Cowan, who had served in the wartime CBC Overseas Unit. Much of Cowan's work concerned the CBC's new short-wave International Service. Liaison with the BBC was also important, particularly securing BBC material for the CBC, and placing CBC material on BBC domestic services.[59] Cowan was often frustrated by what he perceived to be the bureaucratic, pedantic, and superior attitude of senior BBC officers:

> The BBC suffers from what I call institutional arrogance. It is a form of imperialism. They unconsciously feel that they have the best broadcasting system in the world, staffed by the best possible people, producing the best possible programmes in the best

[53] 'Bernard Braden dies at 76', *The Guardian* (London), 3 February 1993.

[54] T. W. Bearup to V. Taylor, 6 April 1947, NLA, T. W. Bearup papers, MS7290, 1/1/2.

[55] 'Transcript of Taped Interview with T.W. Bearup, Recorded 1968–70', ML, Clement Semmler papers, MSS5636/19. Bearup, 'Report on Broadcasting in the Union of South Africa', 5 January 1946, ABC Document Archives, Charles Moses correspondence.

[56] Moses to Bearup, 15 November 1945, NAA, Sydney, SP1558/2, box 76, 'ABC Overseas Representative—policy London Office opening of (Bearup), 1945–46'.

[57] Moses to H. B. Chase, 31 October 1945, LAC, RG41, 35719-12-1/2.

[58] 'CBC–BBC Financial Agreement from 1st April 1949', WAC, E1/524. Press cutting, 'CBC Expands London Operation', *The Star* (Montreal), 19 November 1966, LAC, RG41, 391/20-9/10.

[59] I. Dilworth to A. Cowan, 26 July 1951, LAC, Andrew Gillespie Cowan fonds, 8.

possible way. While they look for an unrestricted export of their culture, they have a high tariff policy towards any incoming material.[60]

Despite such tensions, the establishment of the ABC and CBC London offices brought a marked improvement in broadcasting liaison between Australia, Canada, and Britain.

Similarly, the BBC's own outposts in Canada, Australia, and India, and the BBC's Overseas Liaison Office in London, became vital links with Commonwealth broadcasters, encouraging programme exchange. However, contrary to the urgings of the Dominions Office, which wished 'to ensure that the B.B.C. adequately competes with various forms of United States publicity', the BBC declined to establish similar representation in South Africa or New Zealand.[61] As during the 1930s, the BBC would only go so far in reshaping its organizational structure, or its programming, in order to promote imperial interests.

During the 1950s the pattern of the BBC's connections with overseas broadcasters began to change, as the dependent colonies began to loom larger on the corporation's horizons. By the mid-1950s the BBC was receiving more visitors from the colonies than from self-governing Commonwealth countries, even though the introduction of television was bringing a significant number of CBC and ABC officers to Britain. Between 1952 and 1956 twelve Commonwealth visitors were seconded to BBC External Services for a year or more, sixty attended the BBC Training School, and 131 were attached to various BBC departments.[62] This compared to 125 trainees from the dependent colonies over the same period (including twenty-three from Gold Coast and thirty-eight from Nigeria), and a further 108 colonial attachments to BBC departments (including eighteen from Singapore/Malaya, sixteen from Cyprus, ten from Gold Coast, and nine from Nigeria).[63] Visitors from the dependent colonies to the BBC often reported in a positive manner on their experiences in Britain, although not as unambiguously as did their white Commonwealth counterparts.[64]

The outward flow of officers on secondment from the BBC was similarly directed overwhelmingly towards the dependent colonies rather than the self-governing Commonwealth. By 1953 the BBC had loaned seventeen engineers and nine programme staff to eight different colonial territories: that year, it seconded an additional fourteen members of staff to Nigeria alone.[65] In contrast, only five two-way staff exchanges were arranged between the BBC and the ABC between 1945 and 1952, and only one with the SABC.[66] No further exchanges involving BBC officers and their dominion counterparts were organized until 1956/7, when a BBC–ABC

[60] Cowan to [Neil Morrison], 26 September 1952, Cowan fonds, 16, 'CBC International Service, Correspondence 1947–52'.

[61] R. B. Pugh to Clark, 15 March 1946, WAC, E1/356/1.

[62] 'Commonwealth Broadcasting Conference 1956—BBC Memorandum—Staff Exchanges, Secondment and Training—Agenda Part 1—C/10', Boyer papers, folio box, folio 1.

[63] 'Commonwealth Broadcasting Conference, 1956—BBC Memorandum—C/68—Colonial Guest Students at BBC Training Courses (Agenda Part 1)', Boyer papers, folio box, folio 1.

[64] Potter, 'Colonization of the BBC'.

[65] Watrous, 'Broadcasting in the Colonies'.

[66] 'Commonwealth Broadcasting Conference 1952—BBC Memorandum—Staff Exchanges and Training', LAC, RG41, 367/19-31/5.

exchange was agreed.[67] With more than a hint of paternalism, the BBC's Australian representative claimed that while officers arriving at the BBC from overseas were seldom of 'the same calibre as those we send out from London', the resultant drain on BBC resources was to be accepted as part of the corporation's broader empire and Commonwealth commitment. Effectively, exchanges were viewed as a form of training for the staff of overseas broadcasting authorities, which would work to the BBC's long-term benefit in terms of future collaboration.[68]

Meanwhile, some BBC officers moved overseas on a long-term basis: imperial careers in broadcasting were still a possibility for British broadcasting personnel, made particularly attractive before the mid-1950s by the absence of any domestic alternative to the BBC as an employer of broadcasters. The BBC television producer Harry Pringle thus moved to Australia to become the ABC's Director of Variety, while Neil Hutchison (BBC Australian representative, 1945 to 1949) elected to stay in Australia at the end of his tour of duty, and was appointed ABC Director of Features, despite internal opposition to an 'outsider' taking a senior post. Hutchison's subsequent promotion to Director of Drama and Features, and his candidacy for the post of Entertainment Programme Director, provoked further complaints about jobs being given to 'refugees' from the BBC. Nevertheless, in 1962 he became director of productions, and in 1965 controller of programmes. Patrick Jubb, Hutchison's successor as BBC Australian representative, meanwhile went on to become controller of programmes at the Kenya Broadcasting Service.[69] Broadcasters still moved in an imperial world.

EXTERNAL SERVICES

Short-wave and transcription services had played a dominant role in the BBC's efforts to use broadcasting as a tool of empire during the 1930s, and had expanded significantly during the war. In the post-war years the stated aims of BBC External Services (the overall label now applied to short-wave, transcription, and monitoring operations) were 'the spread of truth through the provision of an accurate, comprehensive and objective service of information [and] the projection of Britain by making known the British ways of life and thought in all their aspects'.[70]

In practice, of course, it remained difficult to separate 'neutral' information and benevolent national projection from official Cold War propaganda. Indeed, the exact nature of the relationship between BBC External Services and the British state

[67] 'Commonwealth Broadcasting Conference 1956—BBC memorandum—Staff Exchanges, Secondment and Training—Agenda Part 1—C/10', Boyer papers, folio box, folio 1.

[68] Jubb to McCall, 23 May 1952, WAC, E1/332/2. On the broader context of media training and British (and American) influence see Jeremy Tunstall, *The Media are American: Anglo-American Media in the World* (London, 1977), 214–18.

[69] *Seventeenth Annual Report of the Australian Broadcasting Commission year ended 30th June 1949* (Sydney, 1949), 22. Hutchison to McCall, 18 January 1949, WAC, E1/356/2. Press cutting of 'Aussies Not Wanted', *Australian Worker*, 13 December 1957 and R. Wood to A. Calwell, 21 August 1959, both in NLA, Arthur Calwell papers, MS4738, 132/43.

[70] *BBC Memorandum—General Survey of the Broadcasting Service*, 39.

remained obscure to outside observers. While the domestic side of BBC operations returned to the inter-war system of funding by listener licence fee, the war-time practice of a Treasury grant-in-aid remained in place to fund External Services. The Foreign Office designated the target countries for BBC broadcasts, and the number of hours to be dedicated to each audience. Broadcasters worked in close daily contact with civil servants from different government departments, and senior BBC officers were granted access to confidential state documents. Yet in theory at least, the BBC retained full 'editorial control' over programme content. In the dominions, public broadcasting authorities tasked with managing overseas services saw all this as a model to emulate, but politicians and civil servants often proved reluctant to grant them the autonomy seemingly enjoyed by the BBC.[71]

During the war the idea of establishing a formal, coordinated Empire Broadcasting Network had made little progress. Governments in Canada, Australia, and India had instead established their own separate short-wave services and, in the early post-war years, Pakistan, South Africa, and New Zealand followed suit. These services tended to target audiences of particular importance to the originating country, but often also discharged a complementary Commonwealth function. Radio Australia's service to Asia and the Pacific sought, for example, to present the broader 'British Commonwealth case', and included regular relays of BBC news, commentaries, and entertainment transcriptions.[72] Practical opportunities for collaboration were frequent, particularly in the context of the western bloc's broader Cold War propaganda effort. Broadcasters cooperated to monitor and coordinate frequency usage, thus minimizing the effects of Soviet jamming, and to some extent arranged their short-wave schedules on a non-competitive basis.[73] Yet such cooperation was ad hoc rather than systematic, and had its limits: for example, while the BBC assisted the CBC by providing it with additional transmitter time for its European services, it was reluctant to carry too many Canadian programmes, for fear that they would compete with its own.[74]

The BBC's short-wave activities generally overshadowed those of other Commonwealth countries. During the early stages of the Cold War the BBC operated thirty-one short-wave transmitters in the UK, controlled the British Far Eastern

[71] Dilworth, 'Appendix to Report of Sub-Committee Appointed by Advisory Committee, CBC International Service', 10 May 1948, LAC, RG41, 988/11. Barry, 'BBC Organisation, with Particular Reference to Programmes', [*c.* February 1949], NAA, Sydney, SP724/1, 25/1/29, 3165637. 'Overseas Shortwave Service—Commission Meeting 26–7 April 1950', Boyer papers, 2/15. Alban Webb, 'Constitutional Niceties: Three Crucial Dates in Cold War Relations between the BBC External Services and the Foreign Office', *HJFRT*, 28/4 (October 2008), 557–67. Errol Hodge, *Radio Wars: Truth, Propaganda and the Struggle for Radio Australia* (Cambridge, 1995). James Larry Hall, *The History and Policies of the Canadian Broadcasting Corporation's International Service* (Univ. of Ohio Ph.D. thesis, 1973). The Suez crisis did see a serious attempt by the UK government to undermine BBC editorial control through threatened funding cuts. See Briggs, *History of Broadcasting in the UK*, v, *Competition* (Oxford, 1995), 73–137.

[72] 'Radio Australia and Programmes', 26 October 1948, University of Melbourne Archives, Tom Hoey papers, 1/11.

[73] Stead to J. A. Terraine, 9 July 1954, WAC, E1/332/3. Minutes, Commonwealth Broadcasting Conference, 30 June 1952, LAC, RG41, 367/19-31-1/2.

[74] Cowan to Jacob, 9 July [1947] and Clark to C. R. Delafield, 5 November 1951, WAC, E1/497/1.

Broadcasting Service (BFEBS) at Singapore, and reserved eight and a half hours of programme time daily on the Ceylonese government's short-wave transmitter (formerly Radio SEAC). While bespoke scheduling on the Pacific Service and the North American Service (NAS) was slimmed down, the BBC's General Overseas Service expanded to carry more programmes of shared interest to listeners around the world. Reductions in the grant-in-aid after 1949 necessitated cuts, but these tended to be made in the BBC's European operations. Services in Afrikaans were one of the few early imperial casualties, and during the late 1950s other services to Africa were actually expanded, with regular broadcasts in Hausa, Somali, and Swahili aimed at maintaining British influence and counteracting hostile radio propaganda from Radio Cairo, which had replaced the tamer Egyptian State Broadcasting Service.[75] The main impact of austerity was the failure to implement plans to build short-wave relay stations in the West Indies, Kenya, or Ceylon.[76] In 1951 a relay station was opened at Tebrau in Malaya; but this had two, rather than the recommended four, transmitters, and was designed primarily to serve Asia, rather than to boost signals to Australia and New Zealand.[77]

While Britain remained a significant short-wave broadcaster into the 1950s, Canada, Australia, and New Zealand became less important short-wave target areas. The war had added a temporary urgency to the flow of news and programme material from Britain, making short wave a particularly valuable channel for communication, with timeliness compensating for poor signal strengths. After the war, however, the usefulness of short wave as a means to reach audiences that already enjoyed well-developed local medium-wave services again came into question. Direct listening in Australia, New Zealand, and Canada was minimal, restricted to 'the fanatical pro-British' and 'inveterate "knob-twiddlers"'.[78] Rebroadcasting by local medium-wave stations (the only way to reach significant audiences) meanwhile declined markedly from its wartime peak. As the BBC's representative in Canada reported, regular rebroadcasting by the CBC fell, partly due to CBC officers' 'renewed sense of their national duty to Canada (interpreting one part of the country to another, and so on)', partly because the CBC wanted to produce its own equivalents of BBC programmes like *Radio Newsreel*, and partly because many BBC programmes simply did not meet the CBC's requirements.[79]

During 1948 the ABC, CBC, and SABC each carried around five hours of BBC rebroadcasts weekly, and the NZBS almost seven. Unsurprisingly, the focus was on topical material: news (including special reports from 'home' for the post-war wave

[75] *BBC Memorandum—General Survey of the Broadcasting Service*, 39–43, 96–7. The BFEBS closed in 1948. Gerard Mansell, *Let Truth Be Told: 50 Years of BBC External Broadcasting* (London, 1982), 211–15. Briggs, *History of Broadcasting in the UK*, iv, 125–47, 463–77.

[76] Clark to H. Bishop, 27 April 1949, WAC, E2/360/1. The idea of a relay station in India had already been dropped, due to fears about the country's future stability and relations with Britain.

[77] Minutes, Commonwealth Broadcasting Conference Technical Committee, 26 June 1952, CBA archive, box 1, C/1/2.

[78] Barkway to MacAlpine, 14 January 1946, WAC, E1/509/3. Hutchison to J. Gough, 22 January 1946, E1/356/1.

[79] Barkway to R. A. Rendall, 3 July 1945, and Barkway to MacAlpine, 3 July 1945, WAC, E1/565/1.

of British migrants to the dominions), current affairs, sports, and, to a lesser extent, farming news, coverage of ceremonial occasions, and political speeches.[80] Unsurprisingly, fewer BBC news bulletins were rebroadcast in the dominions than during the war. Nevertheless, the fact that the rebroadcasting of US news and commentary largely ceased in Australia and New Zealand offered the BBC some consolation. Moreover, although the ABC and CBC continued to expand their own news operations, they were still largely reliant on agencies for international news, and thus often took additional material for their own bulletins from the BBC's growing network of foreign correspondents. The state-controlled NZBS continued to rely on government sources for its domestic news, and took BBC bulletins as a source of 'independent' international coverage. Five BBC news bulletins were rebroadcast daily in New Zealand, as well as *Radio Newsreel*, and regular BBC news commentaries.[81] The authoritativeness of BBC news, and its apparent freedom from government influence, meant that the BBC even managed to emerge from the Suez crisis of 1956 with its reputation largely intact. One of the most traumatic episodes in Britain's end of empire thus did little to disrupt Commonwealth broadcasting collaboration: in Australia, for example, it was rather the ABC's handling of its own commentators that caused problems during the crisis.[82]

The BBC meanwhile took some incoming news from the CBC and ABC for use in its domestic and external services, and in 1959 the BBC and ABC agreed to more fundamental collaboration in covering Asian news. Henceforth, reports filed by ABC journalists at Singapore and Jakarta, and the BBC's correspondents at Delhi and Hong Kong, would be shared by both organizations. If major stories broke, ABC and BBC correspondents would work together as a team. In this instance, Greene (now BBC Director of News and Current Affairs) was willing to put aside the principle of BBC editorial control, which had been so fiercely guarded since the war, because he regarded the BBC and ABC as 'organisations of the same stamp, with similar responsibilities and standards'.[83]

Along with news, sports coverage provided the bread and butter of post-war short-wave Commonwealth broadcasting collaboration: in both cases, the desire

[80] Sloan to A. G. Huson, 22 May 1952, WAC, E1/509/5. Terraine to P. J. Saynor, 26 September 1957 and K. Worsley to Looker et al., 23 April 1958, E1/1675. J. F. Mudie to Looker, 16 May 1958, E1/1636/2. From November 1956, the Australian Macquarie network of commercial stations also carried BBC current affairs material in its magazine programme, *Monitor* (and material from the US National Broadcasting Company programme of the same name, upon which it was modelled). See Mudie, 'The Australian Scene', 28 August 1958, E1/1636/2.

[81] 'Commonwealth Broadcasting Conference, 1956—CBC Memorandum—C/12—News Exchanges (Agenda Part 1)', Boyer papers, folio box, folio 1. *Annual Report of the New Zealand Broadcasting Service, for the 12 months ended 31st March 1947* (Wellington, 1947), 6. Mudie to Looker, 16 May 1958, WAC, E1/1636/2.

[82] Briggs, *History of Broadcasting in the UK*, v, 73–137. Tony Shaw, *Eden, Suez and the Mass Media: Propaganda and Persuasion during the Suez Crisis* (London and New York, 1996). Gary David Rawnsley, 'Cold War Radio in Crisis: the BBC Overseas Services, the Suez Crisis and the 1956 Hungarian Uprising', *HJFRT*, 2/16 (1996), 197–219. For the ABC's handling of Suez see Inglis, *This is the ABC*, 191–2.

[83] Extracts from BBC Board of Management minutes, 25 May and 8 June 1959 and Greene to S. C. Hood and A. H. Wigan, 27 May 1959, all in WAC, E1/1640.

for live and up-to-date coverage overcame the deficiencies of short-wave reception quality. Summaries of, and commentaries on, many different types of sporting events were exchanged, including horse racing, tennis, soccer, boxing, rowing, and athletics. For many listeners in Britain, Australia, New Zealand, and South Africa, cricket and rugby tours were of transcendent interest, and they necessitated almost constant contact among broadcasting organizations. In terms of sports coverage, South Africa was more central to structures of Commonwealth broadcasting cooperation than in most other spheres of collaboration, and Canada (integrated into US sporting circuits) was more marginal. Sport also provided important opportunities for collaboration among the ABC, NZBS, and SABC, without the need for the BBC to act as an intermediary: when a South African soccer team toured New Zealand in 1947, for example, coverage was successfully relayed back to the SABC via Radio Australia.[84] The deeper failure fully to integrate Caribbean, African, and Asian countries into structures of Commonwealth broadcasting was also reflected in sporting broadcasts. Even though some sense of connection was provided through coverage of Caribbean, Indian, and Pakistani participation in cricket, it was, for example, as late as 1959 that an Australian cricket team (accompanied by an ABC commentator) first officially toured India and Pakistan.[85]

Much of the ideological work done by sporting coverage was implicit, through practical cooperation that made the British world seem a familiar and natural on-air presence. During the 1950s the number of tours covered by broadcasting organizations increased substantially as international travel became easier: in 1957/8, for example, seventeen overseas teams toured New Zealand, and nine New Zealand teams toured in other countries.[86] As with earlier war reporting, attempts were made to include dominion voices in BBC sports commentary teams, to satisfy the requirements of audiences overseas, but also to make the British world seem more real to UK listeners. When the New Zealand Army rugby touring team played fixtures in England in 1945/6, it was accompanied by the popular New Zealand commentator Winston McCarthy. The head of BBC Welsh Programmes recalled that the effect of McCarthy's energetic style 'was at first a little startling, but very soon he became quite a favourite with our listeners'.[87] When a British team toured New Zealand in 1950, the BBC asked that McCarthy supply commentaries, and when McCarthy came to Britain with the All Blacks in 1953, the BBC covered part of his costs, and used his commentaries as much as possible.[88]

Every four years, the Empire Games offered opportunities for a more explicit harnessing of sport to the cause of empire and Commonwealth unity. The CBC told the NZBS that the 1950 Auckland Empire Games were 'an occasion for first

[84] J. Shelley to Moses, 20 June 1947, Archives New Zealand (ANZ), AADL 564/89d 1/3/19.

[85] *28th Annual Report and Financial Statement of the Australian Broadcasting Commission Year Ended 30th June 1960*, 25.

[86] *Report of the New Zealand Broadcasting Service for the year ended 31 March 1958* (Wellington, 1958), 11.

[87] Looker to Shelley, 6 November 1945 and A. W. Jones to R. Alston, 22 July 1953, WAC, R30/1904/1.

[88] Jubb to W. Yates, 13 April 1950, ANZ, AADL 564/1c 1/3/27. G. H. Simpson to C. Max-Muller, 1 June 1953, WAC, R30/1904/1.

class Commonwealth publicity', although New Zealand's poor short-wave infrastructure ultimately limited radio coverage overseas.[89] Preparing for the 1954 Empire Games in Vancouver, the CBC produced a special Empire Day programme that included messages from prominent athletes around the empire and Commonwealth. The New Zealand contribution emphasized that:

> Our ties with the Mother Country and our sister Dominions and colonies are very close...The proudest moment of our Empire heritage...is when sportsmen and sportswomen stand on the soil dedicated to the British Empire Games. It is there that we meet all members of the Empire family.[90]

These echoes of the traditional language of imperial unity perhaps reflected the invigorating effects of the coronation and royal tour of 1953–4. The CBC International Service provided full short-wave coverage of the games, tailored to the specific rebroadcasting requirements of organizations around the empire and Commonwealth.[91] One BBC officer argued that it was 'a most successful example of Commonwealth cooperation'.[92]

TRANSCRIPTIONS

For non-topical material, transcriptions became an increasingly successful way to export BBC programmes to the Commonwealth and empire, building on the wartime achievements of the London Transcription Service (LTS), and supported by continuing subsidies from the British government. The BBC's expanded post-war Transcription Service, directed first by a South African, Tom Gale, and then by Malcolm Frost, provided overseas subscribers with a wide range of material on disc. Pre-war attempts to use transcriptions to project a clearly defined set of ideas about Britishness were abandoned, in favour of a wholesale export of British broadcast culture, in almost all its aspects. Initially offering up to seven hours of programmes per week, by the end of 1948 the bulk of the Transcription Service output was being carried on multiple stations in each of the dominions. By 1952 the estimated overall cost of running the service was £245,000, but only around 10 per cent of this was covered by the subscriptions paid by overseas users. The rest was funded by the UK Treasury grant-in-aid.[93] This meant that the service was extremely good value for subscribers: the NZBS, for example,

[89] Bushnell to Yates, 16 November 1949, ANZ, AADL 564/3a 1/3/25 part 1.

[90] Script for 'Saluting the British Empire Games', April 1954, ANZ, AADL 564/2a 1/3/25 part 3.

[91] 'International Service, CBC, Overseas Broadcasts, British Empire and Commonwealth Games, Vancouver 1954', ANZ, AADL 564/1d 1/3/25 part 4.

[92] 'International Service, CBC, British Empire and Commonwealth Games—Extracts of Engineering Correspondence', ANZ, AADL 564/1d 1/3/25 part 4. On the Vancouver Games more generally see Michael Dawson, 'Acting Global, Thinking Local: "Liquid Imperialism" and the Multiple Meanings of the 1954 British Empire & Commonwealth Games', *International Journal of the History of Sport*, 23/1 (February 2006), 3–27.

[93] *BBC Memorandum—General Survey of the Broadcasting Service.* 'Commonwealth Broadcasting Conference 1952—BBC Memorandum—The Transcription Service', LAC, RG41, 367/19-31/5. The transcriptions were also used in the colonies, the US, Europe, and particularly in South America.

was in 1949 paying the almost nominal amount of £15 per programme hour, far less than the BBC had charged for transcriptions before the war.[94] To help tailor transcriptions to dominion requirements, Frost arranged for staff from overseas broadcasting authorities to be attached to the Transcription Service on an annual rotating basis, and offered to include particular programmes on request from subscribers whenever possible.[95]

Features, documentaries, drama, music, and talks (taken from BBC domestic schedules or specially produced for overseas listeners) all figured prominently in the output of the Transcription Service. Australia and New Zealand provided a particularly ready market for talks, documentaries, and features. In these fields, the ABC and NZBS both focused their own scarce resources on local topics. While this struck Hutchison as a 'melancholy' result of 'exclusive' nationalism, it created a niche for the BBC, allowing it to provide coverage of international issues.[96] The ABC and NZBS were also hungry for BBC entertainment programmes. The ABC could not pay top Australian artists the sorts of fees offered by the private networks, and BBC transcriptions offered a means to fill the resultant gap in its schedules. Patrick Jubb argued that even comedy could discharge a useful function if it reinforced the British connection: 'one of the best ways to put across the Commonwealth idea is through light entertainment'.[97] In Canada, the BBC representative maintained that, unless the BBC provided entertainment programmes, American material would dominate completely.[98]

The BBC had already started to provide entertainment programmes during the war as part of the LTS. By 1952 variety programmes constituted almost a sixth of the total programme hours of the Transcription Service.[99] Series originally aired on the domestic services such as *Much-Binding-in-the-Marsh*, *Take it From Here*, *Variety Bandbox*, and *Ray's a Laugh* proved extremely successful overseas. *Much-Binding* became a 'religion' in Australia. The ABC took around 40 per cent of its variety programming from the BBC Transcription Service. Making a virtue of necessity, it argued that such programmes played 'a valuable part in emphasising the bonds between the people of Britain and Australia', and exceeded in popularity any of the US entertainment programmes broadcast during the war.[100] BBC entertainers like Arthur Askey, Richard Murdoch, and Kenneth Horne visited Australia and made special programmes for the ABC.[101] After an 'undistinguished start', *Take it From Here* even built up a loyal following in Canada: one of the first signs that some

[94] T. Gale to Shelley, 2 May 1949, WAC, E17/124/3.

[95] Frost to J. D. F. Green, 10 July 1960, WAC, E14/45.

[96] Hutchison to McCall, 20 April 1949, WAC, E1/356/2. J. H. Hall to Thomas, 19 March 1953, E17/124/3.

[97] Jubb to Clark, 3 October 1952, WAC, E1/311.

[98] Barkway to MacAlpine, 19 February and 21 May 1946, WAC, E1/509/3.

[99] 'Commonwealth Broadcasting Conference 1952—BBC Memorandum—The Transcription Service', LAC, RG41, 367/19-31/5.

[100] 'Extract from the Chairman's Statement', 13 February 1948, NAA, Sydney, SP613/1, 8/1/29 part 5, 3188855. Jubb to Clark, 5 May 1949, WAC, E1/332/1. 'Commonwealth Broadcasting Conference 1952—ABC Memorandum—Program Exchanges', LAC, RG41, 367/19-31/5.

[101] Barry to Moses, 5 January 1950, NAA, Sydney, SP368/1, 7/6/7, 3164541. *Twenty-second Annual Report of the Australian Broadcasting Commission Year Ended 30th June 1954* (Sydney, 1954), 19.

British comedy might be palatable in North America after all.[102] BBC comedy programmes were generally included in the Transcription Service in edited form, to remove risqué jokes and 'honest vulgarity' deemed unacceptable by overseas broadcasters, especially in Australia and New Zealand.[103]

A particularly significant BBC comedy export was the *Goon Show*, which incorporated the absurd and sometime satirical humour of Spike Milligan, Peter Sellars, Michael Bentine, and Harry Secombe. Produced for domestic listeners from 1951, the series made enemies as well as fans, but quickly built up a cult following.[104] First offered on transcription in 1956, the series similarly divided Australian listeners into 'Goonadicts' and those who 'violently dislike the session'. However, while some remained allergic, 'The flood of criticism which followed the first programme gradually changed to a steady stream of keen appreciation.' Listener demand proved so great that the ABC was obliged to repeat the first series as soon as it finished, not once but twice. The Transcription Service happily provided further series.[105] When Milligan visited Australia in 1958 and 1959, he made two hugely popular *Idiot Weekly* series for the ABC.[106] The Goons were also a success in Canada, where devoted listeners complained en masse whenever an episode was rescheduled.[107] In New Zealand, Goon catchphrases provided key elements of the secret vocabulary of broadcasting engineers, who also victimised announcers by playing impromptu snatches from the series at moments timed to cause maximum embarrassment.[108]

The success of the post-war BBC Transcription Service seemed to vindicate the BBC's continuing determination to operate on non-commercial lines in its relations with the dominions, and to prove the wisdom of state funding for BBC External Services. From 1956 programmes were recorded onto vinyl LPs, making it cheaper and easier to distribute programmes. By 1960 over seven hundred different programmes were being issued annually, involving the despatch of some sixty thousand discs.[109] The increased use of BBC transcriptions by the CBC was particularly striking, as Canada had previously remained largely closed to UK entertainment programmes. The cheapness of the service undeniably made it attractive to the cash-strapped CBC. By 1950 the BBC was getting roughly four

[102] 'Rebroadcast Report, May 1950', WAC, E1/565/3. 'Rebroadcast Report, January 1952', E1/565/4.

[103] Lawrence Constable, 'The Programme Doctors', *New Zealand Listener*, 29 June 1956, 7. Report by Barry, 3 December 1957, NAA, Sydney, SP613/1, 8/1/57, 3188949.

[104] Stuart Ward, '"No Nation Could be Broker": the Satire Boom and the Demise of Britain's World Role', in Ward (ed.), *British Culture and the End of Empire* (Manchester, 2001), 93–5.

[105] Barry, 'Goon Show', 18 April 1956; Barry to C. Semmler, 19 September 1956; memorandum by D. Porter, 25 September 1957, all in NAA, Sydney, SP1423/1, R18/4/6 part 1, 3162623.

[106] *Twenty-sixth Annual Report and Financial Statement of the Australian Broadcasting Commission Year Ended 30th June 1958* (Sydney, 1958), 33. *Twenty-seventh Annual Report and Financial Statement of the Australian Broadcasting Commission Year Ended 30th June 1959* (Sydney, 1959), 11.

[107] C. Curran to J. Jackson, 17 April 1957, WAC, E1/1775.

[108] '33 Years Behind the Mast: Recollections of a career in New Zealand broadcasting by Godfrey Gray', Alexander Turnbull Library, Godfrey Gray papers, MS-Papers-6488-1.

[109] 'Commonwealth Broadcasting Conference 1960—BBC paper—Distribution of BBC programmes overseas', LAC, RG41, 369/19-35/1.

times more Canadian airtime than in 1947, and in August 1949 BBC 'station hours' in Canada exceeded those of the US networks for the first time.[110]

Yet the rigorously non-commercial nature of the Transcription Service also had its drawbacks. The BBC was unwilling to allow its programmes to be used by commercial stations overseas in association with sponsorship. This derived partly from fears that copyright holders and artists would demand higher fees if this was allowed, and that overseas artists' bodies would protest about the perceived damage done to local employment opportunities by imported British programmes.[111] The BBC thus placed substantial restrictions on the use of its transcriptions by commercial stations, against the advice of the BBC's Australian and Canadian representatives. Jubb pointed out that even if '95 per cent of the time [Australian private radio] stinks, it has got the major audience, and what is perhaps more important, it has the ears of the people at whom we ought to be getting'. Moreover, he claimed, many of the 'commercial blokes' were just as eager to 'promote pro-British feeling' as their ABC counterparts.[112] Some Australian private stations simply went ahead and broadcast BBC transcriptions in sponsored slots, without telling the BBC. Ian Mackay of the Macquarie Network thought that 'the ties of empire do not appear to have weakened as a result'.[113] The BBC's Canadian representative similarly claimed that BBC transcriptions would reach only a minority audience if the CBC remained their sole outlet, and argued that the CBC possessed no monopoly of pro-British sentiment. The proclaimed dedication of people like Bushnell to 'using broadcasting more vigorously to strengthen the connection between Canada and the United Kingdom' had 'worn thin in my ears with constant repetition', remaining 'unrelated to any practical programme direction'.[114]

While dominions public broadcasters used BBC transcriptions in unprecedented quantities, the determination of the BBC to operate on rigorously non-commercial principles overseas thus still seemed to limit the imperial potential of broadcasting. Valuable opportunities to collaborate with popular private broadcasters were lost, and BBC programmes were largely restricted to public stations that catered only to a small portion of the local audience.

ROYALTY

In the early post-war years the broadcasting of royal ceremonial continued to offer a means of beating the Britannic tribal drum. In 1947 the Royal Family visited South Africa, and in 1951 Princess Elizabeth and Prince Philip visited Canada.[115]

[110] 'Canada—Rebroadcast Report—August 1949', WAC, E1/565/2. Polwarth to Huson, 3 April 1950, E1/585/1.

[111] Clark to Barkway, 30 April 1946, WAC, E1/509/3. The problem did not arise in New Zealand, where the state-owned commercial network was willing to use the programmes in non-sponsored slots.

[112] Jubb to McCall, 4 April and 23 May 1952, WAC, E1/332/2.

[113] I. Mackay to R. J. Wade, 22 October 1953, ANZ, AADL 564/3d 1/3/42.

[114] Barkway to Clark, 11 March 1948, WAC, E1/509/4.

[115] On the royal visit to South Africa see Thomas Hajkowski, *The BBC and National Identity in Britain, 1922–53* (Manchester, 2010), 97–9.

The BBC's Canadian representative urged that, rather than send its own team of commentators, the BBC should leave the job of covering the 1951 tour to the CBC International Service:

> If London has been making comparisons with what happened on the last Royal Tour of Canada [in 1939], then I think one must bear in mind that the CBC has grown up since then . . . As you know, there is a very firm sense of independence here, and when reporting the visit of their future Queen, Canadians will not want to have their programmes, domestic or Overseas, plastered with English accents. We shall have people talking about Colonialism.[116]

Yet although the BBC assured the CBC's International Service that UK listeners would hear plenty of its material, the BBC still planned to send three commentators of its own to Canada, to provide background material 'as seen through English eyes' and news coverage to match the BBC's specific requirements. Subsequently, press rumours of a 'first-class row' between the CBC and BBC had to be denied.[117]

That monarchy remained a potent symbol of identity in the dominions was demonstrated a few months later, on the death of King George VI. Boyer wrote of 'the spontaneous national and personal mourning' that followed among 'we of the British people'.[118] Public radio authorities followed the procedures adopted on the occasion of the death of King George V in 1936. The CBC cancelled all regular programmes on the day of the King's death and all advertising for two days. All soap operas and US programmes were cancelled on the day of the proclamation of the new Queen, Elizabeth II, and programming was again suspended on the day of the King's funeral.[119] The NZBS similarly cancelled all normal programming and advertising to mark the King's death and funeral.[120]

BBC short-wave coverage of the King's funeral, and of the proclamation of Queen Elizabeth II, was widely rebroadcast around the Commonwealth. For the funeral it was decided to include a CBC commentator, Captain W. E. S. Briggs, in the BBC's radio team, and the Australian Chester Wilmot was one of the television commentators. This was not a particularly risky move: Wilmot was well known to UK audiences, and Briggs's accent was more English than Canadian. However, it set an important precedent. Reiterating the lessons of Elizabeth's 1951 tour, Wilmot argued that the BBC should include Commonwealth voices more systematically in its coverage of future royal events:

> The Queen is now the Queen of each Commonwealth country, except India, in her own right and not merely by virtue of being Queen of England. I think, therefore, that it would be appropriate, and would be a reminder to the home audience of what we may call the 'Commonwealth character' of her position, if at the time of the Coronation

116 Thornton to Clark, 14 August 1951, WAC, E1/570/1.

117 Jacob to Bushnell, 29 August 1951, WAC, E1/570/1. CBC press release, 6 September 1951, LAC, RG41, 747/18-16-2-31/1.

118 Boyer, 'The British Monarchy in 1952', Boyer papers, 4/32.

119 I. Richie, 'Network Arrangements on the Death of His Majesty King George Sixth', 18 February 1952, LAC, RG41, 242/11-37-10/2.

120 *Annual Report of the New Zealand Broadcasting Service, for the 12 months ended 31st March 1952* (Wellington, 1952), 9.

> for instance, Dominion broadcasters should give part of the commentary and should not make merely an incidental contribution to it.[121]

De Lotbinière agreed that this would be a 'good Commonwealth gesture': the principle involved was so important, he argued, that the BBC could justify using commentaries that were not 'top line'.[122] As during the 1930s, it was believed that great events of overwhelming sentimental appeal provided one of the few excuses for allowing programme standards to slip in order to serve imperial interests.

The coronation of Elizabeth II on 2 June 1953 and the subsequent royal tour of 1953/4 provided a series of 'media events' that together marked the apogee of Commonwealth broadcasting.[123] Britain and Canada were the only Commonwealth countries in which television was available at the time of the coronation, making this the last great royal ceremony to take place during the empire's radio age. A single seven-hour programme covering the processions and ceremonies was provided for English-speakers in the UK and around the world. Domestic and overseas audiences were thus, momentarily, united. To boost the BBC's short-wave signals, transmitters belonging to Voice of America, the CBC, and Radio Ceylon were used as relay stations.[124] Following the discussions that had attended the 1951 royal tour, the Commonwealth was represented much more fully than ever before in radio coverage of a royal ceremony, and UK audiences were exposed to the accents and observations of broadcasters from other parts of the British world. Briggs was again brought in from the CBC, while the ABC sent Talbot Duckmanton, already marked out by Moses for accelerated promotion. Duckmanton described his role in the coverage as

> the experience of a lifetime, both as a citizen of the British Commonwealth and as a broadcaster. I am, and always will be, grateful for the opportunity given me to take part in what I believe was one of the most important broadcasts in the history of radio.[125]

The tropical colonies were represented by a West Indian producer in the BBC Overseas Service, the Trinidadian Willy Richardson. Wilmot and Braden were included in the television commentary team, in order to bring the Commonwealth aspect of the event home to UK viewers.[126]

BBC domestic, short-wave, and transcription services carried a wide range of coronation-related radio programming. On the day of the ceremony itself an hour-long round-the-Commonwealth feature (*Commonwealth Greetings*) incorporated recorded messages from 'ordinary' people and statesmen, and a message from the new Queen. The programme aimed to show, 'through a diversity of voices and

[121] Wilmot to C. McGivern, 20 February 1952, NLA, Chester and Edith Wilmot Papers, MS8436, 5/43.

[122] De Lotbinière to Max-Muller, 3 June 1952, WAC, T14/841.

[123] Daniel Dayan and Elihu Katz, *Media Events: The Live Broadcasting of History* (Cambridge, Mass. and London, 1992). Wendy Webster, *Englishness and Empire, 1939–1965* (Oxford, 2005), 92–118.

[124] *The Year that Made the Day: How the BBC Planned and Prepared the Coronation Day Broadcasts* (London, [1953]), 46.

[125] T. Duckmanton to Moses, 11 September 1953, NAA, Sydney, SP724/1, 8/17/4, 3160069.

[126] *The Coronation and the BBC* (1953), Wilmot papers, 6/36.

music and experiences, a multi-coloured, multi-racial British Commonwealth, with strength and wisdom, looking towards a bright future'.[127] That evening another feature (*Coronation Night*) used live feeds to tell listeners how the coronation had been celebrated in different parts of the Commonwealth.[128] This was an occasion on which perceived apathy among UK listeners towards the Commonwealth could be expected to abate, and the BBC exploited this by producing a number of special programmes for its domestic services featuring Commonwealth performers, including a 'Gala Commonwealth Variety Programme' starring overseas artists who had made a name for themselves in the UK. On the Third Programme, the BBC broadcast a special concert of music by Handel; the first part was recorded in Canada by the CBC Symphony Orchestra, and the second recorded in Australia and performed by the Sydney Symphony Orchestra. Jennings thought this 'a real prestige job for the CBC'.[129] The ABC took similar pride in its programme *Coronation Music*, which included examples of music played at every coronation since that of King Charles I, and was included in the BBC's domestic, overseas, and transcription services.[130]

Thus, while the coronation was a global media event, with an estimated worldwide audience of more than 150 million people in fifty countries, it also clearly played a more particular function in a Commonwealth context.[131] BBC coverage made it easy for listeners in the UK and other Commonwealth countries to imagine shared membership in an imperial community. The *New Zealand Listener* concluded that

> British people are spread throughout the world, but they are not cut off from the beating of that strong heart in London. On June 2nd, as they gather around their receiving sets, they will feel the almost mystical sense of unity which suddenly takes form and colour, and is comprehensible, when the Crown is placed reverently on the Queen's head.[132]

The CBC similarly anticipated massive audience interest: 'the audience will take all the Coronation programming we can give them'.[133] It produced its own special women's, children's, schools, and music programmes to mark the coronation, and despatched its own team of commentators to Britain several weeks in advance to provide additional background material.[134] In the month of June 1953 the BBC

[127] A. Burgess to I. M. Elford, 12 March 1953, WAC, R47/129/1.

[128] 'Minutes of Coronation Planning Committee Meeting', 8 April 1953, LAC, RG41, 539/11-37-5/2. For more on the Commonwealth aspect of BBC home coverage of the Coronation see Hajkowski, *BBC and National Identity*, 100–3.

[129] Jennings, 'Special Recorded Coronation Programme for BBC', 25 February 1953, LAC, RG41, 539/11-37-5/1.

[130] Jubb to A. N. Finlay, 30 December 1952 and Moses to ABC Commissioners, 11 March 1953, both in NAA, Sydney, SP724/1, 8/17/3 part 1, 3160063.

[131] 'External Broadcasting Audience Research—Reaction to the Coronation Broadcasts Overseas—August 1953', WAC, E3/20.

[132] Editorial, *New Zealand Listener*, 29 May 1953.

[133] '[CBC] Notes on BBC Paper no. 18 of 2 June 1952, on the Coronation of Queen Elizabeth', LAC, RG41, 367/19-31-1/2.

[134] Minutes, CBC Coronation Committee Meeting, 21 November [1952], LAC, RG41, 539/11-37-5/1. Minutes, 'Coronation Meeting', 4 December 1952, WAC, R47/129/1.

achieved its highest ever rebroadcasting figures in Canada.[135] Ira Dilworth, CBC director for Ontario, found BBC coverage of the coronation service itself 'very moving. I was proud to be a citizen of this commonwealth, although it is perhaps rather "corny" for me to set it down as simply as that.' Canadians with access to television sets could also witness the coronation for themselves. Footage was flown to Canada in British jet bombers, allowing the CBC to screen the ceremony the same day. Again, the CBC saw this as a prestige operation, and was particularly happy to beat the US networks: J. P. Gilmore of the CBC thought this had 'underlined the achievements of Commonwealth broadcasting'.[136] According to Dunton,

> many were moved by the feeling of immediacy and I think television succeeded in getting a feeling of more direct connection with the Crown than by any other way...the effect of almost direct participation in the Coronation was very great, and the enthusiasm for a joint Commonwealth effort that came off so well has been enormous.[137]

The CBC also insisted that the BBC provide full radio coverage of the coronation in French on the NAS, and include a French Canadian, Gerard Arthur, in the commentary team. Rejecting earlier BBC proposals for a more modest NAS service in French, the CBC argued that French Canadians had to be treated as 'men from a Commonwealth country', and would not be satisfied with a truncated service prepared for European Francophones. True or not, the CBC clearly had to protect itself and its nation-building image, and ensure that coverage of what was meant to be a Canadian as well as a Commonwealth event was as generous in Quebec as in the rest of Canada.[138] No such demands came from South Africa, where the politics of identity worked differently. Monolingual Afrikaans-speaking monarchists (a rare breed) had to content themselves with a forty-five-minute recorded coronation summary. Nevertheless, it was estimated that up to 69 per cent of the English-speaking South African population and 51 per cent of Afrikaners (many of whom would have tuned in to the English-language service) listened to the BBC coverage.[139]

Collaboration among public broadcasting authorities during the coronation was remarkably smooth, due in part to the willingness of the BBC to include dominion broadcasters in its commentary teams, and to offer opportunities for dominion public broadcasting authorities to participate in 'prestige' special projects. Successful liaison was secured in a similar fashion during the royal tour of 1953/4. The Queen and the Duke of Edinburgh made relatively brief stops in the West Indies, Tonga, Fiji, the Cocos Islands, Ceylon, Aden, Uganda, Malta, and Gibraltar; but the centrepiece of the tour was a stay of over a month in New Zealand, and almost

[135] Rebroadcast report for June 1953, WAC, E1/565/4.

[136] Memorandum by Dilworth, 3 June 1953 and J. P. Gilmore to Clark, 3 June 1953, LAC, RG41, 539/11-37-5/4.

[137] Dunton to G. Barnes, 8 June 1953, WAC, T8/10/2.

[138] 'Coronation Coverage', WAC, R47/129/2. Minutes, CBC Coronation Committee Meetings, 22 October [1952] and 21 November [1952], LAC, RG41, 539/11-37-5/1.

[139] 'Coronation Coverage', WAC, R47/129/2. 'External Broadcasting Audience Research—Reaction to the Coronation Broadcasts Overseas—August 1953', E3/20.

two months in Australia. This was the first time that a reigning monarch had visited either country, and the ABC and NZBS both provided a full service of outside broadcasts and special programmes. Again, this was an opportunity to stress the contribution that public broadcasting could make to both national and Britannic unity. The ABC claimed that the tour illustrated

> the exceptional value of the broadcasting medium in enabling the entire population, even of a country with so widely scattered a population as our [Australian] Commonwealth, to share intimately in a national experience of historic significance.[140]

In New Zealand the tour finally prompted a start to be made on the installation of a national landline network capable of linking up the various NZBS stations.[141]

The tour also provided an opportunity to project Australia and New Zealand to Britain and the Commonwealth, with the help of the BBC. The ABC and NZBS were asked to provide appropriate features for use by the BBC domestic services, and the BBC and ABC cooperated to produce special schools programmes in advance of the tour.[142] While as usual a team of BBC commentators was sent out to cover the tour, it was agreed that Duckmanton should replace the BBC's commentator with the royal party on board the *HMS Gothic* as the ship approached Sydney.[143] With goodwill secured by this diplomatic gesture, both the ABC and NZBS welcomed the BBC correspondents, providing access to studio and other broadcasting facilities, and using much of the BBC correspondents' material as a means of adding 'balance and colour' to services for Australian and New Zealand listeners.[144] The ABC meanwhile helped the BBC get its reports back to Britain by providing airtime on Radio Australia, although the vagaries of short-wave transmission meant that much material had to be despatched by airfreight.[145]

During the later 1940s Christmas features had tended to diverge from the round-the-Commonwealth pattern established before the war. Segments from Europe were often included, along with recordings made in colonies that had previously been excluded due to the lack of radiotelephone facilities. By the early 1950s there was some pressure from the dominion public broadcasting authorities for a return to the older format. Empire Day, so prominent during the 1930s, was disappearing from the broadcasting calendar, and Christmas Day, due to the royal message, became by default the key opportunity for celebrating the Commonwealth connection. Dominion public broadcasters were also eager to use the occasion to exploit the reach of the BBC's short-wave transmitters and have their voices heard overseas.[146]

[140] *Twenty-second Annual Report of the Australian Broadcasting Commission Year Ended 30th June 1954* (Sydney, 1954), 6. On the royal tour and New Zealand more generally, see Jock Phillips, *Royal Summer: The Visit of Queen Elizabeth II and Prince Philip to New Zealand, 1953–54* (Wellington, 1993).

[141] Day, *History of Broadcasting in New Zealand*, i, 312.

[142] Bridson to Stead, 20 October 1953, WAC, E1/401/2. J. Scupham to R. Bronner, 21 July 1953, E1/417.

[143] Jubb to Greene, 3 July 1953, WAC, E1/406/1.

[144] Jubb to D. Fleming, 4 August 1953, WAC, E1/401/1.

[145] Fleming, 'Australia—Communications', [*c.* January 1953], WAC, E1/401/1. Similarly, while the CBC planned to take summaries of the progress of the tour through Radio Australia, poor reception meant that in the end it turned to the BBC for its coverage. See LAC, RG41, 242/11-37-14-1.

[146] Jubb to McCall, 4 April 1952, WAC, E1/332/2.

For Christmas 1953, the Queen's presence in New Zealand provided an irresistible opportunity to transfer responsibility for the broadcast to the dominions, for one year at least. Although the NZBS did not possess the technical facilities to coordinate a round-the-Commonwealth broadcast or feed the programme to the BBC's short-wave transmitters, the ABC and the Australian PMG's Department did. Nevertheless, the Australians were still reinforced by a large BBC contingent.[147] Laurence Gilliam, who had produced almost all of the previous Christmas features, travelled to Australia, accompanied by the BBC scriptwriter Alan Burgess, two BBC effects officers, and a BBC engineer. Gilliam co-produced the programme with the ABC's Director of Drama and Features Neil Hutchison (himself a former BBC producer). Burgess scripted the programme alongside the ABC's Mungo MacCallum and John Thompson.[148] The BBC invasion seemed to generate remarkably little tension, although Gilliam's condescending and eccentric manner did create some uncomfortable social situations.[149]

The aim of the 1953 Christmas programme was to

> fulfil the new concept of the Monarch as Queen of the individual nations of the British Commonwealth by emphasising that wherever Her Majesty is, there, for the time being, is the centre of the Commonwealth.[150]

The resulting programme tracked the progress of the royal tour, taking contributions from along the route in the style of the pre-war Christmas features. Music was provided by the Australian composer John Antill, and narration by Wilmot, emphasizing the fact that this was an Australian production, even if Wilmot's voice would have been very familiar to BBC listeners in the UK.

The programme's script acknowledged the existence of unrest in some colonies, including Mau Mau in Kenya and the guerrilla war in Malaya. However, it emphasized the violence of the rebels rather than that of the colonial state, and generally depicted the empire as a force for peace. Indeed, ultimately even violent struggle in the colonies was portrayed as a positive sign that 'people of many races are being welded into a new nation . . . hearts are high and full of hope'. Old ideas about Commonwealth unity endured: this was still a 'vast family circle', characterized by 'shared responsibility and shared freedom under a young Queen'. The recent ascent of Everest was presented as emblematic of the challenges facing the Commonwealth and its ability to surmount them, and the programme finished with a segment from Norfolk featuring the newly minted hero of empire, the New Zealander Sir Edmund Hillary.[151]

[147] Moses to Barry, 8 April 1953, NAA, Sydney, SP613/1, 8/1/56 part 1, 3188945.

[148] *Twenty-second Annual Report of the Australian Broadcasting Commission Year Ended 30th June 1954* (Sydney, 1954), 12.

[149] Mungo MacCallum, *Plankton's Luck: A Life in Retrospect* (Hawthorn, Vic., 1986), 186–8.

[150] Minutes, 'Meeting to Discuss Christmas Programme for 1953', 23 April 1953, NAA, Sydney, SP613/1, 8/1/56 part 1, 3188945.

[151] Typescript, 'The Queen's Journey' [Christmas Day, 1953], Wilmot papers, 5/30. For a copy of the recording see National Film and Sound Archive, 636774. Wilmot died in an air crash while travelling back to Britain on one of the ill-fated early Comet airliners. On Hillary and coronation year see Peter H. Hansen, 'Coronation Everest: the Empire and Commonwealth in the "Second Elizabethan Age"' in Ward (ed.), *British Culture and the End of Empire.*

Coverage of subsequent royal occasions was not without its tensions. The lesson that dominion royal visits should be covered primarily by dominion public broadcasting authorities was not absorbed by all at the BBC. Senior BBC officers were themselves critical when only limited CBC coverage of the Queen's 1957 visit to Canada was rebroadcast in the UK. Dominion voices were now deemed essential for the sake of UK listeners, who wanted something 'different and new'. Without Canadian accents and perspectives, 'it is the same old round and it might, except for the names of places or people, be Paris, Portugal or even Westminster Pier'.[152] Moreover, as J. B. Clark (now BBC director of external broadcasting) concluded: 'Even though invasion by a BBC commando unit may seem to be welcome, I wonder if it is necessarily right to despatch it, and if there is not likely to be a growing feeling of resentment underlying the atmosphere of goodwill.'[153] When the Queen visited Canada again two years later for the opening of the St Lawrence Seaway, it was agreed that CBC and BBC staff should form a single cooperative 'Commonwealth unit', rather than operate separately as in the past.[154]

CONCLUSIONS

In many ways, the late 1940s and 1950s seemed to mark the culmination of efforts to promote Commonwealth broadcasting cooperation, with the strengthening of collaborative structures that had been built in the 1930s and during the Second World War. Public broadcasting was preserved in Britain and the dominions, and established in embryonic form in the colonies. Haley's updated version of Reithianism was exported overseas. The BBC and the dominion public broadcasting authorities promoted colonial development, and justified colonial control, for the benefit of audiences at home and overseas. Broadcasting personnel moved around the British world in unprecedented number, the ABC and CBC operated permanent offices in London, and the BBC maintained outposts in Canada, Australia, and elsewhere. Periodic Commonwealth Broadcasting Conferences provided further opportunities for personal contact among senior officers. The BBC remained a formidable presence in the world of short wave, while the dominion public broadcasters each managed parallel external services. If rebroadcasting declined in the dominions, then the increasing use of BBC transcriptions helped make up the shortfall. The monarchy was the principal remaining constitutional connection linking the component parts of the British world, and fittingly provided Commonwealth broadcasting with some of its most inspiring media events.

Underlying Commonwealth collaborative efforts in radio broadcasting in the 1940s and 1950s was the principle of free exchange of material, on a non-commercial basis: part of the Reithian heritage. When radio material was exchanged between

[152] J. Snagge to H. R. Pelletier, 15 October 1957, WAC, E29/248/3.

[153] Clark to Curran, 14 November 1957, WAC, E1/1778.

[154] Minutes, 'Royal Visit to Canada, 1959 & St Lawrence Seaway', 14 March 1959, WAC, E1/1777.

Commonwealth public broadcasting authorities, fees were not generally charged, and only reimbursement for costs incurred was sought. Thus, for example, the BBC still allowed the CBC to transmit material to Canada from London using the NAS without payment, and even continued to cover speakers' fees for material included in such transmissions. In return, the CBC paid speakers' fees for programmes made in Canada for the BBC, saving the expenditure of scarce British dollars.[155]

Under this non-commercial regime for Commonwealth exchange, the overall burden of costs was not evenly distributed, for the BBC sent out much more than it received in return. Yet the BBC was willing to write off its net expenditure on Commonwealth commitments as part of its expansive public-service mandate, serving the UK's best interests and contributing to the BBC's own prestige and status. It could afford to do so because such expenditure constituted a relatively minor element of its overall balance sheet. Radio remained a relatively cheap medium, and the Treasury grant-in-aid anyway paid most of the bill for short-wave services and transcriptions. Dominion public broadcasters could be similarly generous in providing the BBC and other Commonwealth radio authorities with reciprocal facilities, *gratis*, because they knew that in practice demand for such privileges would be limited. They gained much more from 'free reciprocity in a context of free and mutual collaboration' than they lost.[156]

Television, however, was a much more expensive proposition: an hour of television cost roughly twelve times more to produce than an hour of radio.[157] As the BBC would soon discover, it was as a result simply impossible to fund the sort of one-way outward flow of recorded material that had become customary in radio, particularly as the UK Treasury proved unwilling to extend the grant-in-aid to the new medium. Despite its successes, Commonwealth collaboration in radio in the 1950s might thus be interpreted as a doomed rear-guard action, fought using the weapons of a bygone age.

[155] Barkway, 'Payment for and Scrutiny of Programmes arranged by BBC's Canadian Office', 17 September 1947, WAC, E1/509/4. 'CBC–BBC Financial Agreement from 1st April 1949', E1/524.

[156] Clark to Hutchison, 29 March 1947, WAC, E1/356/2.

[157] Andrew Crisell, *A Study of Modern Television* (Basingstoke, 2006), 22.

6

Challenges, 1945–59

During the late 1940s and early 1950s William Haley, as director-general, had attempted to restore the pre-war order at the British Broadcasting Corporation (BBC), repairing the damage caused by wartime exigencies, but also reversing some of the innovative, populist initiatives that been introduced to help boost civilian morale. In some ways, his successor Ian Jacob (director-general from 1952 to 1960) seemed to continue this task. The bowler-hatted product of an imperial military family, Jacob had left a senior administrative position in the army (he had been one of Churchill's closest wartime aides) for better career prospects at the post-war BBC.[1] Jacob sometimes seemed out of touch with the changing times. In 1955 he remarked that he had yet to see anything worthwhile on television (the BBC had been producing television programmes continuously for a decade), and on his retirement, on *Woman's Hour* no less, he described the BBC as 'hag-ridden'.[2] Some programme-makers felt that he lacked understanding of the creative challenges they faced. Geoffrey Bridson caricatured him as an old-fashioned imperial commander operating in unfamiliar territory: 'Sir Ian's role with the BBC was to form square in the old manner, and make a firm last stand against the circling hordes of showmanship, commercialism, controversy and wogs.'[3] Perhaps most damagingly, Jacob failed to dissuade a government led by his old chief, Churchill, from introducing commercial competition in the field of television.

During the 1950s the BBC also faced challenges overseas. Once the BBC's domestic monopoly was breached, many thought the future prospects for public broadcasting in the dominions would also be diminished. Commercial broadcasting seemed to be in the ascendant around the Commonwealth, bringing with it the threat of further cultural 'Americanization', particularly in the burgeoning medium of television. In terms of sending television programmes to Commonwealth countries, the BBC still tried to operate on a non-commercial, public-service basis. This proved difficult without a state subsidy to support the

[1] Leonard Miall, 'Jacob, Sir (Edward) Ian Claud (1899–1993)', *Oxford Dictionary of National Biography* (Oxford, 2004; online edn, September 2004) <http://www.oxforddnb.com/view/article/52132> accessed 28 July 2010. General Sir Charles Richardson, *From Churchill's Secret Circle to the BBC: The Biography of Lieutenant General Sir Ian Jacob* (London and Oxford, 1991).

[2] 'Report by Neil Hutchison on Recent Overseas Visits', 21 September 1955, National Archives of Australia (NAA), Sydney, SP724/1, 11/1/3 part 2, 3160816. Leonard Miall, *Inside the BBC: British Broadcasting Characters* (London, 1994), 173.

[3] D. G. Bridson, *Prospero and Ariel: The Rise and Fall of Radio, a Personal Recollection* (London, 1971), 225.

production of television transcriptions, and the BBC fell behind US producers eagerly selling their wares overseas. In the field of radio, traditional Commonwealth exchanges continued. However, the programmes often remained relatively simple and uninspiring, and formats that had once seemed innovative (such as round-the-empire Christmas broadcasts) were now stale. Attempts to build more complex collaborative arrangements seldom succeeded. Moreover, in Britain, Canada, and Australia, as television services were established, so radio audiences dwindled: a pattern that was repeated in other Commonwealth countries over the years to come. The BBC retained imperial mastery of what appeared to be a dying medium.

The BBC's overall response to the challenges of the period thus appeared unadventurous and uninspired. It would be unfair to blame this solely on poor leadership, for it reflected a broader British pattern of post-war relative decline, and the UK's more general failure to provide inspiration to help the rest of the British world resist the onslaught of American popular culture.[4] Moreover, Jacob was in fact more of a pragmatist than his critics allowed. He proved willing to jettison the more elitist aspects of Reithian thinking, in order to make the BBC relevant to the concerns of a changing society. He encouraged alterations in the BBC's approach to programming that enabled the later, more radical, innovations of his protégé and successor as director-general, Hugh Carleton Greene.[5] Nevertheless, there were clear limits to the nature and pace of the changes that Jacob would countenance. At least during the early years of his regime, the BBC continued to deploy the paternalistic language of cultural standards to justify its policies and programming, particularly during debates about television and commercial competition in Britain, Canada, and Australia. By the 1950s this way of thinking seemed badly out of date, and still restricted Commonwealth collaboration by providing an excuse for BBC domestic controllers to reject supposedly inferior offerings from dominion public broadcasting authorities.

PUBLIC BROADCASTING AND PRIVATE COMPETITION

In the early post-war years the BBC and its counterparts in the dominions faced similar challenges, and tended to view those challenges in a broader Commonwealth setting. In 1949, prior to the renewal of the BBC's royal charter, a committee was appointed under the chairmanship of Lord Beveridge to assess the BBC's past record and future prospects. The BBC sought to impress upon the committee the dangers inherent in the introduction of any form of commercial competition, arguing that in broadcasting,

[4] J. M. Bumsted, 'Canada and American Culture in the 1950s' in J. M. Bumsted (ed.), *Interpreting Canada's Past*, ii, *After Confederation* (Toronto, 1986), 399–401.

[5] Michael Tracey, *The Decline and Fall of Public Service Broadcasting* (Oxford, 1998), 70–84. Su Holmes, *Entertaining Television: The BBC and Popular Television Culture in the 1950s* (Manchester and New York, 2008).

> the crucial test is standards... By standards is here meant the purpose, taste, cultural aims, range, and general sense of responsibility of the broadcasting service as a whole. Under any system of competitive broadcasting all these things would be at the mercy of Gresham's Law... The good, in the long run, will inescapably be driven out by the bad... because competition in broadcasting must in the long run descend to a fight for the greatest possible number of listeners, it would be the lower forms of mass appetite which would more and more be catered for in programmes.

The BBC turned to the US to illustrate the problems associated with commercial broadcasting, and to the dominions for evidence of the damage that commercial competition could inflict on public broadcasting.[6] These international comparisons were hotly contested, as advocates of commercial broadcasting (including one member of the Beveridge Committee itself, Conservative MP Selwyn Lloyd) tended to present private broadcasting in the US and the dominions as a positive model for Britain.[7]

In June 1948, before the Beveridge Committee had even been appointed, the BBC Chairman Lord Simon of Wythenshawe asked colleagues what lessons could be learned from broadcasting in Canada.[8] In response, J. B. Clark (who had visited Canada earlier that year) reported that the dominant position of the Canadian Broadcasting Corporation (CBC) within the Canadian 'dual system' had been progressively eroded. Its regulatory powers over private broadcasting had been diluted, and it had itself become overly preoccupied with generating advertising revenue. As a result, programmes had become commercialized, 'sensationalised', and unadventurous, and public-service and nation-building obligations were being neglected.[9] The BBC's Canadian representative submitted a similar report, more positive about the CBC's achievements, but no less condemnatory of private broadcasting.[10] Simon visited Canada and the US himself that year, and returned convinced that 'the sponsoring of broadcasting is fundamentally wrong and even dangerous from the point of view of the strength and development of democracy'.[11] He argued that commercial broadcasters seeking the biggest audiences (in order to attract advertisers and maximize profits) had little incentive to produce programmes with a constructive social or political purpose, or to cater to minority audiences. Instead, they went for simple entertainment, often accompanied

[6] *BBC/27—Confidential—Broadcasting Committee 1949—Monopoly and Competition in Broadcasting* (BBC, April 1950), 2–4, Manchester Archives and Library Services, Lord Simon of Wythenshawe papers, M11/6/5c.

[7] Briggs, *The History of Broadcasting in the UK*, iv, *Sound and Vision* (Oxford, 1995 [1979]), 299–301, 335. Extract of Hansard debate on Beveridge Report, 19 July 1951, Simon papers, M11/6/5d. Michele Hilmes, *Network Nations: A Transnational History of British and American Broadcasting* (New York and London, 2012), 181–6.

[8] Memorandum by I. Jacob, 9 June 1948, BBC Written Archives Centre (WAC), E1/489.

[9] J. B. Clark, 'Some Aspects of Commercial and Public Service Broadcasting', 20 August 1948, E1/489.

[10] M. Barkway to Simon, 'Note on the Canadian Broadcasting Corporation', [October 1948], E1/489.

[11] Simon to A. D. Dunton, 15 December 1948, Library and Archives Canada (LAC), RG41, 988/2.

by sensationalism and violence, which had a 'bad and demoralising' effect on audiences, and especially on children.[12]

A. Davidson Dunton, the CBC chairman, subsequently sent Beveridge a confidential, candid statement which supported many of the BBC's arguments about the damaging effects of commercial competition in Canada. In addition, Dunton explained how private stations affiliated to the CBC constantly worried about their audience share, and how they pressed for a reduction in the proportion of CBC network programming dedicated to 'good music, talks and discussion programs, and serious drama': such programmes attracted few listeners. Dunton claimed that while the CBC still tried to serve 'those who wish to get more serious fare', it was impossible to provide what it deemed to be a proper balance of programmes. Under a dual broadcasting system, Dunton concluded, choice was illusory. Competition lowered the overall standard of programming, and restricted the variety of programmes on offer.[13]

In 1948 Beveridge had himself visited Australia and New Zealand, and he subsequently asked the BBC about the Australian experience of a dual system.[14] Simon had already requested that Robert McCall (now back in Britain and working for the BBC) provide a report on Australian broadcasting. McCall's statement chimed with what others had said of the dual system in Canada. He argued that Australian private stations broadcast a uniformly poor service of repetitious and unadventurous programmes, dictated by the demands of powerful advertising agencies. The Australian Broadcasting Commission (ABC) carried material of the highest cultural standard, but attracted at most only twenty per cent of listeners. The ABC could not follow the BBC's example, and mix entertainment with uplift, because commercial producers could afford to buy up all the most popular artists. Thus, McCall argued, the vast majority of listeners never tuned in to ABC stations, and were never exposed to 'the better-class programmes'.[15]

With Richard Boyer's permission, the BBC meanwhile reprinted in its evidence for the Beveridge Committee some of the ABC chairman's own confidential criticisms of the Australian dual system.[16] Boyer also visited Britain, and gave further evidence to the committee in person, repeating many of the points made by McCall, and arguing in good Reithian terms that competition had debased the tastes of Australian audiences and the cultural standard of programming on ABC stations.[17] William Yates, New Zealand Broadcasting Service (NZBS) director-in-waiting, also visited London in 1949, and advised the BBC that even when commercial broadcasting was administered by a public authority (as it was in New

[12] His report on his trip is reprinted in Lord Simon of Wythenshawe, *The B.B.C. from Within* (London, 1953), 209–36.

[13] Dunton to Beveridge, 26 April 1950, LAC, RG41, 988/2.

[14] 'Dinner at Marsham Court on 15th March 1949, with Lord Beveridge, Lord Chorley, Edward Muir and Desmond MacCarthy—note by S. of W.', Simon papers, M11/6/5b.

[15] R. McCall, 'Broadcasting in Australia', 15 January 1949, WAC, E1/342.

[16] W. Haley to R. Boyer, 20 October 1949 and Boyer to Haley, 2 November 1949, both in NAA, Sydney, SP1489/1, NN, 317542, 'BBC'.

[17] Simon J. Potter, '"Invasion by the Monster": Transnational Influences on the Establishment of ABC Television, 1945–1956', *Media History*, 17/3 (2011), 253–71, esp. 259.

Zealand) competition for listeners still tended to 'debase the programme standards' on both commercial and non-commercial services.[18]

These damning verdicts on the state of public broadcasting outside Britain undoubtedly reflected a certain amount of posturing, designed to impress upon the Beveridge Committee the dangers of introducing commercial competition to the UK. It was not just the BBC that had something to lose. The CBC, and particularly the ABC, both feared that any successful assault on the BBC would ultimately undermine their own positions at home. However, the emphasis on the shortcomings of the dual system also reflected a genuine dislike of the effects of commercial competition in Canada and Australia, and of the US approach to broadcasting. Candid reports designed for internal consumption, written by ABC visitors to North America, provide ample proof of this. In Canada in 1947 T. W. Bearup was struck by the fact that, unlike the ABC, the CBC sold on-air advertising time, and also took many of its programmes from US commercial networks. It was thus 'deprived of the opportunity of acting as a curb on advertising excesses', or of controlling its own programme standards.[19] Frank Clewlow, ABC Director of Drama, visited the US the following year and was astonished by 'the low standard of the material of most programmes'; he thought they were aimed at an audience with an average mental age of eight or nine. To Clewlow, CBC programmes seemed to be conditioned by American approaches, and marked by a 'stereotyped regularity'. Commercial competition had eroded audiences for educational programmes.[20] When listener licences were abolished in Canada in 1953, leaving the CBC dependent on advertising and the goodwill of the government for funding, the BBC's Canadian representative sent a colleague a picture of a CBC announcer tearing up a radio licence on air: 'If you ever thought that Hari Kari or ceremonial suicide was out of fashion, have a look at the enclosed!'[21]

TELEVISION

Debates about the future of public broadcasting and the role of commercial competition were given added urgency by the introduction and development of television services. The BBC had operated a television service between 1936 and 1939, but this was available only to those living in London and wealthy enough to afford a receiving set. Regarded as a luxury, the service was suspended for the duration of the war. It started again in 1946, but even into the early 1950s it was assigned a relatively low priority by BBC managers. Post-war austerity made it difficult to justify investing in new studio and transmitter infrastructure; British manufacturers had limited capacity to produce receiving sets, and dollar shortages made it

[18] Simon to C. Conner, 17 February 1949 and Conner to Simon, 4 March 1949, WAC, E1/1091/2.

[19] 'Report from Mr T. W. Bearup—American Visit—10th April—28th May 1947', NAA, Sydney, SP1558/2, box 83, 'London Representative, 1938–59'.

[20] F. Clewlow, 'American Report' [*c.* December 1948], NAA, Sydney, SP613/1, 11/1/3, 304893.

[21] T. Sloan to I. E. Thomas, 23 February 1953, WAC, E1/493/4.

undesirable and impractical to import receivers from the US.[22] A virtue was made of these necessities. Haley told Charles Moses that while rapid post-war expansion of television in the US would cause problems for the Americans, in Britain gradual development under properly controlled conditions would allow the medium to be disciplined into fulfilling a constructive role in society, as had been the case with radio in the 1920s. 'There is no doubt that to control this infant and to teach him to grow up properly is going to be one of broadcasting's main problems.'[23]

In South Africa and New Zealand, the underdeveloped nature of private broadcasting interests allowed a similarly leisurely approach. The main result of a 1951 investigation by NZBS engineers into broadcasting in Britain, Europe, and the US was a decision to postpone the introduction of television indefinitely.[24] Meanwhile, in India and the tropical colonies, poverty necessitated delay. Most Africans and Asians still lacked access to radio sets, let alone televisions. Indian resources were tied up in expanding the meagre radio broadcasting infrastructure inherited from the raj. A very modest experimental television station was established in Delhi in 1959, with the support of the United Nations Educational, Scientific and Cultural Organization, but did not start to operate on a regular basis until 1965.[25] Even in the Federation of Rhodesia and Nyasaland, where the white minority was wealthy, but tiny, the new medium simply seemed unaffordable. If a television service was introduced, it seemed inevitable that it would be dominated by imported programmes, 'mainly American, [and] limited in quality and variety'.[26]

However, in Canada and Australia the situation was very different. Powerful private media interests were prepared to move into television from an early stage, obliging public authorities to make their own plans. In Canada, reception of transmissions from across the border acted as a further stimulus for action. Thus, in 1947 J. A. Ouimet, CBC assistant chief engineer, and H. G. Walker, manager of the CBC's Dominion Network, visited Britain and the US to investigate television programming and operating costs.[27] Maurice Gorham, formerly Director of the BBC North American Service (NAS) and thus well known in CBC circles, had been appointed Head of the BBC Television Service. He provided Walker with preliminary information by mail before they met up in Britain, and the Canadians were then given a full tour of BBC facilities.[28] Fact-finding visits to European countries and the US followed, and in 1949 Bushnell, now CBC director-general of programming, was sent to Britain to conduct further investigations.[29] Reporting back, Bushnell argued that although the BBC had not yet found answers to all of the questions raised by television broadcasting, '[w]e can take many leaves from

[22] Briggs, *History of Broadcasting in the UK*, iv, 180–244, 892–5.

[23] Haley to C. Moses, 8 October 1948, NAA, Sydney, SP613/1, 12/1/2, 3189891.

[24] Patrick Day, *A History of Broadcasting in New Zealand*, ii, *Voice and Vision* (Auckland, 2000), 14.

[25] H. R. Luthra, *Indian Broadcasting* (New Delhi, 1986), 191–204, 407–9.

[26] 'Report on the Commission of Inquiry into the Organization of Broadcasting within the Federation', 31 May 1955, Bodleian Library, Sir Hugh Carleton Greene papers, box dep.c. 903.

[27] CBC press release, 15 February 1947, LAC, RG41, 339/14-4-12.

[28] J. W. MacAlpine to M. Gorham, 3 March 1947, WAC, E1/581. J. A. Ouimet to H. Bishop, 27 October 1947, E1/550.

[29] 'Statement on Television by the [CBC] Board of Governors', 17 May 1948, WAC, T8/10/1.

their book'. What was at stake was the future shape of Canadian society: Bushnell claimed that the British and American systems 'have a very different effect on the social habits of the people of both countries—social habits that help to form a pattern of living'.[30] Later that year, a further delegation of six senior CBC officers visited Britain, headed by Ouimet, who had since been appointed coordinator of television. Government approval was given for the CBC to establish television production and transmission facilities in Montreal and Toronto, and the directors of the two centres each spent a fortnight training with the BBC.[31]

At the same time as the Beveridge Committee was sitting in Britain, a royal commission chaired by Vincent Massey was investigating Canadian 'arts, letters, and sciences'. Massey appointed Charles Siepmann (professor of communications at New York University, and formerly a senior officer at the BBC) to undertake a preliminary examination of Canadian broadcasting. Siepmann had authored the US Federal Communications Commission's 1946 'Blue Book', which had been highly critical of how entertainment and advertising dominated American broadcasting. Predictably, his preliminary report for the Massey Commission admonished Canadian private stations for neglecting their cultural and national responsibilities, and commended the CBC's record in these areas.[32] The BBC lent the CBC further support, at Dunton's request, in helping to head off any suggestion that a separate public body should be made responsible for television. When one of the members of the Massey Commission visited the UK, BBC officers impressed upon him the need for 'unified' control of public broadcasting in sound and vision.[33]

In due course the Massey Commission offered a resounding endorsement of the idea of public broadcasting. After outlining television developments in Britain, France, and the US, its report argued in strikingly Reithian terms that,

> Recalling the two chief objects of our national system of broadcasting, national unity and understanding, and education in the broad sense, we do not think that American programmes, with certain notable exceptions, will serve our national needs.
>
> Television in the United States is essentially a commercial enterprise, an advertising industry. Thus sponsors, endeavouring to 'give the majority of the people what they want', frequently choose programmes of inferior cultural standards, thinking to attract the greatest number of viewers. And as television greatly intensifies the impact of radio, so television commercials intensify the methods of appeal to material instincts of various kinds, methods which now disfigure many radio commercials...
>
> In Great Britain, it is assumed that the role of television is not simply to reproduce in picture and sound a reflection of contemporary life. One of the members of this Commission learned from the directors of British television that it is part of British policy in television to present programmes which are consciously educational in nature; indeed, the directors of television in Britain refer to their 'moral and cultural responsibilities'. For this reason British television is extremely varied, but possesses

[30] E. Bushnell, 'Report on Television by Director-General of Programmes on a Visit to the United Kingdom, 20 May to 14 June 1949', LAC, Ernest L. Bushnell papers, 2.

[31] Ouimet to N. Collins, 13 February 1950, WAC, T8/10/1.

[32] Paul Litt, *The Muses, the Masses, and the Massey Commission* (Toronto, 1992), 98, 143.

[33] G. Barnes to Dunton, 17 May 1949; Dunton to Barnes, 25 May 1949; Barnes to Dunton, 2 August 1949, all in WAC, E1/511.

> nonetheless a markedly cultural character... It seems to us that British television is attempting to effect a suitable balance in the varied whole of its productions.[34]

The Massey Report thus recommended control of television broadcasting and programming by the CBC, and urged that no private stations should be licensed in cities which already possessed a CBC station, until some form of public service was properly established. Whether its broader recommendations regarding the cultural mission of television were realistic in Canadian conditions was another matter. The BBC's Canadian representative characterized the report's section on the mass media as 'a fine piece of nostalgic writing... its whole content is quite remarkable in its naïve sincerity'. Nevertheless, a House of Commons broadcasting committee endorsed Massey's recommendations, and approved financing for the gradual extension of the CBC television service.[35]

Australians could not receive American television signals, but recorded American programmes could be imported. ABC officers, and the BBC's Australian representative, worried that Australian commercial radio interests would be allowed to branch out into television, and use American material in great quantities. Boyer and Moses thus looked to both the BBC and the CBC to provide supportive statements about the importance of a public broadcasting monopoly. As in Britain and Canada, it was argued that commercial competition would drive down cultural standards and Americanize Australian broadcasting and culture. Boyer kept in close contact with both Haley and Dunton, and in 1950 toured Britain, Canada, and the US to gather information about the financing, organization, and social impact of television. Travel did not broaden his mind, but rather confirmed his belief that television had to be carefully controlled by a responsible public broadcasting authority, or else frivolous and irresponsible programming would have damaging social and cultural consequences.[36]

Boyer was pleased when the UK Beveridge Committee's majority report eventually endorsed the BBC's monopoly. In the past, Boyer had quoted 'the B.B.C.'s position and status in enormous chunks' to bolster the ABC's defences against hostile politicians and commercial broadcasting interests.[37] The ABC now did all it could to disseminate the conclusions of the Beveridge Report in Australia. Boyer told ABC listeners that the publication of the report represented 'perhaps the most important and interesting event in radio policy in the last ten years'.[38] The findings of an Australian delegation to investigate television in Canada, the US, Britain, and Europe (comprising Moses of the ABC, and representatives of the Australian Broadcasting Control Board and the Postmaster-General's Department) seemed

[34] *Report—Royal Commission on National Development in the Arts, Letters and Sciences, 1949–51* (Ottawa, 1951), 42–9, quote at 47–8.

[35] Sloan, 'Canadian Representative's Report', 1 May 1952, WAC, E1/509/5. Frank W. Peers, *The Public Eye: Television and the Politics of Canadian Broadcasting, 1952–1968* (Toronto, Buffalo, and London, 1979), 3–28.

[36] Potter, 'Invasion by the Monster', 259.

[37] Boyer to A. Powell, 3 December 1946, WAC, E1/315/6.

[38] Boyer, script for talk on 2BL, 'The Beveridge Report on Broadcasting', 18 February 1951, National Library of Australia, Sir Richard Boyer papers, MS3181, 4/32.

further to confirm the superiority of the BBC's public monopoly of television, and the well-regulated Canadian dual system, to 'uncontrolled' American commercialism, in terms of the cultural standard of the programmes produced and their social impact.[39]

However, the ABC's case for a television monopoly was significantly undermined by subsequent developments in Britain. Ignoring the findings of the Beveridge Committee, Churchill's new Conservative government published a white paper in May 1952, recommending that some form of competition be introduced in the field of television. How far this would involve private enterprise or on-air advertising remained unclear, but the possible implications were discussed at the 1952 Commonwealth Broadcasting Conference, held in London that summer. Dunton was unwilling to make too partisan an intervention, fearing not only an adverse reaction in the UK, but also that his comments would be picked up and turned against him in Canada. Nevertheless, along with Moses, he did repeat earlier warnings that, if the experiences of the CBC and ABC were anything to go by, the consequences for the BBC of commercial competition would not be good, and the cultural standard of programming would suffer.[40] Gideon Roos, director-general of the South African Broadcasting Corporation (SABC), was more forthright, arguing that commercial competition had to be prevented in Britain, or else it would be introduced across the Commonwealth, due to the tendency of other governments to look to Britain for broadcasting policy models. Roos argued that in Australia private radio stations and networks, and powerful advertising agencies had debased the standard of programmes. He suggested that a better system had been introduced in South Africa, which had learned directly from the failures of private broadcasting in Australia, Canada, and the US, and could in turn provide a model for Britain: control of all sponsored and non-sponsored broadcasting by a single public authority. When preparing evidence for the Beveridge Committee, Simon and others at the BBC had already considered, and rejected, the New Zealand version of this system. However, whereas in New Zealand all broadcasting was under direct state control, in South Africa a BBC-style broadcasting authority under remote state control seemed to have had more success in satisfying while also regulating the demands of advertisers and other private interests. Roos argued that the SABC had been able to offer listeners a choice of services, and businesses an opportunity to advertise, without allowing private commercial interests to dominate or drive down standards.[41]

In a subsequent letter to *The Times*, Simon (who had since retired as BBC chairman) distilled his thinking on overseas models. In the US, he argued, broadcasting was driven primarily by the profit motive, leading to superficiality, sensationalism, and on-screen violence, all in order to secure the largest possible audiences. Private stations in Canada and Australia adopted a similar approach, diverting audiences

[39] Potter, 'Invasion by the Monster', 260–1.

[40] Ibid., 261–2.

[41] 'Interview with Mr Gideon Roos, on 15th July 1952—Note by S[imon]. of W[ythenshawe]'; Gideon Roos, 'Great Britain', August 1952; and [Roos], 'Springbok Radio', [August 1952], all in Simon papers, M11/6/9.

from the CBC and ABC, and bringing down overall programme standards. New Zealand, and particularly South Africa, Simon argued, offered a more attractive model, in which commercial interests were successfully controlled and standards preserved. Simon recommended a thorough investigation of broadcasting overseas, particularly in the dominions, before any private television licences were granted in Britain.[42] In his book *The B.B.C. from Within*, published the following year, Simon restated this argument, and claimed there was no evidence that private competition had anywhere improved the standard of public broadcasting. In his book he also reprinted an article by Boyer, which claimed that only public broadcasting delivered true choice, catering to minorities and experimenting to offer new programme formats.[43]

Simon's departure from the BBC coincided with Haley's resignation. Jacob, the new director-general, initially rejected the idea that the BBC might itself sell airtime to advertisers in order to avoid the introduction of commercial competition. Moses thought this ill-advised, and fatal to his own hopes that the ABC would be given control of commercial television in Australia. Subsequently, Jacob did in fact suggest something akin to the South African model, whereby the BBC would run two separate services, one with advertising, the other without, but this came too late in the day to influence the government's decision.[44] In March 1954 a bill was published providing for the creation of a new Independent Television Authority (ITA), which would supposedly operate as a 'public service', running transmitters while contracting programming out to private companies funded by advertising. The ITA held its first meeting that August. Separate contracts for weekday and weekend services were awarded on a regional basis. Commercial broadcasts to London and the surrounding area began on 22 September 1955. The BBC's domestic monopoly, in television at least, was over.[45]

The details of the new UK framework had still not been decided when, in January 1953, the Australian government announced that the dual system of public and private broadcasting would be extended from radio to television. A subsequent royal commission recommended few of the Canadian- or British-style safeguards that Boyer argued were necessary if the worst excesses of private competition were to be avoided. The ABC would not enjoy an initial monopoly period, and would have to face competition from the outset. Nor was an Australian version of the ITA created, which might have obliged commercial operators to accept some sort of public-service mandate, and would have allowed commercial licences to be revoked if necessary in a relatively straightforward fashion. Instead, private companies were simply licensed to operate their own transmitters and produce programmes, initially in Sydney and Melbourne, and then across the country. Many of the licensees were existing newspaper and radio

[42] Lord Simon, 'Commercial Broadcasting', *The Times* (London), 3 September 1952.

[43] Simon, *B.B.C. from Within*, 257–72, 290–3.

[44] Briggs, *History of Broadcasting in the UK*, iv, 820, 862–70. Transcript of Charles Moses oral history interview, tape 8 (1971), ABC Document Archive.

[45] Briggs, *History of Broadcasting in the UK*, iv, 872–88, 910–19. Bernard Sendall, *Independent Television in Britain*, i, *Origin and Foundation, 1946–62* (London and Basingstoke, 1982).

conglomerates. Minimal regulation was imposed on broadcasting hours, foreign ownership, or the use of imported programmes.[46]

Meanwhile, in November 1952, a few months after the UK government first indicated that the BBC television service would face some form of competition, the Canadian government announced that private stations would be licensed in those urban centres where the CBC had not yet been authorized to establish transmitters of its own. These private stations would be obliged to carry some CBC network programming, but would also be able to schedule other material. As in Australia, licences were largely granted to existing newspaper and radio combines. For the moment, the CBC remained protected from direct Canadian competition (Canadian stations already had to compete with cross-border US transmissions), but its expansion into new provincial centres had been halted. The BBC's Canadian representative worried that the government's decision might herald the eventual breaking-up of the CBC. At any rate, it was clear that more private stations would be licensed in the near future in cities which already had public stations, and that the CBC would soon have to compete with them for viewers.[47]

At the end of 1955, a Canadian royal commission chaired by Robert Fowler (a businessman with a history of public service) was set up to investigate broadcasting, and particularly the question of how to pay for the rising costs associated with CBC television. Fowler made the customary visit to Britain in the spring of 1956: Jacob asked Dunton if there were any points he wanted the BBC to 'insert into [Fowler's] consciousness'. Dunton replied that it would be good to impart

> some idea of the complexities and costs of television and its organization, especially of program production. And perhaps too, you could bring out some of the facts about 'competition' in broadcasting; how it does not necessarily make for better programming but can lead to a working of Gresham's Law, tending to work against the good programs. Emphasis on the need for some assured basis of revenue to provide for long term planning would also be useful. I would appreciate afterward any indication you feel free to give me personally about how the wind is blowing.[48]

Jacob duly reported that the auguries were good: Fowler would likely make recommendations that would assure the CBC's future dominance of broadcasting in Canada.[49] This proved somewhat optimistic. Fowler's report was laudatory of CBC programming, and critical of the failure of private stations to produce original programmes or develop Canadian talent. It suggested funding CBC operations through five-year grants agreed by parliament, allowing long-term planning. However, while it did not want the CBC to become 'more commercial', the report paradoxically recommended that the corporation should seek to maximize advertising revenues. The report concluded that a second set of private television stations should be licensed, thus paving the way for direct competition between Canadian public and private broadcasters. Fowler also called for the CBC's regulatory powers

[46] Potter, 'Invasion by the Monster', 262–3.
[47] Peers, *Public Eye*, 36–45. Sloan to Clark, 21 November 1952, WAC, E1/509/5.
[48] Jacob to Dunton, 2 March 1956 and Dunton to Jacob, 5 March 1956, WAC, E1/1776.
[49] Jacob to Dunton, 15 March 1956, WAC, E1/1776.

over private stations to be transferred to a separate board.[50] Would some sort of Canadian version of the UK's ITA be created, to curb the worst excesses of the proliferating private broadcasters? Or would regulation become less effectual, as in Australia?

Boyer, who had visited Canada just before the Fowler Report was released, was not impressed with its recommendations.[51] Indeed, the progressive erosion of public service principles was remarked upon by other ABC visitors to Canada in this period, and was often contrasted with the UK experience.[52] However, others were not so sure that the BBC could maintain its traditional approach in the face of competition. Returning from his own travels, Neil Hutchison of the ABC worried that the continued commercialization of broadcasting in the US and Canada would spread to Britain and Australia. Hutchison argued that while the requirements of advertising had driven down the cultural standard of US broadcasting to truly abysmal levels, commercial competition had also damaged CBC programming, and would have the same effect on the BBC and the ABC. All would be forced to focus on entertainment in order to retain an audience, or else risk losing any justification for continued funding.[53] Subsequent ABC visitors to Britain were similarly concerned by what they saw: by 1957 Moses thought the BBC was facing a difficult situation, with both its audience figures and the cultural standard of its programmes declining, as its attempts to cater to mass tastes failed to win viewers over from the commercial stations. Moses thought the solution was a Reithian one: the BBC should 'pay more attention, and devote more thought, imagination and energy, to finding more programmes of quality'.[54]

Clearly, debates about public and private television in Britain, Canada, and Australia were closely intertwined. Public broadcasters deployed common arguments to defend themselves, and turned to their counterparts overseas for advice, information, and inspiration. Much investigative travelling was done between the different broadcasting centres of the British world, and more took place as public broadcasters worked to establish their own television services. Yet none of this prevented the emergence of powerful and successful private television lobbies, or the establishment of commercial television stations in private hands.

RADIO: PROGRAMME EXCHANGE

It was not only in the sphere of television that public broadcasters faced obstacles to effective Commonwealth collaboration. As discussed in Chapter 5, during the

[50] Peers, *Public Eye*, 92–114.

[51] Boyer to Dunton, 1 March 1957 and 15 April 1957, and Dunton to Boyer, 3 May 1957, LAC, RG41, 988/1.

[52] Potter, 'Invasion by the Monster', 263–4.

[53] N. Hutchison, 'Report on Recent Overseas Visits', 21 September 1955, NAA, Sydney, SP724/1, 11/1/3 part 2, barcode 3160816.

[54] C. Moses, 'Report on Television Overseas', [received 1 April 1958], NAA, Sydney, C1574 T2, box 2, WOB1, TV2/2/26, 12203197. See also Report from K. Barry, 3 December 1957, NAA, Sydney, SP613/1, 8/1/57, 3188949.

1950s short wave and transcriptions together allowed a substantial amount of radio programme material to flow from Britain to the dominions, even if this material did not always find the most popular outlets. However, the flow of reciprocal programmes back to Britain from the dominions, and among the dominions, remained limited. In private, BBC officers continued to express grave doubts about the cultural and production standards of programmes made by dominion broadcasting authorities. The BBC's response to an early post-war Commonwealth exchange of short stories was somewhat typical. Dismissing one CBC submission, an officer noted that a dramatized version of the same story had already been produced and had been 'the worst ever radio feature... I reject better stories at the average rate of six per working day.'[55] While the BBC offered to distribute reciprocal programmes from the dominions as part of its Transcription Service from as early as 1949, it was not until 1956 that significant progress was made on this front. In the meantime, both the ABC and CBC inaugurated their own separate transcription services.[56]

BBC officers still tended mainly to turn to the dominions for relatively simple programmes, using unadorned 'ordinary' voices to emphasize concrete, workaday connections among the component parts of the British world. One series in this vein was *Namesake Towns*, which built on earlier precedents. During the war some NAS editions of *Radio Newsreel* had included short recorded greetings from people in British towns, aimed at their counterparts in places of the same name in the US and Canada. It was hoped that this would stimulate interest in rebroadcasting by stations in those towns and cities. In 1943 similar programmes were organized targeting places in Australia and New Zealand, and in 1946 the ABC and the Australian Department of Information together prepared a reciprocal programme of messages to support the Australian government's post-war attempts to attract British migrants.[57] Improving on this format, in 1950 the BBC and the ABC co-produced a series of two-way *Namesake Towns* discussion sessions, using the radiotelephone to link contributors in studios in Britain and Australia. The programmes were carried in BBC regional variations, and on the Pacific Service for rebroadcasting in Australia.[58] By the time they reached Australian listeners, the Australian contributions had traversed the globe by short wave twice, and sound quality had deteriorated significantly. Nevertheless, as these were the only programmes jointly produced by the BBC and ABC, ABC stations were asked to carry them whenever they were audible.[59]

[55] J. K. Rickard to G. R. Lewin, 14 July 1947, WAC, E2/91/1.

[56] Barry to Moses, 14 January 1949, NAA, Sydney, SP724/1, 8/1/25, 1360456. R. Stead to Moses, 9 February 1956, C2327 T1, box 70, 25/1/2 part 3, 12386884, 'Exchange of Broadcasting Information between ABC and BBC'.

[57] Conner to G. Cock, 4 March 1941, WAC, E1/191. N. C. Tritton to G. I. Smith, 5 August 1943, WAC, E1/428. Tritton to J. Shelley, 3 February 1944, Archives New Zealand (ANZ), AADL 564/134a 1/8/7. J. Gough to McCall, 31 December 1945 and E. G. Bonney to McCall, 14 January 1946, both in NAA, Sydney, SP613/1, 25/1/23 part 1, 3190684.

[58] B. H. Molesworth to State Directors of Talks, 18 August 1949, NAA, Melbourne, B2112, 7/BBC, 4154389.

[59] Memorandum by Barry, 15 November 1950, NAA, Sydney, SP613/1, 25/1/23 part 1, 'ABC Programmes Prepared for BBC 1945–50'.

The BBC requested that the ABC include 'ordinary people' rather than 'official types' in *Namesake Towns*.[60] This was an extension of the simple, effective principle developed during the 1930s, when messages from 'ordinary people' were used in coverage of royal and ceremonial events to illustrate and emphasize a genuine sense of Britannic unity and shared ideals. This approach had been deployed in new contexts during the war, and was similarly adopted in *Family Gathering*, a series broadcast in 1946 in the BBC North region.[61] *Family Gathering* comprised recorded contributions from the dominions, in which 'ordinary people' who had migrated from the north of England during the 1920s and 1930s sent messages to relatives at home, and discussed their experiences of life and work overseas. Farmers and factory workers spoke in accents characteristic of the English north as well as of the dominions, while members of the professional classes used the 'refined' tones that would have been more usual on the BBC, ABC, or NZBS. By this time the BBC (and the Dominions Office) was aware that migration schemes were not running entirely smoothly, and did not wish to paint too rosy a picture of the British migrant experience. *Family Gathering* went for balance.[62] Many of the New Zealand participants, for example, were middle-aged, and thus able to offer mature reflection on the difficulties of migration (homesickness and the rigours of farming life) as well as the opportunities (social mobility and improved standards of living). The overall tone was nevertheless positive, encapsulated by one Yorkshirewoman's claim that 'so far as working-class people go, I think New Zealand is the place'.[63]

Similarly, for the 1947 round-the-Commonwealth Christmas feature (*Men of Goodwill*) the BBC asked the ABC to provide 'a sincere, realistically hopeful picture of the kind of life prospective immigrants are likely to find in Australia', and the NZBS to 'soft pedal references [to] abundant food and concentrate [on] opportunities for energetic young able men and women'. The overall aim was 'to show how the people of the British Dominions, including new arrivals, are all helping to strengthen the whole community of British peoples, and through them help to bring back stability and prosperity to the world'.[64] Sometimes, the BBC's cautious treatment of migration drifted into outright pessimism. One exchange programme, focusing on the first Scottish migrants to travel to Canada by air, sought to give a balanced account of their experience, 'and not just the usual rosy picture, plus "Can you hear me, Mother?" '[65] However, the end product was perhaps too balanced: afterwards, one BBC programme assistant in Scotland thought

[60] Molesworth to Director of Talks for Victoria, 23 March 1950, NAA, Melbourne, B2112, 7/BBC, 4154389, 'Talks—BBC'.

[61] 'Broadcasting Review', *The Manchester Guardian*, 15 June 1946. H. C. Fenton to Hutchison, 6 March 1946, United Kingdom National Archives (UKNA), DO35/1141.

[62] C. Conner to G. Looker, 27 July 1949, WAC, E1/332/1. Minute by Pugh, 12 April 1946, UKNA, DO35/1141.

[63] A number of the New Zealand *Family Gathering* recordings are preserved in the Radio New Zealand Sound Archive, media MUCDR45, MUCDR47, MUCDR48, and MUCDR49. Quote from MUCDR47 track 6, 'Mobile Unit Smith Family Gathering, 13 July 1946'.

[64] Conner to Moses, 3 November 1947, NAA, Sydney, ST1790/1, box 33, 25/1/25, 12203270. Conner to Shelley, 6 December 1947, ANZ, AADL 564/140a 7/1/2.

[65] A. Stewart to I. M. Elford, 31 March 1948, WAC, E1/530.

that 'many intending emigrants will possibly have cancelled their bookings because of the black picture [the speaker] painted of conditions in Canada'.[66]

Programmes focusing on migration carried significant ideological baggage, but they also had a practical appeal to listeners who had friends and family overseas, or were themselves considering emigration. In much the same way, short contributions to news, current affairs, and other magazine-type programmes seemed a natural way to introduce straightforward and 'useful' material from the dominions to British listeners. St David's Day and St Andrew's Day provided annual opportunities to feed relatively simple programmes from Welsh and Scottish communities in the dominions back home; perhaps due to the complex politics of identity in Northern Ireland, St Patrick's Day does not seem to have been exploited in quite the same way. Regular newsletters, some in Welsh and Scots Gaelic, were also obtained from dominion public broadcasters. Particularly following the establishment of the BBC's *Woman's Hour* session in 1946, programming for women also offered possibilities for Commonwealth exchange.[67] The ABC regularly included items in its women's sessions from Commonwealth countries and the US, and later also from Asian countries.[68] Similarly, in 1957 for example, the NZBS included in its women's sessions material from Britain, South Africa, the US, and Hong Kong.[69] Religious broadcasting proved another natural area for cooperation, particularly given the Commonwealth-wide reach of the Anglican Church. The ABC regularly used scripts and transcriptions of BBC religious broadcasts, and Dorothy L. Sayers' dramatization of the life of Christ (*The Man Born to be King*) proved a particularly popular BBC export. The ABC also broadcast talks by visiting religious dignitaries, including the Archbishop of Canterbury and the Chief Rabbi of the United Hebrew Congregations of the British Commonwealth.[70]

Cooperation in other specialized areas of broadcasting revealed both the potential of, and the problems with, collaboration. Rural broadcasting had been established in Britain and Canada during the mid-1930s. The ABC appointed its first director of rural broadcasts, John Douglass, in 1945, and the NZBS its first agricultural officer in 1950.[71] After the war, the BBC's Agricultural Liaison Of-

[66] Bill Meikle to Elford, 16 August 1948, WAC, E1/530.

[67] Elford to Collins, 12 August 1947, WAC, E1/1097/2; Polwarth to Peggie Broadhead, [*c.* March 1948], E12/100/3. J. Scott-Moncrieff to Moses, 29 June 1956, NAA, Sydney, C2327 T1, box 70, 25/1/2 part 3, 12386884, 'Exchange of Broadcasting Information between ABC and BBC'.

[68] *Twenty-second Annual Report of the Australian Broadcasting Commission Year Ended 30th June 1954* (Sydney, 1954), 24. *Twenty-third Annual Report of the Australian Broadcasting Commission Year Ended 30th June 1955* (Sydney, 1955), 23. *Twenty-seventh Annual Report and Financial Statement of the Australian Broadcasting Commission Year Ended 30th June 1959* (Sydney, 1959), 15. *28th Annual Report and Financial Statement of the Australian Broadcasting Commission Year Ended 30th June 1960* (Sydney, 1960), 19.

[69] *Annual Report of the New Zealand Broadcasting Service, for the 12 months ended 31st March 1957* (Wellington, 1957), 7.

[70] *Nineteenth Annual Report of the Australian Broadcasting Commission Year Ended 30th June 1951* (Sydney, 1951), 14. *Twentieth Annual Report of the Australian Broadcasting Commission Year Ended 30th June 1952* (Sydney, 1952), 10, 15. *Twenty-sixth Annual Report and Financial Statement of the Australian Broadcasting Commission Year Ended 30th June 1958* (Sydney, 1958), 28.

[71] 'Commonwealth Broadcasting Conference 1952—BBC Memorandum—Rural Broadcasting in the English-speaking World', LAC, RG41, 367/19-31/5.

ficer John Green sought 'to establish a common bond of farming interest throughout the English-speaking world' with a *Land and Livestock* session, broadcast on the Overseas Service. This included not only British material, but special contributions on agricultural themes and events in the dominions, such as coverage of the Sydney Royal Agriculture Show.[72] However, given the nature of rural life and work in Canada and Australia, the countervailing influence of US farm broadcasting was strong. Douglass's enthusiasm for things American was well known: after a trip to Britain in 1951, he contrasted the progressive and extensive nature of farm broadcasting in the US with the atrophied state of its BBC equivalent.[73] Nevertheless, the scale of the ABC's rural broadcasting operations meant that Douglass could find a Commonwealth leadership role for his department. The ABC arranged for a subcommittee on rural broadcasting to be convened during the 1956 Commonwealth Broadcasting Conference, which reported on untapped possibilities for Commonwealth programme exchange. It was agreed to start circulating information among the rural officers of Commonwealth and colonial broadcasting authorities, with the ABC playing a coordinating role.[74] From December 1957 the ABC printed a rural broadcasting bulletin for Commonwealth and colonial broadcasting organizations, *Yabba Yabba*.[75] Several round-the-Commonwealth rural discussion programmes were also organized, involving contributions from Australia, Canada, New Zealand, the UK, India, and Ceylon.[76]

Collaboration was also attempted in schools broadcasting, another area that expanded rapidly around the British world after the war. In 1947 Mary Somerville, the BBC's Assistant Controller of Talks and former Director of Schools Broadcasting, visited Canada, Australia, and New Zealand. During her tour, Somerville gave demonstrations and talks to educationalists and broadcasters, and advised the ABC on its plans to reorganize school broadcasting operations. Somerville arranged for several of the schools broadcasting officers that she met in Australia and New Zealand to visit the BBC.[77] She also opened negotiations for programme exchange. In 1948 the ABC duly prepared three schools programmes for the BBC and one for the CBC. The BBC provided a series of twenty-five programmes in English for use

[72] J. Green to W. Skelsey, 25 January 1949, WAC, E1/308.

[73] J. Douglass, 'Report on Overseas Trip', 8 October 1951, NAA, Sydney, SP724/1, 17/1/6, 3161021.

[74] Moses to Ouimet, 4 February 1955, NAA, Sydney, SP987/1, box 6, 21/8/7, 'British Commonwealth Broadcasting Conference, 1956'. 'Commonwealth Broadcasting Conference, 1956—R/7—Report of the Rural Broadcasts Sub-Committee', Commonwealth Broadcasting Association (CBA) archive, box 1, C/1/3. 'Commonwealth Broadcasting Conference 1956—Summary of Decisions and Recommendations—C/78', LAC, RG41, 368/19-32/1.

[75] Douglass to Moses, 22 November 1957, NAA, Sydney, SP724/1, 22/2/24, 3165565.

[76] *Twenty-seventh Annual Report and Financial Statement of the Australian Broadcasting Commission Year Ended 30th June 1959* (Sydney, 1959), 14.

[77] *Annual Report of the New Zealand Broadcasting Service, for the 12 months ended 31st March 1948* (Wellington, 1948), 9. *Sixteenth Annual Report of the Australian Broadcasting Commission Year Ended 30th June 1948* (Sydney, 1948), 14. Simon J. Potter, 'Strengthening the Bonds of the Commonwealth: the Imperial Relations Trust and Australian, New Zealand, and Canadian Broadcasting Personnel in Britain, 1946–52', *Media History*, 11/3 (December 2005).

in Ontario and Quebec, and subsequently in other Canadian provinces and Commonwealth countries.[78]

However, programme exchange required elaborate planning if material was to meet classroom requirements, which diverged significantly around the Commonwealth, and even between local educational authorities in each country. To this end, the heads of schools broadcasting at the CBC, BBC, and ABC met in Toronto in March 1949.[79] They agreed to produce a series of six exchange programmes, involving Britain, Canada, Australia, New Zealand, South Africa, and Ceylon. The BBC would meanwhile continue to include twenty-five of its own schools programmes each year in the Transcription Service.[80] At the 1952 Commonwealth Broadcasting Conference in London, a special subcommittee on schools broadcasting was convened. This provided an opportunity for overseas visitors to observe BBC schools broadcasting in action, and for all to discuss common concerns. The subcommittee's final report presented schools broadcasting as

> a very powerful instrument for strengthening the ties that link the various Commonwealth countries together, since it has an unrivalled capacity for making children aware of conditions and ways of life in other parts of the world than their own, and for appealing to their imagination and sympathies.

However, it was noted that the multilateral exchange agreed on in Toronto had lapsed after only two years, due to difficulties with unfamiliar accents, different approaches to issues of timing and pace, and divergent school curricula. It was thus recommended that portfolios of basic programme material should be exchanged instead: these could be assembled by different broadcasters into programmes suited to local requirements.[81] However, little progress was made on this front either. Instead, the BBC increased the number of schools programmes in its own Transcription Service, and from 1953 the ABC and CBC arranged their own bilateral exchange of schools programmes.[82] The ABC did attempt to resurrect the portfolio idea at the 1956 Commonwealth Broadcasting Conference, offering to act as a clearing house for such material, and subsequently producing portfolios on the

[78] *BBC/1—Confidential—Broadcasting Committee 1949—General Survey of the Broadcasting Service* (BBC, May 1949), 26–7, Simon papers, M11/6/5c. Barkway, 'Semi-annual Canadian report—January to June, 1948', 5 July 1948, WAC, E1/509/4. R. S. Lambert, 'School Broadcasting in Canada', November 1948, E1/521.

[79] A. Frigon to Moses, 9 April 1949, NAA, Sydney, SP341/1, box 19, 'CBC—December 1943–August 1953'.

[80] R. Postgate, 'Report by Head of School Broadcasting on his Visit to Canada and the United States', 13 May 1949, WAC, E1/584/2.

[81] 'Minutes, Reports and Summaries of Meetings of Commonwealth Broadcasting Conference 1952, BBC, November 1952', LAC, RG41, 367/19-31-1/2. See also R. S. Lambert and R. Bronner 'Inter-Commonwealth School Broadcasting: an Experiment', *BBC Quarterly*, 7/4 (winter 1952–3), 210–15.

[82] *Twenty-First Annual Report and Financial Statements of the Australian Broadcasting Commission Year Ended 30th June 1953* (Sydney, 1953), 15. *Twenty-second Annual Report of the Australian Broadcasting Commission Year Ended 30th June 1954* (Sydney, 1954), 16–17. *Twenty-third Annual Report of the Australian Broadcasting Commission Year Ended 30th June 1955* (Sydney, 1955), 14. G. Winter to J. Scupham, 30 June 1954, WAC, E17/124/3.

wool industry and, the following year, the problem of water scarcity.[83] This material was used in India and Pakistan, and All India Radio (AIR) provided a reciprocal portfolio about the Ganges. However, given the failure of other broadcasters to respond, the ABC discontinued the project.[84]

Outside these special areas, the BBC still continued to express reservations about taking material from dominion broadcasting authorities. The ABC had great difficulty selling its own transcriptions to BBC domestic services, prompting the BBC's Australian representative to stress that, even if the BBC did not need to 'plug the Australian way of life [so] as to try and make people believe it is as good as the Australians themselves think it is', it was important to take some transcriptions in order to please the ABC and build up 'an idea of Commonwealth . . . in this respect the views of Domestic Controllers need overriding'.[85] But domestic controllers were powerful people. At the 1956 Commonwealth Broadcasting Conference, it was agreed that the BBC, CBC, ABC, SABC, and NZBS would each offer one play and one feature for exchange every year. However, Val Gielgud, BBC Head of Drama, subsequently protested at the absurdity of 'resolutions affecting programme departments to be agreed at large scale *tamashas* overseas without those who are responsible for putting such agreements into operation being informed'.[86] By March 1959 the BBC had accepted only one feature and one drama from the ABC, and only one CBC offering (a play called *The Man who was a Horse*, which received a poor audience response when included in the Light Programme).[87] The ABC faithfully despatched its own offerings, but received nothing in return.[88] At the 1960 Commonwealth Broadcasting Conference J. B. Clark suggested that bilateral exchanges of programmes might have more success than multilateral ones, and that 'direct contact between the producers of dramatic programmes might yield better results than decisions taken at a Conference where they were not present'.[89] Enthusiasm for major collaborative projects waned as experience developed of the practical difficulties involved.

The superior attitude of the BBC's domestic controllers was not the only obstacle to programme exchange. The CBC was also reluctant to schedule material from other Commonwealth countries, because Canadian listeners were attuned to North American programming styles that differed from those prevailing in other parts of the Commonwealth. Moreover, it could never afford to use Commonwealth material in peak periods, which had to be reserved for the popular American shows

[83] A. N. Finlay to W. Yates, 21 November 1957, NAA, Sydney, SP1687/1, R13/1/5 part 1, 3164180. *Twenty-sixth Annual Report and Financial Statement of the Australian Broadcasting Commission Year Ended 30th June 1958* (Sydney, 1958), 22.

[84] ABC paper for 1963 Commonwealth Broadcasting Conference, 'Exchange of Portfolios of Basic Material for Radio School Programmes', LAC, RG41, 370/19-36-1/1.

[85] P. Jubb to McCall, 4 April 1952, WAC, E1/332/2.

[86] V. Gielgud to M. F. C. Standing, 9 October 1957, WAC, E2/684/2.

[87] Elford to Conner, 16 March 1959 and H. R. Pelletier to R. D'A. Marriott, 30 April 1959 and 6 October 1959, WAC, E2/684/2.

[88] 'ABC—Review of the 1956 Conference and Discussion of Points Arising', [*c.* December 1959], LAC, RG41, 369/19-35-1/2.

[89] Minutes, Commonwealth Broadcasting Conference, 23 January 1960, CBA archive, box 1, C/1/4.

that attracted the most lucrative sponsorship.[90] The ABC, while generally more positive about possibilities for programme exchange, worried about accents. 'Broad' Canadian and Ceylonese accents were deemed unacceptable to Australian listeners, as were the 'clipped' tones of some BBC announcers, who spoke of 'Orstralia' (Australia) and 'gels' (girls).[91] Local differences sometimes amounted to outright prejudice: Hutchison urged that BBC transcriptions should exclude 'pansy' voices, which were 'not... for the Australian air'.[92]

Joint productions offered a means for participating broadcasting authorities to exert more control over the standard of incoming programmes, and perhaps avoid some of the problems associated with exchanges. A successful experiment in this vein was organized by the BBC Features Producer Geoffrey Bridson. After having visited South Africa to record background material for the BBC during the royal family's visit of 1947, Bridson had been sent to Australia and New Zealand for another anticipated royal tour. Although this was cancelled due to the King's illness, Bridson still spent six months with the ABC, advising on the establishment of its new Features Department, and producing programmes that were heard locally and on BBC domestic and overseas services.[93] Bridson's criticisms of South African apartheid, and of Australian cuisine and private radio, raised some hackles; indeed, his feature on South Africa directly jeopardized an £80 million gold loan to Britain.[94] Nevertheless, Bridson's overseas experience made him the obvious candidate to produce a series of programmes to publicize the Colombo Plan, part of the broader support given by the BBC to policies of colonial development. In a paper presented to the 1952 Commonwealth Broadcasting Conference, Bridson advocated the production of 'a series of large scale Commonwealth Feature Programmes', arguing that in radio 'there has been a tendency for world reconstruction and development to be reported with a predominantly American accent'. A series of co-produced features and talks could make listeners aware of what the Commonwealth was doing under the auspices of the Colombo Plan in India, Pakistan, Ceylon, and Malaya. A travelling BBC writer-producer would cooperate with local broadcasting authorities, using local material and technical facilities, thus keeping down costs while encouraging cooperation. The resulting programmes could be distributed through the BBC Transcription Service. The experiment had the virtue of combining a clear practical purpose with the prospect of high production standards and the creation of full collaborative links on the ground, and was approved by the conference delegates.[95]

[90] 'Commonwealth Radio Conference—Notes on Agenda', [*c*.1945], LAC, RG41, 357/19-12/2.

[91] 'Commonwealth Broadcasting Conference 1952—ABC Memorandum—Program Exchanges', LAC, RG41, 367/19-31/5.

[92] Hutchison to Conner, 6 September 1946, WAC, E17/9/1.

[93] Bridson, *Prospero and Ariel*, 133–69. Leonard Miall, 'Bridson, (Douglas) Geoffrey (1910–1980)', rev., *Oxford Dictionary of National Biography* (Oxford, 2004; online edn, September 2010) <http://www.oxforddnb.com/view/article/30851> accessed 27 January 2012. Moses to Haley, 30 September 1948 and Haley to Moses, 8 October 1948, WAC, E1/315/6.

[94] 'Commonwealth Relations Office and the B.B.C.', [*c*. January 1948], UKNA, CAB124/36.

[95] Bridson, 'Commonwealth Broadcasting Conference 1952—Joint Commonwealth Features', LAC, RG41, 367/19-31/5. Minutes, Commonwealth Broadcasting Conference, 9 July 1952, RG41, 367/19-31-1/2.

The first series of Commonwealth features involved the BBC, ABC, AIR, Radio Pakistan, Radio Ceylon, and the Malayan Broadcasting Service.[96] Although the response to the programmes among listeners and newspaper critics was disappointing, Bridson coordinated a second series focusing on colonial development in Africa. He hoped that the programmes might play a moderating role in the troubled arena of contemporary African affairs.

> By showing what has been achieved in West Africa along the path towards self-government, for instance, we should be offering a useful corrective to extremist opinion in the Union [i.e. South Africa], Central and East Africa; similarly, by showing what has been achieved industrially in the Union and Central Africa, we should be offering a useful object lesson to West and East.[97]

However, producing features in Africa was not easy. There were no African members of the Commonwealth Broadcasting Conference apart from the SABC, and colonial broadcasting services could offer only rudimentary production facilities. The SABC thus played a prominent role in devising and producing the series, and the BBC, CBC, and ABC also contributed staff and resources. After his earlier experiences in South Africa, Bridson suggested that the ABC and SABC, rather than the BBC, should collaborate to cover the Union in the series. As Moses noted, the aim was to draw South Africa 'more closely into the Commonwealth': sending an all-BBC team might prove counterproductive, whereas critical comment from an Australian observer would provoke less inherent hostility. The project was seriously delayed by the need to produce features to coincide with the royal tour of 1953/4, but eventually seven hour-long programmes were recorded, and broadcast in 1956.[98] The SABC judged the programme on South Africa (produced in the end by Laurence Gilliam of the BBC and John Thompson of the ABC, working with an SABC crew) 'strikingly successful'.[99]

The Commonwealth Features Project showed how public broadcasting authorities could successfully work together to produce radio programmes. The project's demise also illuminated some of the obstacles to such collaboration. In 1957, for a short, unhappy interlude, Bridson moved from radio to television. This robbed the Commonwealth Features Project of its coordinating personality. Radio collaboration also seemed a less urgent priority by the late 1950s, as Commonwealth countries began to slip into the television age. A few ideas for a third series of Commonwealth radio features were suggested, but with Bridson off the scene little progress was made.[100] Eventually, the expatriated Canadian Rooney Pelletier (now

[96] *Twenty-First Annual Report and Financial Statements of the Australian Broadcasting Commission Year Ended 30th June 1953* (Sydney, 1953), 14.

[97] Bridson to L. D. Gilliam, [*c.* July 1953], WAC, E2/91/2.

[98] Press cutting, '"Africa 1955"... A Study of a Continent in Turmoil', *CBC Times*, 1–7 July 1956, LAC, Matthew Halton fonds, 24. Moses to ABC commissioners, 14 September 1954, NAA, Sydney, SP613/1, 12/3/16, 3189918.

[99] 'Commonwealth Broadcasting Conference, 1956—SABC memorandum—C/53—Programme exchanges (Agenda Part 1)', Boyer papers, folio box, folio 1.

[100] Bridson, *Prospero and Ariel*, 239–89. Conner to C. Jennings, 1 February and 25 April 1957, LAC, RG41, 380/20-3-23. R. M. Marathay to Conner, 18 March 1958, WAC, E2/684/2.

BBC radio's controller of programme planning) took charge. Pelletier's ideas about Commonwealth programme exchange differed from Bridson's, reflecting his own background and interests. The Commonwealth Features Project was transformed into 'Canada Week'. In the first week of June 1959 the BBC domestic and overseas services broadcast a selection of the CBC's best programmes, including a symphony concert; jazz and folk music sessions; a talk on 'nationalism in Canadian history'; discussion, drama, and light entertainment programmes; and a Wayne and Shuster comedy show.[101] Pelletier hoped that the new approach would help 'avoid the stiffness and slight unreality that has hitherto affected most Commonwealth programmes prepared for export'.[102] However, reactions to the experiment were decidedly mixed, with only two programmes scoring above average for listener appreciation.[103] Crucially, Pelletier's narrow focus on Canada effectively destroyed the broad appeal of the Commonwealth Features Project. Although the CBC agreed at the 1960 Commonwealth Broadcasting Conference to revive and coordinate the scheme, it was subsequently quietly dropped.[104]

TELEVISION: EXCHANGE, SALE, OR CO-PRODUCTION?

In part, the problems encountered by the Commonwealth Features Project derived from a much more profound change: the transfer of broadcasting supremacy from radio to television. How did this transition shape prospects for Commonwealth broadcasting collaboration more generally? Television's insatiable demand for content, and the high costs of television production, created from the outset a clear rationale for transnational cooperation. As Haley argued, television would positively require a greater degree of collaboration between different countries than radio ever had.[105] However, upon what basis would cooperation rest? Could the traditional, non-commercial Commonwealth approach be extended to the new medium? Radio programmes had long been exchanged free of charge, or on a costs-only basis. However, the opportunities for applying the same model to television were limited. If past experience in radio was any guide, then the BBC would send out many more television programmes than it brought in, at least as far as the Commonwealth was concerned. This would result in an imbalance of trade which, due to the expense of television production, would be unsustainable. If state subsidies to make up the loss were not forthcoming, and if the BBC thus had to sell its television programmes overseas at market rates, would other Commonwealth

[101] C. Curran, 'Notes on Meeting with CBC Programme Officials—Canada Week Project', 24 April 1958 and CBC Information Services press release, 8 April 1959, LAC, RG41, 380/20-3-23.

[102] Pelletier, 'Canada Week', 10 April 1959, WAC, E2/685/2.

[103] Report on 'Maple Leaf Special', n.d., WAC, E2/685/2.

[104] Jennings, 'Commonwealth Broadcasting Conference—New Delhi—1960', 25 March 1961, LAC, RG41, 380/20-3-23.

[105] Sir William Haley, *Broadcasting as an International Force: The University of Nottingham, Montague Burton International Relations Lecture, 1950–1* (Nottingham [1951]), 17, Simon papers, M11/6/6.

public broadcasters be able to afford them? BBC officers wrestled with these problems throughout the 1950s, and into the 1960s.

'In its first decade United States television was already a world phenomenon', its overseas sales potential fuelled by a domestic 'gold-rush' as commercial stations multiplied and networks expanded to serve the massive US audience. Hollywood's back catalogue of film provided an instantly saleable resource. By comparison, the BBC produced a mere trickle of television programmes for export.[106] Progress with developing UK domestic television services remained slow, transmission and reception equipment scarce. The BBC thus had a weak domestic base from which to break into the US, which was the only sizeable overseas English-speaking television audience in the early 1950s. Technical problems were also significant. As live television was the norm at the BBC, programmes produced for UK viewers had to be specially recorded if they were to be exported overseas. Before the advent of videotape the only way to do this was to produce 'kinescopes' (also known as 'kines' or 'telerecordings') by pointing a specially prepared 35mm film camera at a video monitor linked up to the live camera. This seldom produced a high-quality recording. The alternative was to make programmes especially for export, recording direct to 35mm film. How could this be funded? Following a trip to New York, Greene argued that the BBC should adopt a commercial approach, and focus on selling to the US, to provide a viable foundation for subsequent exports to other countries. 'Only by breaking into the United States market could a television transcription service hope to make both ends meet.' Greene's language was that of the salesman, not the public servant: he argued that the BBC would have to study the demands of American networks and sponsors carefully, and offer them specially produced films designed to suit US requirements.[107]

However, the BBC also claimed that before a commercially run television transcription service could become self-supporting, an initial government subsidy (of the kind that had allowed it to make such a success of its radio transcription service) would be necessary. In May 1952 the BBC asked the Treasury for £250,000 per annum, for three years, to cover a modest annual output of thirty-nine half-hour entertainment programmes, and twenty-six documentaries (so around one programme per week) all specially produced on 35mm film. The programmes would be aimed at the US mass market: televisual equivalents of Third Programme material might be tackled later.[108] However, the Treasury refused to support the BBC's proposals. Thus, when the BBC established a small Overseas Film Unit the following December, it could undertake only a limited schedule of kinescope production. This did not hold out much promise, Greene having already pointed out that kinescopes of programmes aimed at British viewers would be unsuited to US requirements. A pilot scheme of three specially produced half-hour films was un-

[106] Eric Barnouw, *A History of Broadcasting in the United States*, iii, *The Image Empire: from 1953* (New York, 1970), 3, 65. Briggs, *History of Broadcasting in the UK*, iv, 891–2.

[107] Michael Tracey, *A Variety of Lives: A Biography of Sir Hugh Greene* (London, 1983), 139. Hugh Carleton Greene, 'Television Transcriptions: the Economic Possibilities', *B.B.C. Quarterly*, 7/4 (winter 1952/3), 216–21, quote at 217.

[108] Minutes, Commonwealth Broadcasting Conference, 4 July 1952, LAC, RG41, 367/19-31-1/2.

dertaken, aimed at providing background information about the 1953 coronation for American and Canadian audiences, with Pelletier as the presenter. They were sold to the CBC for $1,000 each.[109]

Frustrated by the meagre amount of BBC material on offer, the CBC pressed for a much fuller supply. Dunton argued that it was 'a tragic thing' that US programmes, paid for by a massive US domestic market, were so cheap and plentiful compared to UK material.[110] Successful BBC negotiations with the actors' union Equity reduced the cost of exporting programmes to Canada somewhat, and Bushnell was sent to Britain to arrange for the supply of more kinescopes. However, Dunton also warned that Norman Collins had already visited Canada, offering to sell British commercial television programmes. Collins, a former BBC controller of television, had become a vocal proponent of commercial broadcasting in Britain: the BBC was now facing the prospect of UK-based competition abroad as well as at home.[111] George Barnes, BBC director of television, himself visited Canada in 1953, and concluded that the BBC was missing 'a great opportunity' to export television programmes to North America. In Canada the need for a BBC on-screen presence seemed particularly pressing. 'The C.B.C. and the new Commercial stations want anything we can sell them at a rate that they can pay... Television is one of the few ways left in which the americanisation of Canada can be counteracted.'[112]

Nevertheless, by the end of March 1956 the BBC had only supplied 138 kinescopes and 66 films to overseas television organizations, an average of just over one programme per week. Due to continued difficulties in negotiations with artists' bodies, most of these programmes were either outside broadcasts or televised 'talks', in which neither professional actors nor musicians were involved.[113] Moreover, BBC kinescopes continued to prove expensive compared to other material available to the CBC (particularly old Hollywood movies), and the production and recording quality of BBC offerings left much to be desired. It was not possible for the CBC to audition material before the BBC made the kinescopes, allowing little opportunity for unsuitable programmes to be either improved or rejected.[114] All in all, this did not represent an impressive BBC effort.

During the mid-1950s the establishment of private commercial television stations in Britain and Canada created competitive markets for programme imports, and threatened to drive up prices. The BBC and CBC could possibly generate more revenue from programme exports, but might also have to pay more for

[109] H. C. Greene to Bushnell, 3 December 1952, WAC, T6/72. [A. G. Cowan], 'Three BBC TV films on the Coronation' [*c.*17 February 1953] , LAC, RG41, 539/11-37-5/1. 'The Coronation—A BBC Television Film—The Second Elizabeth', LAC, ISN 12093, Acc # 1986 024.

[110] Dunton to Clark, 3 October 1952, LAC, RG41, 988/2.

[111] Dunton to Barnes, 30 October 1952, LAC, RG41, 988/2. Dunton to Clark, 13 November 1952, RG41, 378/20-3-5/2.

[112] Barnes, 'Report on Visit to the United States and Canada', [*c.* early 1953], WAC, E1/584/1.

[113] 'Commonwealth Broadcasting Conference 1956—BBC memorandum—Television Transcriptions—Agenda Part 1—C/4', Boyer papers, folio box, folio 1.

[114] Memorandum by Sloan, 3 August 1953, WAC, E1/509/6. J. M. Moore to F. Mutrie, 17 November 1953, LAC, RG41, 378/20-3-5/2.

imports. This made it even more difficult to arrange exchanges among public broadcasters on a costs-only basis: the BBC might forfeit substantial revenues in Commonwealth markets, while still having to pay increasing market prices for imports from the US and elsewhere. Ultimately, the BBC decided to tap all available financial resources to preserve its position at home. Thus while it reaffirmed its commitment to 'public service broadcasting on a national basis', it maintained that it needed to 'exploit every market to the utmost', and obtain the maximum price for its exports.[115] BBC overseas operations would henceforth be run, in part, on a commercial basis. It would be difficult to exaggerate the significance of this departure.

Would cash-strapped Commonwealth public broadcasters be able to afford BBC exports sold at market prices? In Canada, the only Commonwealth country with its own television service at the time, the government's 'single station' policy provided a brief respite. Initially, the CBC and private operators would not be competing in the same urban centres, so would not have to bid against each other for programmes. When direct competition was eventually introduced, the BBC agreed to give the CBC first refusal of its programme offerings, and it was anticipated that similar rights would be granted to the ABC.[116] Some sort of preference was thus retained. However, if public authorities could not afford the programmes, then the right of first refusal would become meaningless.

At the 1956 Commonwealth Broadcasting Conference delegates tried to find ways to trade television programmes at low cost in an increasingly commercialized environment. Exchange, joint production, and joint purchasing schemes were suggested. Given earlier debates about the damaging impact of American commercial programming on young viewers, programmes for children and schools were prioritized for urgent action.[117] Subsequently, some special Commonwealth exchanges of rural television material were also successfully organized, and the BBC and CBC tentatively explored possibilities for joint production of films. However, little of substance was achieved.[118] In Canada, low-budget BBC television programmes could not compare with the US commercial product, and were relegated to the status of 'summer fills' and 'occasional placings' in CBC schedules. Charles Curran, the BBC's Canadian representative, thought that little better could be expected

[115] Clark to Sloan, 22 May 1953, LAC, RG41, 378/20-3-5/2.

[116] Jennings, 'Statement on CBC Policy', 3 August 1955, WAC, E1/1782. Conner to Stead, 29 August 1955, E1/1636/1. The CBC similarly offered the ABC first refusal on its own exports to Australia. See G. Rugheimer to T. Duckmanton, 25 October 1956, LAC, RG41, 378/20-3-7/3.

[117] 'Commonwealth Broadcasting Conference 1956—ABC memorandum—Programme Exchanges in Television—Agenda Part 1—C/32', Boyer papers, folio box, folio 1. Minutes, Commonwealth Broadcasting Conference, 12 November 1956; 'Commonwealth Broadcasting Conference 1956—Report of Sub-Committee appointed to discuss Commonwealth Co-operation in the Production of TV Programme Material'; 'Press Release—Commonwealth Broadcasting Conference', 29 November 1956; and 'Commonwealth Broadcasting Conference 1956—Summary of Decisions and Recommendations—Paper C/78', all in LAC, RG41, 368/19-32/1.

[118] *28th Annual Report and Financial Statement of the Australian Broadcasting Commission Year Ended 30th June 1960* (Sydney, 1960), 15. 'Commonwealth Broadcasting Conference 1960—ABC paper—Joint Production—TV Film Programmes', LAC, RG41, 369/19-35/1. R. Waldman to J. F. Mudie, 5 May 1958, WAC, E1/1636/2.

without substantial British government subsidies for television transcriptions.[119] However, in 1957, as a result of an official inquiry into UK overseas information services, the idea of state-funded BBC television transcriptions was finally and definitively rejected. Instead (perhaps unsurprisingly, given the BBC's independent-minded behaviour during the Suez crisis) the British government opted to distribute television material through the more easily controlled Overseas Information Service.[120]

In Australia, despite warnings to the contrary, even up until the eve of the inauguration of its television service the ABC expected BBC recorded programmes to be readily and cheaply available.[121] Subsequently, it rapidly became clear that this would not be the case. Instead, US companies became a major supplier of imported television programmes, particularly to the new Australian commercial stations. At the ABC, the good Reithian Clement Semmler shuddered at the American programmes now occupying Australian screens, and particularly at the plentiful supply of Westerns.[122] As the BBC's Australian representative, James Mudie, put it,

> In 300,000 homes in Sydney and Melbourne the Waggon Trains [*sic*—a reference to the popular US Western TV series, *Wagon Train*] move steadily towards an endless horizon. Restless guns blaze at a rate that one would envy in any automatic and bodies litter the approach to the controls of the set. But like old soldiers they always live to fight another day, same channel, same time slot, twenty-three of them in a week. Through the bull dust of this stampede there appears dimly and as fleetingly as the Angel of Mons a vision with the wings of War In The Air [an acclaimed BBC documentary series], a trumpet from the Sydney Symphony Orchestra and holding high a heraldic device bearing the letters ABC. But she is a frail undernourished creature, the object of concern to her creators who struggle with the contrast control to try and brighten the vision... The truth is that for a capital expenditure far in excess of the commercial stations, the ABC is commanding less than 10% of the audience.[123]

Perhaps the most, or indeed only, significant achievement in terms of making television a medium of imperial mass communication in the 1950s was the establishment of a Commonwealth organization for the exchange of newsfilm. Initially, the BBC, CBC, and ABC all relied on newsfilm agencies such as United Press Movietone Television (UPMT) and freelance cameramen for world coverage. Exchanges of newsfilm among the three public authorities were limited in scope.[124] Position

[119] Curran to E. Barraclough, 9 April 1958, WAC, E1/1784.

[120] Gerard Mansell, *Let Truth Be Told: 50 Years of BBC External Broadcasting* (London, 1982), 236–7. Linda Kaye, 'Reconciling Policy and Propaganda: the British Overseas Television Service, 1954–1964', *Historical Journal of Film, Radio and Television*, 27/2 (June 2007), 215–36.

[121] Potter, 'Invasion by the Monster', 264.

[122] C. Semmler to Clark, 21 August 1958, WAC, E1/1667.

[123] Mudie, 'The Australian Scene', 28 August 1958, WAC, E1/1636/2.

[124] P. H. Dorté to Jubb, 5 October 1950, WAC, E1/425. Clark to Winter, 15 September 1955, E1/1782. 'Commonwealth Broadcasting Conference, 1956—CBC memorandum—C/12—News exchanges (Agenda Part 1)' and 'Commonwealth Broadcasting Conference 1956—ABC memorandum—News Exchanges—Agenda Part 1—C/16', Boyer papers, folio box, folio 1. Duckmanton, 'Report no. 12', 28 November 1955, NAA, Sydney, C1574 T1, box 9, TV2/2/3 part 2, 11029325.

papers prepared for the 1956 Commonwealth Broadcasting Conference suggested only a modest revision of these arrangements: the BBC proposed marginal increases in the exchange of film to supplement agency sources, while the ABC argued for cooperation in special assignments, 'a step forward towards the ideal—of the Commonwealth reporting the Commonwealth to the Commonwealth'.[125] However, at the conference itself, Jacob proposed more significant change. Arguing that agencies such as UPMT were 'all controlled by United States interests' and provided material unsuited to Commonwealth audiences, and claiming that existing Commonwealth exchanges of newsfilm were too slow to act as a viable substitute, Jacob suggested the establishment of 'a British Commonwealth agency for the supply of news film . . . controlled by the users . . . ownership would be confined to Commonwealth organisations'. Both the ABC and CBC were supportive, as was the SABC (in anticipation of the day when it would have its own television service) and, in more guarded fashion, AIR.[126]

Together, the BBC, the Rank Organisation (a major UK film and entertainment group), the ABC, and the CBC went on to establish the British Commonwealth International Newsfilm Agency (BCINA, also known as 'Visnews'), aiming to provide a newsfilm service with world coverage that was 'politically independent and free from bias of any kind'. The two UK partners together put up the bulk of the necessary funds. The ABC provided film not only from Australia but also, through its Singapore office, from south-east Asia.[127] Some attempt was made to give the service a Britannic flavour:

> The service is not a propaganda service in the sense of being government controlled, but it includes in its world-wide coverage such Commonwealth events as Royal Tours and ministerial visits, elections, sport and the general visible process of building democratic development as well as the violence and conflicts which accompany it.[128]

The establishment of BCINA was a major achievement, allowing the extension of Commonwealth cooperation on a non-commercial basis into this vital area of television. However, the application of public-service principles to the exchange of newsfilm was not unqualified. After a disappointing uptake in the first year, an exchange agreement was signed with an American commercial network (the National Broadcasting Company) to allow better coverage of US affairs.[129] Similarly, although the ABC terminated earlier exchange agreements with the UK's

[125] 'Commonwealth Broadcasting Conference 1956—BBC memorandum—News exchanges in sound and television—Agenda Part 1—C/11' and 'Commonwealth Broadcasting Conference 1956—ABC memorandum—News Exchanges—Agenda Part 1—C/16', both in Boyer papers, folio box, folio 1.

[126] Minutes, Commonwealth Broadcasting Conference, 8 November 1956, LAC, RG41, 368/19-32/1.

[127] *Twenty-sixth Annual Report and Financial Statement of the Australian Broadcasting Commission Year Ended 30th June 1958* (Sydney, 1958), 11. John Tebbutt, 'News from Asia: a Transnational History of Australian Television', *Media History*, 17/3 (August 2011), 289–303.

[128] [Tangye Lean], 'The Impact of Broadcasting: Radio and Television in the Commonwealth', *Round Table*, 50/198 (March 1960), 126–33, quote at 133.

[129] 'BCINA—Special Report to the Trustees', 8 July 1958, NAA, Sydney, C1574 T2, box 16, 11/6/9, 'BCINA—Director's Special Report'.

commercial television news service (Independent Television News (ITN)) the CBC continued to supplement Visnews film by exchanging material with UPMT, ITN, and other commercial agency sources.[130]

CONCLUSIONS

During the 1950s public broadcasters continued to operate in a British world. In Britain and the dominions, debates about the introduction of television were informed by a shared Britannic perspective and set of reference points. Ideas and broadcasting models travelled around the British world, shaping policies and approaches in each of its component parts. Yet this was not sufficient to prevent the inexorable erosion of the status and role of public broadcasting, as policymakers in Britain, Canada, and Australia granted increasing freedom to private commercial operators. Public broadcasters felt each blow inflicted upon their counterparts, and worried about the damage that events overseas might do to their own status, particularly when the BBC's own monopoly of broadcasting was finally breached.

Public broadcasting authorities were not well served during the 1950s by their belief that they had a mission to preserve 'cultural standards'. Policymakers largely ignored the argument that 'Gresham's law' could be applied to broadcasting, that bad commercial programmes designed purely for profit would drive out good material informed by public-service principles. The language of cultural standards even acted as an obstacle to successful Commonwealth collaboration in radio, restricting programme exchange. By the late 1950s even senior BBC officers had come to admit that adaptation was necessary, that the old language of cultural standards was now redundant, and that the purpose of public broadcasting had to be reimagined. In Britain the 1960s would see profound changes at the BBC, as it began to engage in open competition with its commercial rivals. Overseas, a variant of this approach was also deployed, as the BBC began to operate on an increasingly commercialized basis, in order to compete with foreign, and particularly US, rivals.

[130] W. Hamilton to Manager, ITN, 24 December 1957, NAA, Sydney, C1574 T1, TV11/3/12 part 1, 1930850. 'BCINA Management Report no. 15', 28 May 1958 and copy of Ouimet to T. Hole, 3 October 1958, both in NAA, Sydney, C1574 T2, box 15, 11/6/4 part 2, 12203200.

7

Disintegration? 1960–70

During the 1960s, with Hugh Carleton Greene as director-general (1960–69), the British Broadcasting Corporation (BBC) jettisoned the old Reithian language of cultural standards, and adopted a new attitude towards its commercial rivals. While Ian Jacob had argued that the public broadcaster should seek at most a third of the television audience (thus maintaining its commitment to 'quality' and protecting its distinctive cultural contribution) Greene insisted that the BBC must compete outright and secure at least an equal share of the audience. It could do this, he proposed, by producing the same types of programmes as the commercial stations, but doing a better job. Greene also thought that the BBC could attract viewers by taking programming into challenging new areas where commercial producers, scared of the response of advertisers, would be unwilling to tread. In the short term, this new approach seemed a great success, bringing large audiences back to BBC television, and making possible the production of some groundbreaking programmes. The BBC seemed in tune with the sixties, capable of mediating the decade's characteristic challenges: 'youth in revolt, relaxed manners, political radicalism, and puritanism repudiated'.[1] However, in the long term the effects of Greene's regime were more ambiguous. By narrowing the gap between public and private broadcasting, and by operating its overseas services on an increasingly commercial basis, the BBC reduced its ability to defend the distinctive role of public broadcasting, particularly against the free-market orthodoxy of the late twentieth century.[2]

Developments in the sphere of overseas operations provided a hint of what would come at home in subsequent decades. Denied a government subsidy, the BBC had to work on a commercial basis if it wanted to export television programmes. By embracing this option, the BBC helped ensure that the UK remained a partner in the broader 'Anglo-American' alliance that would subsequently exert such a powerful influence over much of the world's mass media.[3] Yet the UK was clearly the junior partner, and British broadcasters and policymakers did not necessarily relish the prospect of playing second fiddle to US media interests. The BBC was not openly 'anti-American': it had, since the 1930s, sought to bring its

[1] Brian Harrison, *Seeking a Role: The United Kingdom, 1951–1970* (Oxford, 2009), 472.

[2] Michael Tracey, *The Decline and Fall of Public Service Broadcasting* (Oxford, 1998), 85–97. Tracey, *A Variety of Lives: A Biography of Sir Hugh Greene* (London, 1983), 155–8, 194–6.

[3] Jeremy Tunstall, *The Media are American: Anglo-American Media in the World* (London, 1977). Jeremy Tunstall and David Machin, *The Anglo-American Media Connection* (Oxford, 1999).

audiences what it regarded as the best products of US culture and broadcasting.[4] However, at the same time, reflecting the enduring distaste for commercial mass culture nurtured by members of the British cultivated elite, many BBC officers deplored the mainstream products of the American media industry, and worried that they would eventually displace British programmes overseas and perhaps even in the UK.

Such dilemmas reflected the broader problems faced by policymakers dealing with Britain's continuing adjustment to imperial decline, and attempting to seek an accommodation with the American superpower that would allow the preservation of key British overseas interests. During the late 1950s the British government began to weigh up the potential benefits and costs of empire, and the possible advantages of decolonization. Subsequently, plans for the gradual extension of self-government to colonies in Asia and Africa were abandoned in favour of rapid, even precipitate, transformation. Policymakers hoped to run before the nationalist wind of change, and win sufficient goodwill to keep newly independent Commonwealth countries friendly and aligned with the Western bloc. By the mid-1960s, almost all of the British Empire that could be abandoned, had been abandoned. It was at this terminal point that the fact of imperial decline belatedly began to register in British popular and elite culture.[5] At the same time, facing the sudden collapse of the old Britannic foundations of settler identities, white English-speaking communities overseas found themselves wrestling with the difficult, perhaps impossible task of creating 'new nationalisms' for themselves.[6]

In each part of the British world, the unravelling of the Britannic connection had its own particular dynamic. As Britain pushed towards black majority rule in its African colonies, South African governments dominated by Afrikaner nationalists retreated into apartheid and isolation. In 1961 South Africa became a republic, and left the Commonwealth before it could be expelled. Meanwhile, Canada's 'flight from Empire' accelerated, as sections of the country were drawn further into the orbit of the US economy, and as federal authorities struggled to contain separatist forces in Quebec. Perhaps most significantly, the UK itself began to disengage from the wider British world, and reorient itself towards Europe. The consequences of this were particularly serious for Australians and New Zealanders, still geared primarily towards the UK, economically and culturally.[7] The 1962 BBC television documentary *Commonwealth Crisis: Britain and the Old Dominions* (produced with the assistance of the Canadian Broadcasting Corporation (CBC), Australian Broadcasting Commission (ABC), and New Zealand National Film

[4] Kelly Boyd, 'Cowboys, Comedy and Crime: American Programmes on BBC Television, 1946–1955', *Media History*, 17/3 (August 2011), 233–51.

[5] Ronald Hyam, *Britain's Declining Empire: The Road to Decolonisation, 1918–1968* (Cambridge, 2006), 241–326. John M. MacKenzie, 'The Persistence of Empire in Metropolitan Culture' in Stuart Ward (ed.), *British Culture and the End of Empire* (Manchester, 2001), 32–3.

[6] James Curran and Stuart Ward, *The Unknown Nation: Australia after Empire* (Carlton, Vic., 2010).

[7] Phillip Buckner, 'Canada and the End of Empire, 1939–1982' in Buckner (ed.), *Canada and the British Empire* (Oxford, 2008). Buckner, 'The Long Goodbye: English Canadians and the British World' and Stuart Ward, 'Worlds Apart: Three "British" Prime Ministers at Empire's End' in Buckner and R. Douglas Francis (eds), *Rediscovering the British World* (Calgary, 2005).

Unit) explored the Britannic consequences of the UK's new European leanings in some detail. The writer and narrator, Robin Day, concluded that the economic impact of UK membership of the European Economic Community would be manageable in Australia and Canada, but potentially catastrophic in New Zealand. The political impact remained unclear: according to Day, everything depended upon what type of union Europe eventually became.[8]

Those worried by Britain's swing to Europe were also frustrated by continued public apathy in the UK towards the empire and Commonwealth: a condition that the BBC had long attempted to alter but without perceptible success. Broadcasting on the BBC Home Service, the future Conservative MP Angus Maude, who had recently returned to England after three years' editing the *Sydney Morning Herald*, claimed that,

> The British do not really care about Australia, whereas the Australians care deeply about Britain... The Empire, the Commonwealth, Australia, New Zealand—admirable things, jolly good ideas. Let's be nice to these people when they come over; and indeed we are delightfully kind to them. But how many of us are interested enough to want to know, to find out, what it is all about? They want to hear whether anyone really knows or cares—in terms of hard figures—what is going to happen to their wheat and metals and beef and butter and eggs. It happens to be a matter of life and death to them. We cannot defend them any more. We cannot even help them much. But we might at least *care*.[9]

Meanwhile, Australia and Canada were becoming noticeably less British. In 1963 the BBC's Canadian representative argued that Canada's Britishness could no longer be taken for granted, given the varied ethnic origins of inward migrants, and economic and cultural integration with the US. Demand for BBC programmes from Canadian broadcasters would, he thought, decline unless the BBC adopted a 'fresh approach'.[10] More optimistically, the BBC's director of television pondered the potential effects of British programming on European migrants to Australia. 'In a continent which is always vaguely, and often openly, "anti-Brit", is it not agreeable to think that many of the new Australians may learn to speak English like the Controller of BBC-2 [David Attenborough], or Sir Michael Redgrave, or Terry Scott (or Steptoe)?'[11] Attenborough and Redgrave spoke in tones that had been familiar to BBC listeners for many years: Scott (a well-known comedian, whose voice was distinctive but did not evoke privilege) and Steptoe (the fictional London rag-and-bone man played by Wilfrid Brambell in the BBC comedy *Steptoe and Son*) did not. Perhaps the New Australians would become more British than the old ones, with the help of a BBC less insistent upon the virtues of received pronunciation and elite culture?

[8] Script for *Commonwealth Crisis: Britain and the Old Dominions*, BBC Television, transmission 4 September 1962, BBC Written Archives Centre (WAC), TV Registry talks scripts 1936–64. See also T32/506/1.

[9] Angus Maude, 'Through Australian Eyes', *Listener*, 28 September 1961.

[10] D. Russell, 'Report on BBC in Western Canada', 16 July 1963, WAC, E1/1738/3.

[11] Kenneth Adam, 'Television across the World—Broadcasting in Australia', *Listener*, 9 June 1966.

The nation-building function of public broadcasting seemed particularly valuable at this time of demographic, social, and cultural change, and offered a useful way of justifying the continued existence of embattled public authorities. However, this way of thinking could also tempt public broadcasters to acquiesce more readily in the demands of government. In Canada the CBC yielded to pressure from federal politicians and Francophone Liberals to combat separatism (which some felt was marked within the ranks of Radio Canada in Quebec), drawing the corporation deep into the controversies that followed the 'Quiet Revolution'.[12] The South African Broadcasting Corporation (SABC) was meanwhile increasingly subordinated to the requirements of the apartheid state. The Chairman of the SABC Board of Control from 1959, Dr P. J. Meyer (an admirer of fascism during the 1930s), also chaired the Afrikaner *Broederbond* secret society. Following Gideon Roos's retirement in 1961, Meyer abolished the position of director-general, increased his own executive authority, and placed fellow *Broederbond* members in key SABC positions. A 'Radio Bantu' network was established for African listeners, to support the state's policy of 'separate development'.[13] Both the BBC and the New Zealand Broadcasting Service (NZBS) declined invitations to attend the official opening of the SABC's segregated FM radio services in 1962.[14] In 1966 the SABC opened a new external service, Voice of South Africa, to put across to outsiders the state's case for apartheid.[15] For public broadcasters in Commonwealth countries, finding a way to deal with racial policies that were abhorred by most listeners, and by most broadcasting officers, caused significant soul-searching and some controversy.[16]

COMMONWEALTH BROADCASTING?

Like the broader relationships that constituted the British world, collaborative Commonwealth broadcasting structures did not suddenly collapse during the 1960s. Rather, they assumed a new guise: whether the end result would be total disintegration was not entirely clear. Britannic sentiments remained in only vestigial

[12] Marc Raboy, *Missed Opportunities: the Story of Canada's Broadcasting Policy* (Montreal and Kingston, 1990), 137–80.

[13] Graham Hayman and Ruth Tomaselli, 'Ideology and Technology in the Growth of South African Broadcasting', 58–63, in Ruth Tomaselli, Keyan Tomaselli, and Johan Muller (eds), *Broadcasting in South Africa* (Bellville, 1989). Patrick J. Furlong, *Between Crown and Swastika: The Impact of the Radical Right on the Afrikaner Nationalist Movement in the Fascist Era* (Hanover, New England and London, 1991), 94–6. Charles Hamm, '"The Constant Companion of Man": Separate Development, Radio Bantu and Music', *Popular Music*, 10/2 (May, 1991), 147–73.

[14] J. B. Clark to G. Stringer, 1 May 1962 and Stringer to Clark, 18 May 1962, WAC, E1/2259/1.

[15] Julian Hale, *Radio Power: Propaganda and International Broadcasting* (Philadelphia, 1975), 87.

[16] See for example N. Hutchison, 'Films Dealing with Apartheid', 4 March 1971, National Archives of Australia (NAA), Sydney, SP1299/2, TV28/5/20 part 2, 3161216. T. Richards, Halt All Racist Tours (HART) press release, [*c.* October 1971] and P. A. Fabian to Richards, 28 October 1971, both in Alexander Turnbull Library, Trevor Richards papers, MS-Group-0860, 99-278-02/09. Howard Smith, 'Apartheid, Sharpeville and "Impartiality": the Reporting of South Africa on BBC Television, 1948–1961', *Historical Journal of Film, Radio and Television* (*HJFRT*), 13/3 (1993), 251–98.

form, and seemed uncomfortably old fashioned. In 1966 Charles Curran, the BBC's director of external broadcasting and a former BBC Canadian representative, rather self-consciously declared himself 'a Commonwealth man, however blown-upon that phrase may seem to be today' (Curran later succeeded Greene as director-general).[17] However, practical connections tended to survive the erosion of old ideological certainties, and Commonwealth cooperation endured, recast as desirable on practical grounds, or as a means to encourage wider international understanding. Technological change meanwhile continued to make collaboration easier, further compensating for the dissolution of the sentimental connection. Trans-oceanic cables capable of carrying sound feeds became available during the early 1960s, most notably the Commonwealth Pacific cable (Compac) in 1963, reducing reliance on unpredictable and unsatisfactory short waves. Communications satellites were also put into orbit during the 1960s, allowing direct television links to be established among Commonwealth countries. From 1962 television signals could be relayed across the Atlantic by satellite, and in 1969 Australians were able to watch live coverage of the investiture of the Prince of Wales at Caernarfon Castle.[18] New technologies might yet breathe life into invented traditions.

Similarly, as international air travel became almost commonplace, staff mobility continued to strengthen connections among the Commonwealth's public broadcasting authorities. Trips by junior staff for training purposes continued, with London remaining the destination of choice for many. The BBC Transcription Service still provided a temporary home each year for an officer from an overseas broadcasting authority, to offer advice on local tastes. Staff exchanges among Commonwealth broadcasting authorities also continued: for example, seconded BBC officers played a major role in the creation of a new broadcast news service for New Zealand between 1964 and 1969.[19] Production crews travelled the world with increasing ease: in the first ten months of 1969, six British television producers spent a combined total of some fifteen weeks in Australia.[20]

Senior officers also made more frequent overseas trips. Short fact-finding missions to gather information about issues of special interest became reasonably common: in 1963, for example, Bill Armstrong visited Britain to examine BBC management structures in anticipation of reform at the CBC; and the following year the chairman and director-general of the newly established New Zealand Broadcasting Corporation (NZBC) visited to study coverage of the British general election.[21] Face-to-face contact made collaboration easier, and officers continued to adapt what they had seen overseas to local requirements. When for example the ABC's Director of Talks, Alan Carmichael, came back from an overseas tour at

[17] C. Curran to Stringer, 20 December 1966, WAC, E1/2259/2.

[18] L. Miall to C. Semmler, 1 July 1969, NAA, Sydney, SP1687/1, R6/8/13 part 1, 3162931.

[19] Simon J. Potter, 'The Colonization of the BBC: Diasporic Britons at BBC External Services, *c*. 1932–1956' in Marie Gillespie and Alban Webb (eds), *Diasporas and Diplomacy: Cosmopolitan Contact Zones at the BBC World Service* (forthcoming). WAC, E14/74/1.

[20] L. A. Woolard, 'D.P.A.'s Annual Report to the Board of Governors—Australia and New Zealand', 27 November 1969, WAC, E10/9.

[21] S. W. Smithers to D. Stephenson, 25 September 1963, WAC, E1/1739/4. 'NZBC Chairman on TV Policy', *New Zealand Listener*, 13 November 1964.

the end of 1962, he proposed a significant restructuring of production practices to give producers the overall control of programmes that their counterparts enjoyed in Britain and the US. Carmichael also recommended that the ABC launch Australian versions of three BBC television programmes: *Tonight* (current affairs), *Choice* (consumer affairs), and *What the Papers Say* (a talk on the contents of the week's newspapers). For radio, Carmichael suggested the introduction of a breakfast-time news magazine similar to the BBC's *Today* programme, and a media highlights session along the lines of the BBC's *Pick of the Week*. It took five years, however, before the ABC finally created its own versions of *Today* and *Tonight*.[22]

The growing ease of international travel also allowed the Commonwealth Broadcasting Conferences to be held more frequently, and in new places. Conference membership continued to be restricted to the national public broadcasting authorities of Commonwealth countries. Private stations were still excluded, as were regional broadcasting authorities established in countries such as Nigeria. The UK's Independent Television Authority (ITA) was similarly kept out, despite its protests, and it was not until 1968 that broadcasting authorities from non-self-governing territories were admitted as associate members.[23] Reflecting the composition of the senior ranks of the member organizations, the gatherings were still overwhelmingly male-dominated (delegates were seldom accompanied by their wives) and could become staggish. At the 1968 conference in New Zealand Clement Semmler discovered Greene drinking Black Velvets in a hot-pool at 2 a.m., along with several of his senior BBC colleagues and 'the same number of the most attractive young ladies from the hotel staff'. At this particular moment of Commonwealth unity, the surplus ABC delegate was decidedly unwelcome.[24]

However, one obvious sign of change was the substantial and growing number of African, West Indian, and Asian delegates attending the conferences, as decolonization accelerated and newly independent countries joined the Commonwealth. In the 1950s the only new public broadcasting authorities eligible to attend were Radio Pakistan and Radio Ceylon. During the 1960s the balance of membership shifted more rapidly, even if the expense involved meant that eligible authorities could not always send a delegate to each meeting. The Ghana Broadcasting System (in former Gold Coast) and Radio Malaya both sent delegates to the 1960 conference in India. At the 1963 conference, held in Canada, new members included the Sierra Leone Broadcasting Service and the broadcasting corporations of Nigeria, Tanganyika (later Tanzania), Cyprus, and Jamaica. Due to South Africa's departure from the Commonwealth, the SABC was excluded from this and future meetings. At the fifth conference, held in Nigeria in 1965, first-time attendees included the broadcasting corporations of Zambia and Malawi, Radio Uganda, the Voice of

[22] A. Carmichael, 'Report by D/Talks following Overseas Tour', 20 December 1962, NAA, Sydney, C1574 T2, box 2, WOB1, TV2/2/37, 'Alan Carmichael, 1962'. K. S. Inglis, *This is the ABC: The Australian Broadcasting Commission, 1932–1983* (Carlton, Vic., 1983), 302.

[23] R. Fraser to J. A. Ouimet, 18 April 1963, Library and Archives Canada (LAC), RG41, 372/19-36-5/8. Minutes, Commonwealth Broadcasting Conference Heads of Delegations Meeting, 5 March 1968, Commonwealth Broadcasting Association (CBA) archive, box 1, C/1/7.

[24] Clement Semmler, *Pictures on the Margins: Memoirs* (St Lucia, Queensland, 1991), 117–18.

Kenya, and the Malta Broadcasting Authority. Radio and Television Singapura attended for the first time in 1968, when the conference was held in New Zealand. Broadcasters from Swaziland and around the West Indies were able to join the 1970 session in Jamaica.[25]

The changing balance of membership significantly altered the nature of the conferences. Increasingly, an attempt was made to present the meetings as practical affairs, during which the older members of the conference would offer training, advice, and information to the newer recruits. As a consequence, the gatherings were treated less as an occasion for the BBC, CBC, ABC, and NZBC to discuss their own common concerns, and more as a chance for them to shape the activities of others. In planning the 1963 session, the CBC gave an early lead in this respect, proceeding 'on the basis that the older and more advanced organizations will be willing to forego some areas of discussion on highly sophisticated broadcast matters in favour of more time being devoted to the basic essentials'.[26] Later, it was agreed to set up a study group to survey the Commonwealth's broadcasting training resources and requirements, and produce a five-year overall training plan (although an attempt to establish a pilot training scheme in Ghana 'met with an astonishing lack of enthusiasm and support', and was delayed until 1970).[27] The impact of decolonization on the composition and nature of the Commonwealth Broadcasting Conference can be related to the broader transformation of the Commonwealth in these years, and to the changing role of organizations such as the United Nations Educational, Scientific and Cultural Organization (UNESCO) as newly independent countries gained a voice.[28]

Some of those attending the broadcasting conferences hoped that the Commonwealth connection might form a bulwark against growing American cultural influence in the Caribbean, Africa, and Asia. Richard Cawston's BBC documentary *Television in the World* was screened at the 1963 session, so that delegates from countries that did not yet have television could see what they would eventually face. Speaking at the conference, Cawston emphasized that the film was 'anti-American TV', rather than 'anti-American'. Delegates discussed not only the flood tide of US programming, but also their belief that many US programmes would be unacceptable to Africans and Asians 'because of the alien mores and culture'.[29] However, while the fear of American penetration of precarious post-colonial cultures may have been near-universal, it was not clear that Commonwealth countries shared enough in common to present a united alternative front, or that all of the delegates wished to harness the Commonwealth connection in such a way. Neither

[25] *Commonwealth Broadcasting Co-operation* (London, 1965 and 1967 editions). 'Commonwealth Broadcasting Conference—New Zealand 1968—Press Release', 6 March 1968, LAC, RG41, 374/19-40/4.

[26] W. T. Armstrong to F. W. Peers, 12 June 1962, LAC, RG41, 372/19-36-5/2.

[27] 'The Commonwealth Broadcasting Conference Secretariat—Handing-over Notes', n.d., CBA archive, box 1, CBA/1.

[28] W. David Mcintyre, 'Commonwealth Legacy', in Judith M. Brown and Wm. Roger Louis (eds), *Oxford History of the British Empire*, iv, *The Twentieth Century* (Oxford, 1999). Anthony Smith, *The Geopolitics of Information: How Western Culture Dominates the World* (New York, 1980), 61–4.

[29] Minutes, Commonwealth Broadcasting Conference, 4 June 1963, LAC, RG41, 371/19-36-3.

Britain nor even Canada was necessarily viewed as a benign potential ally: A. L. Hendriks, general manager of the Jamaica Broadcasting Corporation, stressed that he had no desire to see small countries become 'another little England or little America or little Canada'.[30]

During the 1960s public broadcasters increasingly moved away from attempts to organize ambitious multilateral Commonwealth programme exchanges in either radio or television, even in areas where cooperation was well established, such as schools or rural radio broadcasting. Administrative complications, funding shortages, and divergent programme requirements and 'standards' had all acted as major obstacles to such schemes in the past.[31] The increasing diversity of the Commonwealth Broadcasting Conference's membership, and the poverty of many of the new recruits, meant that collaboration was unlikely to become any easier. It was agreed instead that 'Program interchange should continue to grow by means of individual arrangements made between Commonwealth broadcasters.'[32] Experience had taught that bilateral projects were ambitious enough.

The perception that public-service values were not deeply rooted in newly independent Commonwealth countries also worked against more intimate broadcasting collaboration. This was particularly apparent in the field of news. Earlier conferences had provided for the exchange of reports from 'own correspondents'. With due acknowledgement and regard for editorial integrity, any news department could use the reports gathered in any Commonwealth country by any of its Commonwealth counterparts. By 1962, however, the BBC was worried that the 'comparatively young' African broadcasting authorities were using its reports too frequently, without proper editorial treatment, thus exposing the BBC to criticism.[33] More broadly, it feared that the political independence of many African and Asian public broadcasters was limited. Some African states were beginning to curtail press freedoms and treat broadcasting as an organ of government information and propaganda, and a serious disagreement with the Indian authorities over editorial freedom had already resulted in the abandonment of a major BBC television project. White settler leaders in Central Africa could be just as determined in this regard.[34] Although the BBC had faced political pressure in the past from the Canadian and Australian governments (exerted on correspondents on the ground, and through the high commissions in London), it was felt that the problem was most difficult in relation to Asian and

[30] 'Opening Addresses to the Commonwealth Broadcasting Conference', 27 May 1963, CBA archive, box 1, C/1/5.

[31] '1963 Commonwealth Broadcasting Conference—ABC Paper—Exchange of Radio Programmes', LAC, RG41, 370/19-36-1/1.

[32] 'Commonwealth Broadcasting Conference 1963—Summary of Conclusions', LAC, RG41, 370/19-36/2. A subsequent attempt by the ABC to arrange a multi-lateral rural broadcasting programme exchange proved a failure. See 'Commonwealth Broadcasting Conference—New Zealand 1968—P/I/2—ABC Paper—Rural Programme Exchange', RG41, 373/19-40/1.

[33] B. Moore to R. W. P. Cockburn, 14 June 1962, WAC, E1/1754/1.

[34] D. W. Willis to A. H. Wigan, 26 June 1962 and A. Milne, 'Commonwealth Censorship', [*c.* July 1962], WAC, E1/1754/1. On later clashes between the BBC and the Government of India, see Alasdair Pinkerton, 'A New Kind of Imperialism? The BBC, Cold War Broadcasting and the Contested Geopolitics of South Asia', *HJFRT*, 28/4 (October 2008), 537–55.

African countries. 'Canadians will only go so far to get their point across and can be resisted, while Ghana [for example] will go the whole hog.'[35] It could no longer be assumed that members of the conference were working under even remotely similar conditions.

However, criticism was not all one-sided. At the 1965 conference some African delegates asked for their countries' affairs to be reported with greater 'sympathy and understanding'. More tellingly, others insisted that reports by 'lightning-tour experts' from overseas could be highly misleading, but the suggestion that 'stories must be cross-checked with officials and other responsible people' was not well received by the older public broadcasting authorities.[36] Debate became particularly intense during the 1968 conference, when delegates were told by the director-general of the Nigerian Broadcasting Corporation, in the midst of the Nigerian civil war, that his corporation's 'responsibility to truth has been discharged by reporting only what we are told by the Federal Military Government'.[37] The BBC and the Nigerian Broadcasting Corporation subsequently fell out badly over the former's coverage of the war. Conference delegates were unable to agree on any formula for a new resolution on objectivity in news reporting or freedom of debate. As P. S. Raman, Singapore's director of broadcasting, noted, 'Broadcasting... is closely connected with the state of the country in which it operates, and liberal ideas cannot exist in abstraction.'[38] Echoing the timidity with which broadcasters had approached coverage of imperial affairs in the interwar years, at the 1970 conference it was recommended that programme exchanges should avoid controversial political topics.[39]

Nevertheless, positive steps were taken to ensure that the conferences survived as a meaningful arena for cooperation. At the 1960 meeting in India, Radio Ceylon suggested the establishment of a 'Commonwealth Broadcasting Relations Centre', to promote collaboration and 'to present the Commonwealth as a whole to non-Commonwealth nations'.[40] Although this suggestion was rejected by the conference delegates, the following year Donald Stephenson, the BBC's Head of Overseas and Foreign Relations, noted that the ABC had advanced a further argument for closer liaison. Private broadcasting interests in Britain, the US, and the Commonwealth were forming intimate connections with one another. Public broadcasters needed to draw together in the face of this transnational challenge. Stephenson

[35] Wigan to D. I. Edwards, 29 June 1962, WAC, E1/1754/1.

[36] Minutes, Commonwealth Broadcasting Conference Programme and Administrative Committee, 16 September 1965, CBA archive, box 1, C/1/6. For broader criticisms of western journalists reporting on Third World affairs see Smith, *Geopolitics of Information*, 93–110.

[37] 'Commonwealth Broadcasting Conference—New Zealand 1968—Inaugural Meeting', 20 February 1968, CBA archive, box 1, C/1/7.

[38] Minutes, Commonwealth Broadcasting Conference Programme and Administrative Committee, 23 February 1968, LAC, RG41, 374/19-40/4. On press freedom in the Commonwealth more generally see Derek Ingram, 'The Commonwealth and the Media', *Round Table*, 93/376 (September 2004), 561–9.

[39] 'Commonwealth Broadcasting Conference—Jamaica 1970—Report of the Sub-committee on Programme Interchange', LAC, RG41, 565/30/2.

[40] 'Commonwealth Broadcasting Conference 1960—Radio Ceylon Paper—A Commonwealth Broadcasting Relations Centre', LAC, RG41, 369/19-35/1.

suggested that a modest standing administrative body might stimulate cooperation, and ensure that more of the resolutions from the conferences were actually implemented. This proposal was presented to other members in time for the 1963 session, and it was duly agreed to establish a standing secretariat on an experimental basis. Perhaps inevitably (and with the support of the African and Asian delegates) London was chosen as the most convenient location. The BBC agreed to provide staff and accommodation, and Michael Stephens, formerly BBC Assistant Head of Overseas Talks and Features, was appointed as the first secretary.[41] It was agreed at the 1965 conference to establish the secretariat on a permanent basis, and to fund it through subscriptions from member organizations, levied on a sliding scale (this was in the same year that the Commonwealth established its own secretariat).[42] In 1968 Stephens was succeeded as secretary by John Akar, the former director of the Sierra Leone Broadcasting Service, who was in turn succeeded by Alva Clark from St Lucia, formerly a producer in the BBC's Caribbean Service.[43]

The new emphasis on advice and training at the Commonwealth Broadcasting Conferences reflected the continuing attempts of public broadcasters in the British world to exercise a leadership role in a wider Commonwealth context, and the ongoing 'development' thrust that accompanied the end of empire. Such work continued in the sphere of radio, and expanded to encompass television. During the 1960s television services were established in many African and Asian countries, though often as an emblem of modernity rather than a means to serve large numbers of viewers. Poverty meant that audiences expanded only slowly: even in India, only 248,300 television licences had been issued to viewers by 1970, and it was not until 1972 that a station opened outside Delhi.[44] The former imperial power meanwhile continued to play a development role, with the BBC providing continued assistance with training and technical matters. From 1962 the corporation also offered studios and other facilities to the Centre for Educational Television Overseas, a non-governmental organization that supported children's and adult educational television in developing countries.[45]

Eugene Hallman, CBC director of English radio networks, thought that the CBC should similarly support educational radio and television in new Commonwealth countries, and thus prevent broadcasting being 'devoted to the purely commercial purposes of private American and British capital or [becoming] an instrument of

[41] Stephenson, 'A Permanent Secretariat for the Commonwealth Broadcasting Conference', 22 August 1961, WAC, E1/1754/1. '1963 Commonwealth Broadcasting Conference—BBC Paper—Examination of the Case for a Permanent Secretariat for the Commonwealth Broadcasting Conference', LAC, RG41, 370/19-36-1/1. Minutes, Commonwealth Broadcasting Conference Program and Administrative Committee, 11 June 1963, RG41, 371/19-36-3. 'Board of Management Minute—The Commonwealth Broadcasting Conference', 24 June 1963, WAC, E1/1754/2.

[42] Minutes, Commonwealth Broadcasting Conference Meeting of Heads of Delegations, 27 and 28 September 1965, CBA archive, box 1, C/1/6.

[43] T. L. Laister to D. N. Brinson, 19 June 1970, United Kingdom National Archives, FCO26/539.

[44] P. C. Chatterji, *Broadcasting in India* (New Delhi, Newbury Park, and London, 1991, 2nd edn), 51–7. [Tangye Lean], 'Africa Listens In: the Impact of Broadcasting', *Round Table*, 55/219 (June 1965), 234–41, esp. 238.

[45] *Commonwealth Broadcasting Co-operation* (1970 edition).

Government propaganda'.[46] Playing such a role also enhanced the CBC's domestic prestige and international profile. CBC officers devoted particular attention to Ghana and Malaya/Malaysia, providing advice and training, and their overseas counterparts were brought to Canada for courses and visits to CBC facilities. Some funding for this was provided through the Colombo Plan, and the CBC also supported UNESCO courses for broadcasters in Africa.[47] Reflecting Canadian regional interests, the CBC loaned staff to the Jamaica Broadcasting Corporation to assist with the establishment of television, and became involved in plans to establish a 'Commonwealth Caribbean-Canada Broadcasting Centre' in the West Indies.[48]

The ABC and NZBC meanwhile continued to exert influence overseas, particularly in the Pacific and Asia, but also further afield. The ABC loaned officers to Malaya and Singapore, Talbot Duckmanton visited both Pakistan and Tanganyika in an advisory capacity, and in 1963 John Douglass was seconded to the United Nations Food and Agriculture Organization (FAO) for six months to direct communications workshops, focusing on the Middle East.[49] Broadcasters in Africa and Asia also began to offer one another assistance and, in an unprecedented collaborative effort, the Commonwealth Broadcasting Conference secretariat organized a combined advisory mission to Sierra Leone in 1969, involving the BBC, CBC, and the Ghana Broadcasting Corporation.[50] Rural broadcasting proved a particularly important area for cooperation between new and established Commonwealth broadcasters. In the early 1960s the CBC ran a successful drive, through its *Farm Forum* listening groups, to supply Indian villages with radio sets and thus disseminate knowledge of new agricultural techniques.[51] With the assistance of the FAO, the Colombo Plan, and the Special Commonwealth African Assistance Plan, the ABC meanwhile ran three-month rural broadcasting training courses for visitors from Africa, Asia, and Papua New Guinea.[52]

Through such initiatives, could Commonwealth broadcasting remain a meaningful concept? Or would transformation become disintegration, even total collapse? The signs were mixed. Added confusion was caused by the broader erosion of the position of public broadcasters during the 1960s, simultaneously in many different

[46] E. S. Hallman to C. Jennings, 24 August 1959, LAC, RG41, 369/19-35-1/1.

[47] Jennings to vice-presidents and regional directors, 16 December 1960; J. W. R. Graham to Jennings, 31 July 1961; and Peers to A. K. Morrow, 24 October 1962, all in LAC, RG41, 380/20-4-4. Armstrong to D. Ramli, 28 January 1964 and A. L. Hendriks to Armstrong, 8 August 1963, RG41, 373/19-36-5/11. CBC press release, 24 August 1962, WAC, E1/1739/3. *Commonwealth Broadcasting Co-operation* (1965 edition).

[48] Ouimet to CBC staff, 18 October 1966, WAC, E1/1739/5.

[49] *Twenty-ninth Annual Report of the Australian Broadcasting Commission Year Ended 30th June 1961* (Sydney, 1961), 4–5. 'Report on a UNESCO Mission to Tanganyika to Investigate the Establishment of Television in the Country', [*c.* November 1962], National Film and Sound Archive, Talbot Duckmanton papers, 0510311. '1963 Commonwealth Broadcasting Conference—CBC paper—Radios for India Project', LAC, RG41, 370/19-36-1/2.

[50] Minutes, Commonwealth Broadcasting Conference Programme and Administrative Committee, 11 June 1970, LAC, RG41, 565/30/2.

[51] R. G. Knowles to J. C. Mathur, 9 December 1960, LAC, RG41, 379/20-3-9-2. Knowles, 'Radios for India', 10 January 1962, LAC, RG41, 227/11-24-4-7/1.

[52] C. Moses to Clark, 9 November 1960, WAC, E2/684/3. R. G. Thompson to Knowles, 25 June 1962, LAC, RG41, 227/11-24-4-7/1.

parts of the Commonwealth. In Africa and Asia the assault by governments upon the political autonomy of public broadcasters was particularly obvious. In the British world the threat was more subtle, but serious nonetheless. It generally took the form of a continued drive towards increased commercialization.

PUBLIC BROADCASTING IN THE SIXTIES

During the 1950s and 1960s broadcasting in the British world became an ever more complex patchwork of private and public, commercial and non-commercial, operations. In the UK the BBC's domestic broadcasting monopoly was breached in 1954 with the establishment of the ITA. In 1962 the report of the Pilkington Committee offered supporters of public broadcasting some consolation: it condemned the ITA for failing sufficiently to regulate the programme policies of the private stations, and recommended that the BBC be granted permission to establish an additional national television network (BBC2 duly began broadcasting in 1964) and a new local radio service (eventually inaugurated in 1967). Nevertheless, the BBC faced continued assaults upon its domestic position. During the 1960s its non-commercial radio service lost many listeners to unlicensed offshore 'pirate' broadcasters, obliging the BBC to devote a whole network (Radio 1) to pop music as part of a general reclassification of radio networks by genre. The Reithian plan of educating individual listeners by exposing them to material that might alter their established preferences, partially discarded under William Haley, was now largely abandoned.[53]

In Canada, Australia, and New Zealand supporters of public broadcasting continued to look to the UK for inspiration. The Melbourne *Age*, for example, argued that Australia's broadcasting regulators could learn much from the Pilkington Report.[54] However, tighter regulation of private broadcasting in Australia was a remote prospect, and during the early 1960s the Menzies government licensed additional private television stations in the Australian state capitals.[55] At least the ABC continued to be seen as 'a social necessity, even if, in the cruder fibre of Australian political life, it was not something to be cherished or very much understood'. Politicians might not appreciate the highbrow programmes carried by the ABC, or believe its claims to political neutrality, but they still saw it as a useful counterbalance to the powerful media interests that controlled many private stations and newspapers.[56]

In Canada, by contrast, antagonism to the principle of public broadcasting was implacable in some quarters, and support for private enterprise entrenched.

[53] See Briggs, *The History of Broadcasting in the United Kingdom*, v, *Competition* (Oxford, 1995), 257–308 for Pilkington, and 502–15 and 571–81 for pirates.

[54] Press cutting, 'Warning to Commercial Television Here', *The Age* (Melbourne), 3 July 1962, WAC, E1/1650.

[55] Inglis, *This is the ABC*, 227.

[56] Report by J. Green, 21 January 1962, WAC, E1/1657. For a nice illustration, see John Gorton's views on American TV Westerns as quoted in Curran and Ward, *Unknown Nation*, 113.

Between 1957 and 1963, John Diefenbaker's Progressive Conservative government licensed new private television stations to compete directly with the CBC, and established a new regulatory body, the Board of Broadcast Governors (BBG), to which both the CBC and the private stations would be subordinate. The CBC's own governors were replaced by a board of nine government-appointed part-time directors, and the position of chairman was abolished. The general manager became the president, carrying an unwieldy burden of increased responsibilities, and lacking the protection against direct political pressures previously provided by the chairman. Funding was now allocated by parliament on an annual basis, with little guarantee of stability. The CBC lost much of its authority over networking, paving the way for the creation of the privately owned Canadian Television Network. Intensified competition, combined with a desperate shortage of funds, obliged the CBC to schedule more 'popular', and more American, programmes in order to attract viewers and commercial sponsors.[57]

Graham Spry, one of the key lobbyists for Canadian public broadcasting in the 1930s, and subsequently Saskatchewan's agent-general in London, sought to defend the CBC in the face of this onslaught. Spry also tried to strengthen the BBC's hand during the Pilkington Committee's investigations by publicizing the CBC's plight.[58] He visited Canada at the same time as did the committee, and submitted written evidence: an article he wrote on 'The Decline and Fall of Canadian Broadcasting' may also have carried some weight with Pilkington.[59] J. Alphonse Ouimet, the embattled CBC president, subsequently drew some solace from Pilkington's criticisms of commercial broadcasting in the UK, and argued that Canada should move towards the simpler division between public and private broadcasting that prevailed in Britain and Australia. If the CBC became fully public, shedding its commercial affiliates and dependence on advertising revenue, then the BBG would need only concern itself with the private sector, leaving the CBC to its own devices.[60]

In 1963 Diefenbaker's Conservatives were defeated by Lester Pearson's Liberals, who asked Robert Fowler to undertake another inquiry into Canadian radio and television (for Fowler's earlier report see Chapter Six above).[61] After meeting Fowler during his 1964 fact-finding mission to Britain, Greene warned the CBC that Ouimet's proposals were unlikely to find favour with the committee.[62] However, Greene did what he could to buttress the CBC's position, following the BBC tradition of giving Canadian public broadcasting behind-the-scenes support. Fowler had a number of lengthy meetings with Greene and other senior BBC

[57] Frank W. Peers, *The Public Eye: Television and the Politics of Canadian Broadcasting, 1952–1968* (Toronto, Buffalo, and London, 1979), 123–81, 232–47.

[58] David James Smith, 'Intellectual Activist: Graham Spry, a Biography' (D.Phil. thesis, York University, Ont., 2002), 415–45. Spry had established a Canadian Broadcasting League in 1958.

[59] G. Mosley to M. G. Farquharson, 2 May 1961 and Mosley to L. G. Thirkell, 19 April 1962, WAC, E1/1780. Graham Spry, 'The Decline and Fall of Canadian Broadcasting', *Queen's Quarterly*, 68/2 (Summer 1961), 213–25.

[60] Ouimet, 'Address to the Canadian Club of Ottawa', 5 December 1962, WAC, E1/1772.

[61] Peers, *Public Eye*, 263–4, 277–82.

[62] H. C. Greene to Ouimet, 1 September 1964, WAC, E1/1734/2.

officers, asking questions about a wide range of issues, including the nature of the BBC's executive and administrative structures, financing, programme policies (particularly the relationship between administrative and programme staff, and with the government, both sources of continued tension at the CBC), and relations with private broadcasters. Greene meanwhile kept Ouimet informed about Fowler's activities and changing views.[63] When at one point Fowler began to doubt the wisdom of using the CBC as a bulwark against Americanization (if this meant that programmes of lesser quality were broadcast on the grounds that they were Canadian), Greene and Curran were quick to remind him that Canadian public broadcasting had to play a nation-building role. Greene argued that a quota to restrict imports of the worst American products, and a strong CBC to act as a patron of Canadian culture, were both essential.[64] As in earlier decades, British influence on Canadian broadcasting debates tended, perhaps paradoxically, to support the Canadian nation-building project. This did not strike contemporaries as 'cultural imperialism'.

Fowler's final report roundly condemned the amount of US programming carried by Canadian private stations and by the CBC, but failed to recommend any reduction in the CBC's dependence upon advertising. Greene told Fowler that this was a mistake. If the CBC had to generate commercial revenue, then it had to schedule American programmes, which were the only means of attracting lucrative sponsorship.[65] Fowler's report also ignored Ouimet's proposals that the BBG's authority over the CBC be reduced. When a parliamentary committee subsequently considered how best to reform existing legislation, Ouimet repeated his proposals for the complete separation of public and private broadcasting, and both Greene and the head of the ITA travelled to Canada to explain to the committee how this worked in Britain. However, the new legislation (which was finally passed in 1968) brought little change. The BBG was abolished, but the Canadian Radio-Television and Telecommunications Commission established in its place had much the same regulatory authority over both the CBC and the private stations.[66]

In New Zealand and South Africa non-commercial and commercial radio services continued to be operated by public broadcasting authorities. South Africans lived without television until 1975, largely due to fears about the impact of the new medium on Afrikaner culture and the apartheid order.[67] In New Zealand, after a brief dalliance with the idea of introducing some form of competition, in 1960 the government awarded the NZBS a monopoly over

[63] 'A note of a meeting between Sir Hugh Greene and Mr R. M. Fowler', 1 September 1964 and 'A note on the Fowler Committee on Broadcasting in Canada' [*c.* November 1964], both in WAC, T8/111/1. Greene to Ouimet, 18 December 1964, E1/1734/2.

[64] R. M. Fowler to Greene, 28 December 1964; note by Curran, n.d.; Greene to Fowler, 11 January 1965, all in WAC, E1/1734/2.

[65] Greene to Fowler, 11 October 1965, WAC, E1/1734/2.

[66] Peers, *Public Eye*, 292–6, 317–20, 368–412.

[67] Experimental services operated in 1975, and a regular service was established at the beginning of 1976. Stanley Uys, 'TV Brings New Fears to S Africa', *The Guardian* (London), 5 January 1976.

both commercial and non-commercial television. It was hoped that unified control would ensure coverage across the country.[68] In 1962, the NZBS became the NZBC. Although modelled in part on the BBC and CBC, the autonomy of the new corporation remained limited, particularly in terms of finance and employment policies. Short-wave broadcasting remained under direct state control.[69] Meanwhile, pirate attacks on the NZBC's monopoly of domestic radio led to the creation of a new regulatory body: in 1968 the New Zealand Broadcasting Authority (NZBA) was established by act of parliament. Potentially, the NZBA combined many of the functions of the British ITA, the Australian Broadcasting Control Board, and the Canadian BBG. Crucially, in 1970 it issued the first new private radio broadcasting licences.[70] It also launched an ultimately inconclusive inquiry into the proposed inauguration of a second television network and the introduction of colour. Witnesses included Ouimet, now retired from the CBC, and Huw Wheldon, the BBC's director of television.[71] Wheldon, using many of the arguments that had been rehearsed in Britain before the Beveridge and Pilkington committees, made an impassioned statement in favour of NZBC control of the second network. He maintained that this was the only way to offer meaningful choice and to prevent private competition lowering standards. Wheldon concluded in strongly Reithian terms that, if meddlesome governments and greedy commercial interests could be resisted, public broadcasting could still provide

> standards of truth and accuracy; hours of delight, and pleasure, and contentment; and moments of insight and splendour which between them have added and not detracted from the quality of life and the processes of civilisation.[72]

Despite such stirring rhetoric, the commercialization of broadcasting continued across the British world. By the early 1960s private television companies were beginning to form transnational business structures that crossed Commonwealth borders, paralleling collaborative connections among public authorities. Links were established between private broadcasting companies in Australia and Britain at an early stage, and British television companies like Granada and Associated Television developed significant interests in Canada. UK companies including Associated Rediffusion and the Thomson Group meanwhile maintained a power-

[68] Patrick Day, *A History of Broadcasting in New Zealand*, ii, *Voice and Vision* (Auckland, 2000), 20–31. Experimental services had preceded the establishment of the full NZBS service, which was inaugurated on a non-commercial basis, but soon included advertising.

[69] Ibid., 32–40. The BBC was concerned that the British system had been misconstrued in the debates surrounding the creation of the new authority, and its Australian representative was briefed to clear up misunderstandings. Clark to D. Fleming, 27 February 1962, WAC, E1/2260/1.

[70] Day, *History of Broadcasting in New Zealand*, ii, 154–60.

[71] *Report of the New Zealand Broadcasting Corporation for the year ended 31 March 1971* (Wellington, 1971), 7.

[72] 'Evidence of Huw Wheldon', WAC, R78/2467/1. The second network eventually began operations in 1975, by which time both the NZBC and the NZBA had been abolished. They were replaced by four separate public authorities, one to manage radio, one for TV1, one for TV2, and a broadcasting council to oversee the entire system. This arrangement did not last long: in 1977 a unified Broadcasting Corporation of New Zealand was created.

ful presence in broadcasting in Britain's dependent colonies and in newly independent Commonwealth countries.[73] American companies also established a role: in Nigeria, television services were pioneered by a US National Broadcasting Company (NBC) subsidiary, NBC International.[74]

Public broadcasting authorities also themselves succumbed to increased commercialization. At the CBC, reliance on commercial sponsorship grew, and at the NZBC the principle of publicly controlled commercial broadcasting was extended from radio to television. In India the state radio authority also began to rely on advertising revenue.[75] The need to compete with private stations also exerted a subtle influence over the programmes produced by public broadcasters. At the 1960 Commonwealth Broadcasting Conference, delegates agreed that public broadcasters needed 'to make a special effort to provide distinctive and better quality light entertainment' if they were to win viewers back from private stations.[76]

At the BBC, Greene gave a new generation of producers free rein to innovate. One of those who provided additional encouragement under the new regime was Sydney Newman, who had left the CBC to work in UK commercial television, and had subsequently been appointed Head of BBC Television Drama.[77] Many of the groundbreaking BBC programmes produced in this new era were admired by broadcasters overseas, but proved difficult to import or emulate without similarly adventurous leadership. In 1963 CBC officers watched a special screening of the pioneering BBC satire show *That Was the Week that Was* (*TW3*) with 'hilarious incredulity'. Few thought the CBC had '[the money], the writers, the talent, [or] the nerve' to produce anything along the same lines.[78] Their counterparts at the ABC similarly believed that Australian audiences possessed insufficient 'broad-mindedness and tolerance' to stomach such a show, brilliant though it was.[79] The

[73] Cockburn to R. H. Waldman, 11 July 1960, WAC, E1/1782. *Commonwealth Broadcasting Co-operation* (1967 edition). Mark Hampton, 'Early Hong Kong Television, 1950s–1970s: Commercialisation, Public Service and Britishness', *Media History*, 17/3 (August 2011), 305–22.

[74] P. O. Elegalem, 'Economic Factors in the Development of Mass Communication in Nigeria' in Frank Okwu Ugboajah (ed.), *Mass Communication, Culture and Society in West Africa* (Munich, 1985), 70–1.

[75] H. R. Luthra, *Indian Broadcasting* (New Delhi, 1986), 369.

[76] 'Commonwealth Broadcasting Conference 1960—Summary of Conclusions', LAC, RG41, 369/19-35/3. As well as competition, there were at least token attempts to forge collaborative links among private and public broadcasters: the Intertel project involved public broadcasters in the US, Canada, and Australia cooperating with the UK commercial company Associated Rediffusion to produce documentaries. See Inglis, *This is the ABC*, 247.

[77] Simon J. Potter, 'Strengthening the Bonds of the Commonwealth: The Imperial Relations Trust and Australian, New Zealand, and Canadian broadcasting personnel in Britain, 1946–52', *Media History*, 11/3 (December 2005), 193–205. Michele Hilmes, 'The North Atlantic Triangle: Britain, the USA and Canada in 1950s Television', *Media History*, 16/1 (February 2010), 31–52 and *Network Nations: A Transnational History of British and American Broadcasting* (New York and London, 2012), 259–65.

[78] Russell to Waldman, 12 March 1963, WAC, E1/1739/3.

[79] Bert Read to Greene, 9 April 1963, WAC, E1/1629/2. See also Sandra Hall, *Supertoy: 20 Years of Australian Television* (Melbourne, 1976), 66, 138.

BBC itself doubted whether *TW3* could ever be cleared for export, and indeed ran into problems with the series at home.[80]

A version of *TW3* was eventually produced by NBC in the US and, despite initial reservations, the CBC commissioned a series of its own inspired by *TW3*, the purposefully controversial *This Hour Has Seven Days*, an uneasy mixture of public affairs and satire.[81] The BBC's Canadian representative thought *Seven Days* 'a very poor copy' of *TW3*, but even this pale imitation provoked massive criticism from politicians, and provided the occasion for a serious disagreement between CBC managers and producers. Attempts to kill the programme spawned a major news story, public demonstrations of support for the producers and presenters, a strike threat, and hearings before a parliamentary committee. The programme's producer told the committee that the CBC's timid approach to controversy compared poorly with Greene's bold regime at the BBC. The crisis was resolved with a spate of dismissals and resignations, and the effective cancellation of the programme. Ouimet left the CBC the following year.[82]

Could public broadcasters henceforth justify their funding by developing a distinctive role as providers of adventurous, creative programming that commercial stations could not risk producing? Or, by courting controversy and abandoning old ideas about cultural standards, was the BBC setting in train a process that would eventually destroy the corporation's own distinctive role as a provider of sweetness and light?[83] Gilbert Stringer, director-general of the NZBC and a self-confessed 'Edwardian', was firmly against Greene's policy.[84] ABC and CBC officers sat on the fence. They were particularly uncomfortable with some of the 'earthy' BBC television comedies of the 1960s, such as *Steptoe and Son* and *Till Death Us Do Part*. These programmes proved extremely popular with Australian audiences, but ABC commissioners judged them guilty of lamentable lapses of taste. Due to its jokes about homosexuality, *Up Pompeii!* was deemed particularly offensive to Australian sensibilities, although Hutchison managed to override objections and have the show broadcast.[85] In the early 1970s *Monty Python's Flying Circus* also caused some soul-searching. The ABC's Head of Audience Research R. F. Newell felt that the series might prove as innovative and popular as the *Goon Show* had two decades earlier. Yet he struggled when dealing, for example, with the 'Bruces' sketch, set at the University of Woolloomooloo: did it satirize 'the Australian atti-

[80] Russell to Hallman, 28 January 1963, LAC, RG41, 378/20-3-5/3. On *TW3* in a broader post-imperial context see Stuart Ward, '"No Nation Could be Broker": the Satire Boom and the Demise of Britain's World Role', in Ward (ed.), *British Culture and the End of Empire* (Manchester, 2001), 100–7.

[81] On NBC's *TW3* see Jeffrey S. Miller, *Something Completely Different: British Television and American Culture* (Minneapolis and London, 2000), 113–23.

[82] Smithers to K. Adam, 5 October 1964, WAC, T8/120. Smithers to Adam, 2 May 1966, E1/1783. Knowlton Nash, *The Microphone Wars: A History of Triumph and Betrayal at the CBC* (Toronto, 1994), 328–66.

[83] Tracey, *Decline and Fall*, 97.

[84] Day, *History of Broadcasting in New Zealand*, ii, 56.

[85] *Thirty-first Annual Report of the Australian Broadcasting Commission Year Ended 30 June 1963* (Sydney, 1963), 16. *Thirty-fifth Annual Report of the Australian Broadcasting Commission Year Ended 30th June 1967* (Sydney, 1967), 22. T. Duckmanton to Hutchison, 17 August 1970 and memo by Hutchison, 14 September 1970, both in NAA, Sydney, SP1299/2, TV28/5/9 part 8, 3160800.

tude towards free thought and homo-sexuality', or British stereotypes about intolerant Australians? Newell worried that if the ABC declined to show the programme, it would be suspected of 'conforming to the stereotype that is being lampooned'.[86] In Canada the first series of *Monty Python* was dropped by the CBC on grounds of poor taste, but then reinstated following organized protests from viewers. One fan thought the programme appealed to 'the basic anti-English sentiment inborn in each man', and that its cancellation reflected the conservatism of an 'over-civilized, stuffed-shirt, Masonic—English no doubt—CBC bureaucracy'.[87] Ideas about Britishness (and Englishness) were clearly changing. Old imperial values and badges of loyalty were being challenged as part of the broader revolt against what, in Britain, was now called 'the Establishment'.

SELLING TELEVISION

During the later 1950s it had become clear that transnational flows of television programmes would be organized on a commercial basis, even among Commonwealth countries, despite the precedent set in radio for exchange on a free or costs-only basis. Greene had been one of the first to embrace the new spirit of commercialism and in 1961, under his leadership, the BBC established Television Enterprises, headed by Ronald Waldman, to sell British programmes overseas (in 1968 it was renamed BBC Enterprises and given responsibility for non-Transcription Service radio exports as well). Television Enterprises became one of the world's largest programme exporters. The effects of the BBC's accelerated drive towards commercialization of programme exports on its relationship with other Commonwealth public broadcasting authorities were complex. On one hand, Commonwealth countries proved an important and growing market. By the mid-1960s the BBC was selling some five thousand television programmes to former colonies each year, and many also to Canada, Australia, and New Zealand, generating nearly £3 million worth of revenues annually by the end of the decade. UK private television companies were also successfully exporting their wares to the Commonwealth, allowing Britain to become a significant on-screen presence.[88] This was particularly striking given the failures of the 1950s. On the other hand, although the value of sales to the Commonwealth grew, they were outstripped by revenues generated in what, by the end of the 1960s, was the BBC's single most important overseas market: the US.[89] In the commercialized mind of Television Enterprises, it was thus increasingly likely that the requirements of American viewers would loom larger than those of Commonwealth audiences.

The BBC also sold its programmes to private stations and networks in Commonwealth countries, further complicating its relationship with Commonwealth public broadcasters. It did continue to offer public broadcasting authorities the right of

[86] R. F. Newell to Assistant Federal Director of Television, 3 October 1972, NAA, Sydney, SP1299/2, TV28/5/9 part 9, 3161211.

[87] Press cutting, *McGill Daily*, 27 January 1971, LAC, RG41, 830/178.

[88] Kenneth Adam, 'Television across the World III—Commonwealth Broadcasting from Singapore to Antigua', *Listener*, 23 June 1966. Briggs, *History of Broadcasting in the UK*, v, 712.

[89] Hilmes, *Network Nations*, 244–7.

first refusal of BBC programmes; this was a significant concession in an increasingly commercialized business.[90] 'Cousinly' television exports on a 'definable extra costs' basis also continued to be organized for coverage of key British public ceremonies, and initially also for relatively inexpensive productions such as interviews and panel discussions.[91] However, by 1963, the BBC was charging near-market rates for the latter type of programme as well, claiming that it could not otherwise cover the costs involved in fulfilling the increasing number of requests it received. There was insufficient demand at the BBC for similar material from Commonwealth countries to make any reciprocal, costs-only exchange feasible.[92] While it could still use BBC radio facilities freely, the CBC's London office now had to pay on 'a hard business basis' for access to similar television services, and so generally used cheaper freelance cameramen. In 1966 it moved off BBC premises to offices of its own.[93] Newsfilm seemed to be one of the few areas of television where a joint venture was possible, with the British Commonwealth International Newsfilm Agency continuing to operate on a cooperative Commonwealth basis.[94]

Between 1959 and 1961 the CBC spent seven times more on BBC programmes than it generated from selling its own material to the BBC. The BBC's Canadian representative noted that it was thus unrealistic to talk in terms of an 'exchange of programmes' in television: the flow was overwhelmingly one way. There nevertheless remained limits on what the CBC would buy from the BBC. Public affairs, sports, and children's programmes were deemed most acceptable, but in the same period only one BBC entertainment series (*Hancock's Half-Hour*) and a few dramas, were purchased. Many of the Canadian criticisms that had been levelled against BBC radio transcriptions in the 1930s were now applied to BBC television exports. Much of the BBC's output was deemed too local in its points of reference to appeal to Canadian audiences, and the irregular running times of BBC dramas were not compatible with the CBC's commercial schedules. UK private television companies seemed to produce material more in line with the requirements of Canadian commercial sponsors, but US exports were even more appealing in terms of content and price. Ouimet estimated that US television programmes were sold in Canada for between 5 and 8 per cent of their original production cost. When Canada followed the US, and introduced colour television before the UK did, the fact that British colour programmes were simply not available meant that BBC sales dropped again.[95]

[90] Armstrong to vice-presidents and general managers, 22 November 1963, LAC, RG41, 378/20-3-5/3.

[91] D. Bennett to W. Weston, 11 June 1960, LAC, RG41, 378/20-3-5/2.

[92] Waldman to Adam, 19 March 1963, WAC, E1/1740/1.

[93] Spencer Moore, 'CBC Overseas', *CBC Inside*, 20 January 1961. Press cutting, 'CBC Expands London Operation', *The Star* (Montreal), 19 November 1966, LAC, RG41, 391/20-9/10.

[94] The NZBC paid a subscription upon the inauguration of its own television service, and in 1965 took up a substantial shareholding in the company. By this time, Reuters had also become a shareholder. See *Report of the New Zealand Broadcasting Corporation for the Year ended 31 March 1966* (Wellington, 1966), 8.

[95] M. Sadleir to H. G. Walker, 31 October 1961 and memo by Walker, 14 May 1962, both in LAC, RG41, 725/10. [Russell], 'Notes on Some Functions of BBC Representative in Canada' [*c.* September 1963], WAC, E1/1738/3. Smithers to Adam, 1 April 1966, E1/1739/5. Smithers to Stephenson, 28 November 1967, E1/1739/6.

Australia meanwhile remained one of the best customers for both BBC and UK private television exports. By 1968 it was the BBC's most important Commonwealth market.[96] Yet the very popularity of BBC programmes with Australian viewers, and the massive scale on which they were imported, caused problems for the relationship between the BBC and the ABC. To Charles Moses' consternation, the ABC took enormous amounts of material from the BBC, paying market prices, but the BBC bought almost nothing in return. A visit by Moses to London only made matters worse: he reportedly returned to Australia harbouring a 'pathological dislike' of Waldman, and a belief that the BBC was staffed by 'rapacious incompetents'.[97] J. B. Clark, the BBC's director of external broadcasting (whose own relationship with Moses and the ABC dated back to the 1930s) thought the BBC should still be 'playing ball to the fullest possible extent with the ABC both within a Commonwealth context and also as a blood-brother in relation to our respective national purposes'. However, sentiment only went so far: Clark noted that 'this relationship cannot be exclusive of everyone else, nor can it bend over too far for reasons of philanthropy rather than those of real mutual interest'.[98]

The ABC had from the start underestimated how much imported programmes would cost, and Moses' resentment of BBC pricing policies derived in part from a general ABC over-spend on overseas material, which obliged the embarrassed commission to seek additional funding from the Australian Treasury.[99] However, in terms reminiscent of earlier suspicions about sales of BBC radio transcriptions in the 1930s, Moses also believed that BBC prices in Australia were inflated, and that the ABC was being milked to subsidize BBC television exports to other countries. Clark thought the ABC wanted to have its cake and eat it, to 'stand on their dignity (as a vigorous and creative young nation)' when it suited them, but still to ask the BBC 'to make concessions as to a junior or poor relation'.[100] Doug Fleming, the BBC's Australian representative, sought to convince J. R. Darling, the ABC's new chairman, that the BBC was not seeking to make heavy profits from the ABC, but merely wanted to sell at market rates. Darling was sympathetic, but emphasized that 'both sides should do more to preserve the dwindling Commonwealth link'. However, playing the Commonwealth card now had little effect on BBC officers: Fleming maintained that 'the ABC's attitude... must be governed by its wish to show this material, combined with its ability to pay the price'.[101] Greene sent Kenneth Adam, the BBC's director of television, to Australia to try to improve relations between the two organizations, and a few months later Darling met with

[96] *Combroad*, 4 (January 1968).

[97] Moses to I. Jacob, 11 November 1959 and Jacob to Moses, 15 December 1959, both in NAA, Sydney, ST1790/1, box 51, 10/6/1 part 1, 'Film Sources, BBC—June 1955 to January 1961'. J. F. Mudie to Clark, 5 and 20 October 1960, WAC, E1/1629/1.

[98] Clark to Mudie, 26 October 1960, WAC, E1/1629/1.

[99] Moses to Semmler, 11 and 24 January 1962, NAA, Sydney, ST3051/2, TV28/1/6 part 1, 1886243.

[100] [Moses] to J. Darling, 23 July 1962, NAA, Sydney, SP1489/1, box 8, 'BBC'. Clark to Fleming, 19 January 1962, WAC, E1/1650.

[101] Fleming to Clark, 21 December 1961, WAC, E1/1650.

senior BBC officers in London.[102] However, these diplomatic missions did little to improve Moses's mood. In protest at BBC television export prices, Moses arbitrarily cut the amount that the ABC paid for BBC radio transcriptions by more than a quarter. The BBC retaliated by revoking the ABC's exclusive rights to transcriptions, and selling some to Australian private radio stations.[103]

Meanwhile, progress with television co-productions, which offered a potentially more equitable way to pool available programme resources, proved painfully slow. The BBC and CBC had some early success when they cooperated with an independent film company in the production of a series about the Royal Canadian Mounted Police, which was also offered for sale to the ABC.[104] However, subsequent joint ventures proved problematic. Towards the end of 1962, Clement Semmler (ABC assistant general manager for programmes), Stuart Hood (BBC controller of television programmes), and Doug Nixon (CBC director of programming, English networks) agreed that their organizations would together produce a television series called *Commonwealth Jazz Club*. Each would record and contribute four half-hour programmes to the series, and each would then screen all twelve programmes.[105] While both the ABC and CBC were eager to move ahead with this relatively simple project, nothing happened for two years. After a 'running battle' with Hood's successor, the most Semmler could get from the BBC was an agreement to go ahead with a truncated, six-episode series. Semmler commented: 'If this is all Commonwealth co-operation can achieve, God help the Commonwealth!' The series was not ready for screening until mid-1965. As Nixon remarked: 'It is truly astounding how time-consuming and how complex even the most simple project can become when two or three organizations in different parts of the world get involved.'[106]

A 1966 ABC–BBC drama co-production called *Kain* proved a similarly sobering experience. The BBC flew a three-man production crew out to Australia, along with three lead actors, two of whom were Australians resident in London. The ABC provided all other facilities, including those required for filming sequences in the outback, and covered around 40 per cent of the total costs. The venture was rushed through by its champions, without the full agreement of senior officers at either the ABC or BBC. The script, written by another London-based Australian, was deemed by Hutchison to be passable at best. Production proved costly, and both sides concluded that more time needed to be spent in planning and discussing any future co-productions.[107] Two years later the ABC rejected a BBC proposal

[102] Greene to Moses, 27 February 1962, NAA, Sydney, ST3051/2, TV25/4/2 part 2, 1886236.

[103] D. M. Hodson to Clark, 27 November 1964, WAC, E1/1647.

[104] Mudie to C. Conner, 18 March 1958, WAC, E1/1636/2. Paul Rutherford, *When Television was Young: Primetime Canada, 1952–1967* (Toronto, 1990), 377–9.

[105] Semmler to D. Miley, 7 December 1962, NAA, Sydney, C2218/6, 18/2/12, 12203775.

[106] Semmler to J. D. Nixon, 3 December 1964 and Nixon to Semmler, 9 December 1964, NAA, Sydney, C2218/6, 18/2/12, 12203775. P. A. Boswell to Armstrong, 18 August 1965, LAC, RG41,378/20-3-7/3.

[107] Memo by D. Goddard, 21 February 1966; D. Stone to Semmler, 11 March 1966; memo by Hutchison, 28 March 1966, all in NAA, Sydney, SP1299/2, TV11/2/3 part I, '"Kain"—ABC–BBC Co-production February 1966–July 1966'. D. G. Scuse to Duckmanton, 29 March 1966 and S. Newman to Semmler, 24 April 1967, NAA, Sydney, C1574 T1, box 34, TV10/10/15, 'ABC–BBC Production "Kain"'.

to co-produce a drama series based on the experiences of British migrants working on the Australian Snowy Mountains hydro-electric project. Hutchison feared that the commission would end up subsidizing a series made primarily for British viewers, which it could buy much more cheaply at market rates if the BBC simply made it on its own.[108]

Relations between the BBC and ABC improved following Moses' retirement in 1964. His successor as general manager, Talbot Duckmanton, was worried by the ABC's loss of control over some of the BBC's more popular programmes, particularly given the licensing of additional private television stations in the major Australian cities. Duckmanton was determined to buy all repeat rights to key BBC television programmes, and to regain exclusive Australian rights to BBC sound transcriptions. The BBC was meanwhile in talks with Australian private radio stations, to provide them with a specially produced radio news programme. Humphrey Fisher (the BBC's Australian representative) and Malcolm Frost (still Head of the Transcription Service) both recommended that the BBC should maintain its historic partnership with the ABC. Fisher based his case not on the importance of the Commonwealth connection, but rather on the need to support the principles of public broadcasting overseas.[109] A compromise was eventually reached: Duckmanton agreed to pay a higher price for BBC radio transcriptions and television exports, in return for exclusive rights to the former, and stricter limits on sales of the latter to private stations.[110] The BBC's trade with New Zealand meanwhile followed a somewhat different course: an ever-increasing amount of BBC material was sold, but at lower prices, given the absence of competition. By 1970 around 40 per cent of NZBC television programming came from the BBC.[111]

Yet the BBC did not have Australian and New Zealand television screens to itself. In the case of the NZBC, the total amount of material purchased from other overseas producers more or less matched that supplied by the BBC. This included material produced by British private companies, but also American imports. In Australia by January 1965 ABC television filled almost 40 per cent of broadcast hours with its own programmes (the NZBC managed only 23 per cent in the same year), more than 40 per cent with British material (from the BBC and private companies), and only 20 per cent with American programmes. However, ABC television won a mere 15 per cent of the total audience. Private stations attracted the rest, and they scheduled American programmes in vast quantities.[112] Trade in

[108] H. Wheldon to Hutchison, 26 January 1968 and Duckmanton to Wheldon, 9 February 1968, both in NAA, Sydney, SP1299/2, TV30/2/111 part 1, 3162358. Hutchison to Duckmanton, 5 February 1968 and memo by B. D. Sands, 1 May 1970, both in NAA, Sydney, C1574 T1, box 34, TV10/10/10, 'Coproduction ABC/BBC'.

[109] H. Fisher to G. Steedman, 6 November 1964, WAC, E1/1647.

[110] Greene to H. J. G. Grisewood and O. J. Whitley, 3 December 1964, WAC, E1/1647.

[111] Fisher to Stephenson, 13 October 1967 and Woolard to Stephenson, 9 April 1968, WAC, E10/9. Minutes, 'Television Weekly Programme Review', 2 December 1970, R78/2467/1.

[112] P. J. F. Lord, 'Notes on the Australian Broadcasting Commission', January 1965, WAC, E1/1678. *Report of the New Zealand Broadcasting Corporation for the Year ended 31 March 1966* (Wellington, 1966), 9.

television programmes among dominion broadcasting authorities hardly acted as a counterweight. While the CBC had modest success in exporting its programmes to other Commonwealth countries, the quantities involved remained insignificant compared to British and especially American overseas sales.[113]

By the early 1960s the ABC was being criticized for failing to prevent 'American cultural colonialism', and for allowing the spread of American popular culture and social mores in Australia.[114] Even if the ABC minimized its own use of American imports, it tempted few viewers away from the US programmes shown by private stations. BBC visitors to Australia were aware of the ABC's dilemma but, without some form of state subsidy for British television exports, no obvious solution presented itself. John Green lamented that while '[n]o section of the British people is nearer in sympathy to Lancashire variety or more naturally estranged to improvised [American] spectacle than Australians and New Zealanders,' the affinities between UK and Australian popular culture were being eroded by the flood tide of US programming.[115] Kenneth Adam thought that 'Australia was being Americanised at an alarming rate and that the main instrument was commercial television.' By 1966 the threat seemed even more marked in other parts of the Commonwealth, such as the West Indies, where the US presence amounted to 'neo-colonialism'. Adam argued that the failure of the British government to keep colonies and former colonies 'out of the hands of the commercial broadcasters', or to subsidize BBC television exports, had made this possible.[116]

A RADIO REVIVAL?

Neither commercialization nor Americanization had the impact on radio in the British world that they had on television. This was largely due to the dwindling resources allocated to sound broadcasting and the erosion of radio audiences, as more and more listeners became viewers. The BBC was able to retain domestic and overseas supremacy in what was essentially a declining medium.

In Canada the changing status of radio was particularly apparent. As advertisers rapidly redeployed their resources into the new medium, CBC English-language radio was badly affected, and in 1962 the Dominion Network, established at the end of the war to maximize commercial revenues, was closed down. Private stations meanwhile began to follow the path pioneered by US radio, and focus on news, recorded music, and local community service. All this reduced demand for BBC transcriptions, and some heavy promotional work by the BBC's Canadian

[113] See for example H. Salmon to B. Zimmerman, 13 February 1968, LAC, RG41, 378/20-3-5/3.

[114] Press cutting, Max Harris, 'Is Americanism a Threat?', Sydney *Bulletin*, 1 June 1961 and press cutting, A. G. Lowndes, 'Is Americanism a Threat?', *Bulletin*, 8 July 1961, WAC, E1/1636/2.

[115] Report by J. Green, 21 January 1962, WAC, E1/1657.

[116] Press cutting, 'Americanised Australia—TV to blame', [*c.*1962], LAC, Graham Spry papers, 120/5. Kenneth Adam, 'Television across the World III—Commonwealth Broadcasting from Singapore to Antigua', *Listener*, 23 June 1966. See also Eric Barnouw, *A History of Broadcasting in the United States*, iii, *The Image Empire: from 1953* (New York, 1970), 112–15.

office was necessary to get programmes heard.[117] However, the BBC also faced much less competition for the Canadian radio audience: there were now few major sponsored programmes coming from the American networks, and Canadian producers were themselves increasingly starved of resources. The BBC also benefited from the fact that Commonwealth material counted toward the 'Canadian content' quotas required of broadcasters by the BBG.[118]

There was scope for the BBC in Australian radio, too. As in Canada, private stations were beginning to adopt the American format of 'relaxed and non-continuous listening'. Struggling to survive, the old commercial networks attempted to maintain a role for themselves by offering affiliates new types of programmes: the Macquarie Network, for example, turned to the BBC for help with a new current affairs session, *Monitor*.[119] Meanwhile, guided by Semmler, a committed Reithian, the ABC sought from 1963 to refocus its radio offerings for urban listeners. On the assumption that those who wanted lighter material had largely turned to television, radio could now be used 'to meet the needs of those—and there are many—who continue to rely largely on radio for serious music and discussions and talks and plays'. One ABC radio network would henceforth carry light material, making way for broadcasts from Canberra when the federal parliament was in session. The second network would focus year-round on serious material, providing 'something midway between the B.B.C. Home and Third'. Country dwellers would have their own, mixed network.[120] In New Zealand the NZBC followed a similar path, networking its non-commercial stations into a single National Programme, carrying a range of different genres of programming, and repositioning its commercial stations as a local, 'community' service.[121]

There thus remained a significant role in Canada, Australia, and New Zealand for BBC radio programmes. Indeed, some argued that CBC and ABC radio were becoming too British in tone. One Canadian critic claimed that

> In an effort to escape the corrosive elements of American culture, Canadian [radio] programming has submerged its once-distinctive qualities under a flood of British programs, British topics, and British accents . . . network programs appeal consistently to a mere handful of listeners who—either because they lack a television set—or because they are imbued with an overwhelming desire to be bombarded with knowledge—enjoy the very British ultra-conservative, ultra-Victorian aspects of CBC radio.[122]

In Australia (referring both to recruitment of former BBC staff and use of BBC transcriptions) another critic complained that '[t]here'll Always be an England, while there's an ABC'.[123] Yet whenever they took BBC programmes off air, public

[117] Curran to Stephenson, 28 March 1958, WAC, E1/1775. R. St Clair, 'Duty Visit to CBC Stations Western Canada', 6 October 1964, E32/11. Curran to Cockburn, 22 July 1958, E1/1738/2.

[118] L. M. Stapley to Cockburn, 18 November 1959, WAC, E1/1738/2.

[119] Mudie, 'The Australian Scene', 28 August 1958, WAC, E1/1636/2.

[120] Semmler to Fleming, 6 May 1963, WAC, E1/1629/2.

[121] Day, *History of Broadcasting in New Zealand*, ii, 3.

[122] Extract from television script, Joseph M. Mauro, 'Critically Speaking—Radio Review', 19 May 1963, WAC, E1/1739/3.

[123] Press cutting, Alexander Macdonald, 'Around the Dial', *The Mirror* (Sydney), 19 July 1963, WAC, E1/1629/2.

broadcasters in Canada, Australia, and New Zealand could expect significant criticism. When the CBC dropped the BBC radio soap opera *The Archers* in 1968, listeners' letters accused it of hostility to 'anything English' and 'things British', and criticized its provision of 'hours of U.S. football, U.S. baseball, U.S. politics, and other idiotic programmes'.[124]

At the BBC, during the 1960s at least, Radio Enterprises failed to make much of an inroad into the fiefdom of External Services. Here, the Treasury grant-in-aid provided a space in which commercialization could be avoided, most obvious in the continued success of the heavily subsidized BBC Transcription Service. By the mid-1960s more than a thousand separate programmes were being provided on disc each year.[125] For the CBC, BBC transcriptions were a source of music in the daytime, and 'serious' programmes at night. For the ABC and NZBC, supplies of BBC variety and comedy programmes remained the top priority.[126]

However, in the field of short-wave broadcasting, the future of BBC External Services seemed less certain. During the 1960s rebroadcasting of short-wave transmissions in countries with their own well-developed medium-wave services continued gradually to decline, partly due to the enduring and inescapable problem of poor sound quality. The amount of BBC short-wave material aimed at Canada, Australia, and New Zealand also dwindled, in response to the decline in rebroadcasting, but also due to continued cuts to the Treasury grant-in-aid. Remaining resources were focused on direct listeners in Africa and Asia, where the 'Transistor Revolution' (which made available cheap receivers that did not require a mains power supply) was delivering an expanding audience, and where British policymakers wished to perpetuate some influence over areas that had once been part of Britain's formal and informal empire. The specialized African Service thus endured, and a new Arabic Service was created. The BBC's General Overseas Service was reoriented explicitly to serve 'the listener who understands English but is not of British descent', and renamed the World Service in 1965. Such listeners were clearly also prioritized when money was invested in overseas transmitter infrastructure projects, as in the case of the transmitter built on Ascension Island in 1967 to improve services to West Africa. By 1972 the Head of the BBC's African Service estimated that there were 15 million radio receivers in sub-Saharan Africa, and that between a third and a half of Africans listened to radio daily.[127] With its new focus on African and Asian audiences, the BBC World Service became a conspicuous success. Even at the end of the century it still enjoyed perhaps the largest aggregate short-wave audience of any broadcaster.[128]

[124] 'Comments from Listeners' Letters Concerning Cancellation of "The Archers"', n.d., LAC, RG41, 378/20-3-5/3.

[125] Briggs, *History of Broadcasting in the UK*, v, 693.

[126] 'Minutes of Transcription Subscribers Conference held in Kensington House, London', 25 September 1968, NAA, Sydney, SP1423/1, R26/1/12 part 1, 3162540. Miley to Duckmanton, 15 July 1969 and 'Minutes of Second Transcription Liaison Committee, held in Toronto', 4 September 1969, NAA, Sydney, C2327 T1, box 70, 25/1/48, 'BBC Transcription Conference, Toronto 1969'.

[127] J. F. Wilkinson, 'The BBC and Africa', *African Affairs*, 71/283 (April 1972), 176–85.

[128] Briggs, *History of Broadcasting in the UK*, v, 679–717. Gerard Mansell, *Let Truth Be Told: 50 Years of BBC External Broadcasting* (London, 1982), 244–9. James Wood, *History of International Broadcasting*, ii (London, 2000).

Meanwhile, 'British whites', who had been the sole target of the Empire Service in the 1930s, were no longer to be served primarily by short wave. The Pacific Service was pruned back, and in 1962 the North American Service (NAS) was closed down altogether. To senior BBC officers this was particularly galling given the lavish external broadcasting efforts of Britain's Cold War enemies.[129] Yet we should be wary of attributing cuts in short-wave funding entirely to the inevitability of imperial decline, or to the reduced importance of the British world to the UK government. For the scaling down of short-wave services came at a time when new technologies became available, with the potential to improve the international reach of radio. Indeed, with hindsight, it could even plausibly be argued that the BBC had taken a wrong turn in the 1930s, devoting significant resources to short wave, only to discover that the sound quality of rebroadcasts was cripplingly poor in comparison with what was produced locally. In countries with good medium-wave services of their own, direct listening to short wave was the pastime of a tiny minority. Except during the Second World War, when topicality overrode almost all other concerns, short wave had never provided a particularly satisfactory link with the British world. During the 1960s alternative possibilities were explored.

News bulletins and sports coverage (two of the long-term staples of external broadcasting) now started to be transmitted by cable rather than short wave, bringing considerable improvements in sound quality. Most CBC stations continued to carry a daily BBC news bulletin at noon, despite doubts about whether this was appropriate given the 'apparently prevailing temper of [the] country' (a reference to contemporary debate over a new national flag for Canada) and the sense that the bulletins still 'smacked rather of the Empire Service'.[130] BBC officers themselves admitted that the World Service had failed to keep up with the changing format, focus, and presentation style of North American news bulletins, and sounded 'more like a daily religious service, with the Collects intoned by the reverend newsreader, followed by a sermon by a lay-preacher from the London School of Economics'. An attempt was made to offer the news in a more attractive package, and to render it interesting by presenting it explicitly as a British perspective on world affairs.[131] Nevertheless, in 1971 CBC National Radio Program Director Peter Meggs argued for the bulletin to be dropped. He claimed this decision was not 'anti-British', but rather a response to Canada's demographic shift 'away from a predominantly Anglo-Saxon population base'. More prosaically, Meggs worried that if the CBC did not drop the bulletin soon, then 'the BBC will be with us forever no matter how bad it gets'.[132] However, when the bulletin was duly cancelled,

[129] Tangye Lean, 'BBC Lunchtime Lecture—The Revolution Overseas', 12 March 1963, LAC, RG41, 389/20-9/3.

[130] L. Wilson to W. Y. Martin, 15 July 1964, LAC, RG41, 906/PG10-9/3. Steedman to Moore, 13 August 1964, WAC, E1/1775. Smithers to Walker, 9 May 1966, LAC, RG41, 378/20-3-5/3. Smithers to Stephenson, 18 August 1967, WAC, E1/1739/6. On the flag debate see José E. Igartua, *The Other Quiet Revolution: National Identities in English Canada, 1945–71* (Vancouver, 2006), 171–92.

[131] Steedman to K. Fairfax, 24 October 1967 and Smithers to Steedman, 18 December 1967, WAC, E1/1739/6.

[132] P. Meggs to J. Craine, 19 October 1971, LAC, RG41, 771/GM7-1-6/1.

the scale of the hostile reaction from listeners prompted the CBC president to reverse the decision.[133]

Director of the CBC's English Radio Network Jack Craine meanwhile used cable sound feeds to forge a direct link between CBC and BBC domestic services, thereby circumventing a World Service that seemed increasingly oriented to Asian and African listeners, and out of touch with Canadian needs.[134] From 1966 Craine experimented with carrying BBC Third Programme news bulletins on the CBC's FM network.[135] In October 1967 the CBC FM network carried twelve continuous hours of BBC programming in a single day, to mark 'British Week' (an export promotion event in Toronto).[136] The CBC also mined the BBC's topical airmail tapes service (established in 1962) for items for its own magazine programmes.[137]

In Australia Wally Hamilton, the ABC's Controller of News, had long wished to take full responsibility for all news broadcasts carried by the commission's stations, and tried to eliminate straight BBC rebroadcasts. 'I do not think we should ever broadcast B.B.C. news. We have a news service of our own and Australian listeners should get their news through our own channels. We don't need to and should not depend on the B.B.C.'[138] However, Hamilton's declaration of independence was neutralized by a new BBC news programme, *World Round-up*. This had initially been planned in 1963 as a cable sound-feed tailored to the requirements of Canadian private radio stations. Although it was subsequently broadcast on short wave, Australian private stations were eager to pay for a high-quality cable feed of their own.[139] Hamilton did not want *World Round-up* for the ABC, but neither did he want his rivals to have it. Eventually, after much 'tough and frank speaking', and as part of its post-Moses rapprochement with the BBC, the ABC agreed to pay to take an exclusive cable feed of *World Round-up* for itself.[140]

Short-wave transmission facilities in the dominions were now well developed, and sporting coverage could be exchanged freely between South Africa, Australia, and New Zealand, without relying on intermediate BBC transmitters.[141] For the 1962 British Empire and Commonwealth Games in Perth, the ABC had access to multiple overseas radiotelephone lines, and four Radio Australia short-wave transmitters, allowing it to service the extensive requirements of different overseas

[133] Memo by G. F. Davidson, 22 November 1971, LAC, RG41, 906/PG10-9/3.

[134] Smithers to Stephenson, 25 August 1964, WAC, E1/1767. Smithers to Stephenson, 18 August 1967, E1/1739/6.

[135] Smithers to Greene, 5 October 1966, WAC, E1/1739/5.

[136] Smithers to Stephenson, 17 October 1967, WAC, E1/1745. Smithers to Stephenson, 28 November 1967, E1/1739/6.

[137] St Clair, 'Canada—NAS Rebroadcast Report—May 1964', E1/1775.

[138] W. Hamilton to Moses, 28 January 1964, NAA, Sydney, C2327 T1, box 58, 15/3/8 part 2, 'From 1959—Rebroadcasts of BBC News Bulletins'.

[139] Russell to Cockburn, 17 January 1963, WAC, E1/1733. Smithers to Clark, 26 August 1964 and Steedman to Hodson, 15 September 1964, E1/1729. Smithers to C. Bell, 7 January 1966, E1/1738/4. C. Edwards to Duckmanton, 6 January 1966, NAA, Sydney, SP1423/1, R26/3/7 part 1, 3162542.

[140] Fisher to Steedman, 13 and 26 August 1964; Hodson to Clark, 12 October 1964; Fisher to Stephenson, 16 December 1964, all in WAC, E1/1647.

[141] 'Commonwealth Broadcasting Conference 1960—SABC paper—Exchange of material for Newsreels', LAC, RG41, 369/19-35/1.

broadcasters.[142] As with news, however, cable feeds began to replace short wave for sports broadcasting, with the NZBC for example arranging coverage via Compac of the 1965 New Zealand Rugby League tour of England.[143] Commonwealth cooperation meanwhile continued even for coverage of non-Commonwealth sporting events, such as the Olympic Games. Working together made it possible to share the cost of expensive new transmission technologies. The ABC thus acted as the main Commonwealth representative in negotiations prior to the 1964 Olympics in Tokyo, coordinating the requirements for all Commonwealth broadcasters, and organizing a common radio coverage pool that included correspondents from all the participating Commonwealth broadcasting organizations (except the CBC).[144] Similarly, in 1968 the ABC and NZBC worked together to secure television coverage of the Mexico Olympics.[145]

The Compac cable allowed radio producers to link studios in different countries with results that far surpassed earlier short-wave efforts such as *Namesake Towns*. Although the cable was expensive to use, the BBC, CBC, ABC, and NZBC cooperated to make a number of radio quiz programmes, with teams able to participate without leaving their home countries.[146] Further use of the cable was planned during the 1968 Commonwealth Broadcasting Conference in New Zealand, when Frank Gillard, BBC director of sound broadcasting, and the CBC's Gene Hallman contemplated creating 'an English-speaking Radio Union'. Discussions continued when Craine subsequently visited London, and the ABC and NZBC were invited to join in.[147] At the end of September the first informal conference in a projected annual series was held in London, attended by Gillard and other senior BBC sound officers, Craine, Darrell Miley (ABC director of radio), and Lionel Sceats (NZBC director of sound broadcasting). It was agreed to form a loosely constituted 'Radio Projects Group', liaising to exchange ideas and produce joint programmes.[148]

The members of this group tried to avoid some of the problems that had undermined earlier attempts at multilateral radio cooperation. Craine, with the support of Miley and to some extent of Sceats, emphasized that the group 'should not become one more instrument for the domination of the BBC', or habitually hold its meetings in London.[149] It was further agreed that the word 'Commonwealth' would not be used in the group's name: neither would the group be linked formally with the Commonwealth Broadcasting Conference. This was partly in order to keep membership open to the new US Corporation for Public Broadcasting, but it

[142] A. N. Finlay to J. Schroder, 4 February 1960, Archives New Zealand (ANZ), AADL 564/23a 1/3/56.

[143] ANZ, AADL 564/25b 1/3/74.

[144] Memo by Moses, 25 July 1963, ANZ, AADL 564/22b 1/3/57 part 1.

[145] B. F. Kerr to L. Cross, 3 September 1968, ANZ, AADL 564/25a 1/3/72.

[146] *Report of the New Zealand Broadcasting Corporation for the Year ended 31 March 1968* (Wellington, 1968), 7. *Report of the New Zealand Broadcasting Corporation for the Year ended 31 March 1969* (Wellington, 1969), 7.

[147] Craine to Mutrie, 15 March 1968 and Craine to Stephenson, 3 June 1968, WAC, E1/1753.

[148] Minutes, 'Meeting of Representatives of ABC, CBC, NZBC, and BBC to Discuss Radio Projects, Broadcasting House, London, 26th and 27th September 1968', 29 November 1968, NAA, Sydney, SP1423/1, R26/1/12 part 1, 3162540.

[149] J. A. Camacho to B. S. G. Bumpus, 12 February 1969, WAC, R51/1070/2.

also allowed the exclusion of African, West Indian, and Asian broadcasters. This was something of a return to the old, white, English-speaking world as the basic unit for cooperation. There was considerable enthusiasm for the scheme, particularly at the ABC, where Hutchison and Miley hoped the group might re-energize radio by providing stimulating, high-quality, timely coverage of international affairs.[150] However, support at the BBC was more qualified. Gillard thought that 'with such a strong initiative coming from the Commonwealth countries we must swallow any misgivings we may have and go into this project with enthusiasm and energy'.[151] J. A. Camacho, Head of Radio Talks and Current Affairs, noted that, '[r]ightly or wrongly we decided to give the idea, as it were, a run for its money'.[152]

The BBC did pay for its Chief Producer of Documentary Programmes Bob Cradock to visit Canada, Australia, and New Zealand, to liaise with programme staff. Cradock was briefed to let the CBC, ABC, and NZBC take the lead in discussions.[153] After some delay, in June 1969 it was agreed to go ahead with a discussion programme on *The Future of Cities*, using cable feeds and coordinated by the BBC.[154] This was not judged a great success. One critic characterized the programme as wordy, 'civilised', and 'rather boring', and the ABC was not happy about the dominating presence of one of the Canadian participants:

> [Marshall McLuhan] had so many apparently novel observations to make that the programme became not so much a consideration of the future of cities but an extension of McLuhan the man into a 'global theatre' of his own making.[155]

A second discussion programme (*East and West across the Pacific*) was coordinated by the ABC, and took place without the participation of the CBC. Again, the discussion was deemed too polite to make for an interesting broadcast.[156]

A second group conference was held in Toronto in September 1969. Semmler reported that

> we were all reasonably unanimous that unless the situation could show some material dividends in this coming year there wasn't much point in pursuing it. On the other hand—here again Gillard and Craine especially emphasized this—so little seems to come out of an organisation like the Commonwealth Broadcasting Conference, that if this Group can achieve something worthwhile, something tangible, in joint programme production, they believe it's well worth going ahead with.

At Gillard's suggestion, it was agreed to supplement future discussion programmes with a monthly light 'conversation piece', harnessing the talents of transnational

[150] Hutchison to Duckmanton, 11 July 1968 and memo by Miley, 28 September 1968, NAA, Sydney, SP1423/1, R26/1/12 part 1, 3162540.

[151] F. G. Gillard to R. D'A. Marriott, 18 July 1968, WAC, R51/1070/2.

[152] Camacho to Bumpus, 12 February 1969, ibid.

[153] Camacho to Gillard, 13 February 1969, ibid.

[154] R. J. C. Cradock to Miley, 11 June 1969, NAA, Sydney, SP1423/1, R26/1/12 part 1, 3162540.

[155] J. Newsom to Miley, 7 August 1969 and press cutting, *Sunday Telegraph*, 10 August 1969, ibid.

[156] Hutchison to Semmler, 28 August 1969, ibid.

teams of witty, entertaining, well-known speakers.[157] However, although the NZBC and ABC remained enthusiastic, the CBC did not like this idea.[158] Semmler despaired of 'the great difficulty of getting four organizations to do something about anything'.[159] While the ABC and NZBC remained enthusiastic, the BBC seemed 'lukewarm' and the CBC 'cool to cold'.[160] When another meeting was held in London in October 1971, the new BBC controller of radio programmes claimed that it was too difficult to find 'genuinely common ground for programming', and that there was insufficient air time on Radio 4 (the BBC's spoken-word network) to accommodate any programmes produced by the group. The CBC admitted that it was more interested in collaboration with National Public Radio in the US. The Radio Projects Group was duly terminated.[161]

The demise of the Radio Projects Group reflected a number of practical issues: the difficulties of getting four large and complex organizations to work together; the problem of producing lively programmes that would be of interest to audiences in all four countries, but that would not offend local sensibilities; and the disruption caused by changes of personnel that robbed projects of their patrons and left them in the unsympathetic hands of their successors. However, the changing transnational priorities of the BBC and CBC also undoubtedly played a significant role in the failure of the scheme.

There were other indications of this changing climate. During the 1950s many rebroadcasters had expressed growing dissatisfaction with the round-the-Commonwealth Christmas feature, including the fact that the programme continued to be coordinated from London. However, the 1953 Christmas programme, produced in Australia to coincide with the royal tour, proved something of a throwback, disappointing expectations of a new pattern. Indeed, the whole idea of the Christmas feature seemed increasingly dated. In the early 1930s a live round-the-world link-up had been a novelty. By the mid-1950s it seemed too commonplace to rouse listeners in Britain from post-prandial stupors, and only those Australians and New Zealanders with 'very strong personal ties with Britain' stayed up into the early hours to tune in. Moreover, in the newly multiracial Commonwealth, many potential listeners did not even celebrate Christmas.[162] Yet no clear alternative to the established pattern emerged, and for the rest of the decade the Christmas feature limped on, generally remaining a BBC responsibility. In 1959

[157] Report from Semmler, 5 September 1969, ibid.

[158] Cradock to Semmler, 22 April 1970, NAA, Sydney, SP1687/1, R26/1/12 part 2, 'Radio Projects Group (ABC/CBC/BBC/NZBC) 1970–73'.

[159] Semmler to Cradock, 5 May 1970, WAC, R51/1070/2.

[160] Miley, 'Radio Projects Group', 19 May 1970, NAA, Sydney, SP1687/1, R26/1/12 part 2, 'Radio Projects Group (ABC/CBC/BBC/NZBC) 1970–73'.

[161] Craine to Semmler, 22 October 1971 and Miley to Craine, 16 March 1972, ibid.

[162] R. McCall to Jubb, 12 May 1952, WAC, E17/9/3. Minutes, Commonwealth Broadcasting Conference, 8 July 1952, LAC, RG41, 367/19-31-1/2. 'Commonwealth Broadcasting Conference 1956—ABC Paper—Christmas Commonwealth Programme—Agenda Part 1—C/18', NLA, Sir Richard Boyer papers, MS3181, folio box, folio 1. Minutes, Commonwealth Broadcasting Conference, 12 November 1956, LAC, RG41, 368/19-32/1.

the ABC was asked to produce the programme, but its feature on *The Young Commonwealth* recycled what had over the years come to seem like stale platitudes about 'family' unity, freedom, and peaceful multiracial cooperation. CBC officers thought the programme had little appeal: like BBC offerings in previous years, it seemed 'to have lost track of Christmas in its efforts to wave the flag and propagandize Commonwealth relations'.[163] Laurence Gilliam, for many years the producer of the Christmas feature, himself thought the fixture had tended to become 'a sermon wrapped up as a geography lesson—a sort of multi-racial welfare workers' orgy'.[164]

Finally, in 1965 the abolition of the BBC's own Features Department acted as a catalyst for change. At the Commonwealth Broadcasting Conference that year, it was agreed to drop the Christmas feature. It was now proposed instead that 'Commonwealth Day' would be celebrated each year with a special documentary, with members taking it in turns to produce a programme about their own country. Tellingly, the organizers first had to deal with the problem that Commonwealth Day was held on different dates in different countries, and was not marked at all in some member states.[165] It also seemed uncertain whether the new approach could resolve the difficulties associated with the old Christmas features. The CBC judged the first Commonwealth Day programme, produced by Radio Malaysia, to be 'as propagandistic as it is dull', and declined to use it.[166] Similarly, when the ABC was asked to make the documentary for 1972, the producer was reluctant to conform to the expected 'quasi-eulogistic style'.

> The fundamental question facing the Commonwealth is its relevance in today's world; and we could not start making a documentary, for example, on the assumption that the Commonwealth is relevant. We would also have to take into account the fact that the Westminster model of parliamentary democracy has not been successfully transplanted to many Commonwealth countries. We could not ignore such problems as political prisoners in Singapore, communal clashes in Malaysia, the Quebec liberation movement in Canada, the Ulster crisis, the situation of Cyprus and of Malta, the rise of one-party states in Africa, the recent civil war in Nigeria and the war between Pakistan and Bangla Desh.[167]

Eventually, the ABC decided to produce a documentary on the safely non-controversial Anglo-Australian telescope project.[168]

During the 1960s the BBC's Radio Talks and Current Affairs Department continued to sprinkle domestic schedules with its own offerings addressing Commonwealth themes. However, reflecting doubts that had endured since the 1930s, few believed

[163] 'The Young Commonwealth—ABC Xmas Programme 1959, recorded in Sydney', RNZSA, TX/29. F. R. Halhed, 'Christmas Commonwealth Radio Broadcast', 6 January 1960, LAC, RG41, 369/19-35-1/2.

[164] L. Gilliam to K. H. Funnell, 6 July 1962, ANZ, AADL 564/302e 1/9/44.

[165] G. Mansell to Gillard, 31 March 1966, WAC, R51/1070/1. Most broadcasters had ceased to mark Empire Day during the 1950s.

[166] P. Garvie, 'Commonwealth Broadcast ex Malaysia', 20 May 1966, LAC, RG41, 921/PG18-64.

[167] A. Ashbolt to Miley, 7 and 20 March 1972, NAA, Sydney, SP1687/1, R6/7/4 part 2, 3164917.

[168] Semmler to Hutchison, 29 August 1972, NAA, Sydney, C2327 T1, 8/12/6, 12203661.

that audiences were particularly interested. The Commonwealth still had to be covered craftily, by springing stand-alone offerings on surprised listeners, or by infiltrating material into general series such as *World of Books* and *The Critics*. BBC officers doubted whether programmes about the Commonwealth could ever generate much of an audience: 'The Commonwealth hardly evokes popular passion, except in a form very near the knuckle, e.g. coloured immigration.'[169] Major series for home audiences on empire and Commonwealth themes, such as Marjory Perham's 1961 Reith Lectures on *The Colonial Reckoning*, were exceptional.[170] Camacho, the department's head, assured the Royal Commonwealth Society that the BBC remained committed to encouraging greater knowledge and understanding of the Commonwealth, but emphasized that radio now had few opportunities to achieve this goal. Long documentaries or features no longer attracted audiences, and series were generally 'interpreted as public relations' and were thus 'likely to engender listener resistance'. Brief spots in magazine programmes such as *Today* and *Woman's Hour*, related to newsworthy Commonwealth events, were the best way to reach a broad audience.[171] Camacho did attempt to follow this up, but was soon frustrated by the seeming inability of the ABC or CBC to supply appropriate material in any quantity (the BBC had abolished the post of sound assistant in its Sydney office, so could no longer produce such reports for itself in Australia), and the poor quality of what was sent by the NZBC. 'New Zealand', Camacho concluded, 'is just an unbelievably dull country.'[172] BBC officers themselves clearly felt some of the apathy towards the Commonwealth of which they had long accused UK audiences.

CONCLUSIONS

In 1926, Ernest Fisk of Amalgamated Wireless (Australasia) Ltd had sought to predict how broadcasting would be used in 1976. He prophesied that

> We shall be able to sit in our homes in any part of Australia and see and listen to any of the important happenings of the world that we may choose as and when they occur. If our fancy lies in the direction of Transatlantic air ship races, which should be a common feature then, we shall be able to watch the competitors as they rush at 200 miles an hour... Since conservative ideas and the love of old things will still exist in some people, the ancient ceremony of the Oxford and Cambridge Rowing Race on the Thames will also find many spectators in Australian homes. When the leaders of thought in politics, science or art have occasion to speak to the public they will not, as they do now, make themselves uncomfortable by going on a platform in a public building, but they will speak from their own homes and will be seen and heard by a world-wide audience also, in the majority of cases, sitting in their own particular

[169] C. F. O. Clarke to Camacho, 12 March 1964, WAC, R51/783/1.

[170] Perham's lectures were reprinted in the *Listener*, November to December 1961, and also published in expanded form as *The Colonial Reckoning* (London, 1963).

[171] Camacho to Marriott, 15 January 1964, WAC, R51/783/1.

[172] Camacho to J. K. Rickard, 26 March 1968, WAC, R51/1070/1.

> homes. In short, the ether waves will draw aside the curtain of darkness and distance which today separates man from man and nation from nation...
>
> By far the greatest benefit will come from the annihilation of distance, both in transport and communication, and the resultant far better understanding among mankind generally, the steadily decreasing Tower of Babel and the consequent increasing tower of mutual understanding and human freedom which will result from the development and application of organised knowledge, which is commonly described by the word 'Science'.[173]

Fifty years later, cable and satellite technology had indeed made it possible for Australians to hear and watch the boat race and political speeches from around the globe, even if the promise of transatlantic airship races and mutual world understanding had not been fulfilled. What Fisk did not predict, understandably, was the radical transformation in the broader international context which occurred over the intervening half century, and which reconditioned how Australia was plugged into global media circuits.

In the 1920s it was not clear that the British world-system was in terminal decline. By the 1960s it was obvious. Commonwealth broadcasting collaboration had survived, but in a rapidly changing context in which prospects for the future seemed uncertain. Royal tours continued to be covered on radio and television, but were no longer presented as a means to reinforce Britannic or imperial unity. Indeed, even their nation-building function was now open to question. In Canada, newspapers were critical of how the CBC suppressed dissenting voices during the 1964 royal visit to Quebec City, which had been accompanied by separatist demonstrations. Some senior officers within the corporation were also unhappy with this policy: 'we were too timid in our approach to the general atmosphere, the emotional setting in Quebec City... mention might have been made by our commentators which would have indicated the sense of strain behind the set piece ceremonials'.[174] In the light of events in Quebec, and of the assassination of President Kennedy in the US, the CBC even felt the need to devise contingency plans to deal with 'emergency situations' during future visits by the Queen or other dignitaries.[175]

Other 'media events' were similarly drained of the imperial flavour that had been so apparent in previous decades. The death and state funeral of Winston Churchill in 1965 was covered extensively around the Commonwealth but, with some minor exceptions, was not generally an occasion for the expression of vestigial Britannic identities. Churchill was presented as the man who had led Britain through the war, and a great parliamentarian, but not as a hero of empire or of the 'British race' (although he was, after all, half-American).[176] Two years later,

[173] E. T. Fisk, typescript article for 'Fifty Years Ahead' series, *The Herald* (Melbourne), 1926, Mitchell Library, MSS 6275, E. T. Fisk papers, vol. 25.

[174] L. B. McIlhagga, 'Royal Visit (1964)', 4 December 1964, LAC, RG41, 248/11-37-14-10/5. Peers, *Public Eye*, 331.

[175] McIlhagga, 'Emergency Broadcasts during visits of Royalty and VIPs—Centennial Year', 30 November 1966, LAC, RG41, 602/373.

[176] Wendy Webster, *Englishness and Empire, 1939–1965* (Oxford, 2005), 182–94.

the centenary of Canadian confederation and the Montreal Expo provided opportunities for carrying some reciprocal programming from Canada. The BBC decided to take much of its coverage from the CBC, for the sake of both economy and goodwill, and the World Service arranged a 'Canadian Week' to mark the occasion.[177] Again, however, there was little that was imperial or British about how this anniversary, of an event which certainly could have been interpreted as a key marker in imperial history, was represented in either Canada or Britain. The British world had itself been consigned to the past.

[177] 'Expo '67—Minutes of a meeting with the Canadian Representative', 24 November 1965, WAC, E1/1753. Smithers to R. E. Gregson, 15 July 1966, E1/1760/1.

Conclusions

In 1950, a British Broadcasting Corporation (BBC) expert was brought to Australia by Amalgamated Wireless (Australasia) Ltd (AWA). He gave its highly experienced production unit a lecture covering things they already knew about. During the course of his talk, the BBC producer fiddled with a gold pencil attached to a long, gold chain swinging from what one witness described as 'his scrawny neck'. At the end, in an exaggerated BBC accent, the producer asked if there were any questions. One of the members of the audience, who was in fact also an Englishman, replied in an equally exaggerated Australian accent, 'Yeah . . . if I was to pull that chain would you flush?'[1] As this anecdote illustrates, the story of broadcasting collaboration in the British world was one of failure as well as success, of miscommunication as much as communication, of common ways of doing things and divergent perspectives and attitudes. In radio programmes, and in AWA's conference rooms, contemporaries could hear all of this in the difference between 'Oxford' and 'ocker' accents.

It is tempting for the historian to take episodes of conflict and tension, line them up in a neat chronological progression, and point to this as evidence for a steady, inevitable growth to maturity of autonomous national broadcasters in Canada, Australia, and New Zealand throwing off the shackles of BBC empire-building. It is also tempting to present this as a reflection of the broader twentieth-century progress of the dominions towards full national independence. Yet this is not the story that has been told here. Rather, I have argued that collaborative relations among public broadcasting authorities developed according to a more complex chronology: largely non-existent in the 1920s; generally dysfunctional in the early 1930s; improving somewhat in the later 1930s; and meshing together as never before during the war. The pattern then becomes more complex in the 1950s with the introduction of television: continued successful 'Commonwealth' collaboration in radio in the 1950s, although at a time when the significance of sound broadcasting as a mass medium was clearly declining; overall failure to achieve successful collaboration in the new medium of television during the same period; and the establishment of a more successful export trade in BBC television programmes during the 1960s, albeit at the price of a new commercialism that transformed the entire basis for cooperation. By the end of the 1960s close practical links were still pervasive, but the Britannic sentiment that had previously accompanied them had drained away.

[1] 'John' to MacKay, 6 April 1950, Alexander Turnbull Library, Ian Keith MacKay papers, box 3.

All parts of the former British world, including Britain itself, seemed to be facing absorption into a new American media empire.

In this overall picture, I have generally presented miscommunication and tension as an integral but not particularly damaging aspect of a set of close and enduring collaborative relationships. The significance of particular episodes of conflict should be evaluated in the context of other personal, organizational, and technological factors, that often rendered conflicts too complex to be construed as the result of any direct clash between 'nationalism' and 'imperialism'. Reith and the BBC's initially outdated conception of empire; the social agendas of the cultivated elites who provided the Canadian Radio League with much of its support base; the organizational failings of the Canadian Radio Broadcasting Commission; the personalities and beliefs of particular individuals; the unsatisfactory nature of short-wave technology; and the introduction of new communications technologies that made joint projects easier to plan and execute: such complications undermine the validity of any simple, evolutionary account of change over time.

Crucially, after the difficulties of the early 1930s, Britannic solutions were devised that gave dominion broadcasting authorities a considerable degree of autonomy, and at least the illusion of equality with the broadcasting behemoth that was the BBC. This was important, for it allowed dominion broadcasters to embrace collaboration rather than to resist it, and later to assume leadership roles of their own within a Commonwealth context. At the Commonwealth Broadcasting Conferences of the 1950s and 1960s, the Australian Broadcasting Commission (ABC) for example repeatedly came out as the champion of joint broadcasting ventures, and was just as often disappointed by the institutional torpor and limited resources of other broadcasting authorities. Nevertheless, it persisted in its efforts. For broadcasters around the British world, the long-term aim was continued and improved voluntary collaboration that would bring good things from overseas, and allow broadcasting authorities to project themselves and their countries around the Commonwealth, accumulating programmes and prestige. The goal was not separation: the Britannic connection continued to represent the key link with the outside world, the main means to access outside English-language programme resources without resorting to American commercial culture, and the way to find a voice overseas.

It was only in the early 1930s that contemporaries accused the BBC of 'imperialism' with much conviction. Paradoxically, this was at a time when BBC influence in the dominions was at its weakest, and when collaborative structures were non-existent or malfunctioning. Critics were not protesting against any flood of UK programmes crushing the diversity of dominion productions: BBC exports were minimal. What was really being attacked was the centralizing ambition of the BBC, its early belief that it should coordinate all empire broadcasting from London. When the collaborative connection became stronger, few people spoke of 'imperialism', although they might dwell on the wonders of a 'Fourth British Empire' based on reciprocity and mutual discussion of shared interests. The UK presence in dominion broadcasting was largely perceived as benign, and compatible with domestic nation-building agendas. It offered at least some sort of counterweight to the much more powerful potential influence of US commercial culture.

Thus, in the sphere of broadcasting at least, cultural imperialism seemed to be American, not British.

Granted, flows of reciprocal material from the dominions to Britain seldom proved extensive. This was a lasting source of resentment for the Canadian Broadcasting Corporation (CBC) and particularly for the ABC. Yet complaints generally reflected a desire in the dominions for closer links with Britain and the Commonwealth, not for a severing of ties. More significant in terms of 'cultural imperialism' was the aim of some BBC officers to shape the tastes of target audiences around the world, even if, for diplomacy's sake, this policy generally had to be concealed. At times the BBC intervened directly (as in Canada in the 1930s or New Zealand in the 1960s) or indirectly (as in Australia in the 1930s or Canada in the 1960s) in local broadcasting debates, to export the BBC model or support the general idea of public broadcasting. BBC officers also sought to export BBC programmes. Sometimes, the aim was cultural uplift, an extension of the domestic role of BBC experts into some sort of Britannic civilizing mission. At other times, with comedy programmes for example, the role of exports was to sustain broad cultural sympathies between Britain and other countries, but also just to make people laugh. We might now call this 'cultural diplomacy': contemporaries certainly recognized that it was not in itself a neutral act. Sometimes it seemed to dominion observers to be good for the unity and cultural well-being of the 'British race', but at other times it smacked of unwelcome missionary activity.

Some even argued that the basic BBC model was an unwelcome imposition on the dominions. The implication was that public broadcasting would not have emerged naturally as a native growth of the local soil. Ian Mackay, whose career took him from commercial broadcasting in New Zealand and Australia to public broadcasting in Nigeria, claimed that the BBC model was unsuited even to the requirements of the UK's most willing Britannic collaborator. New Zealand's small and isolated population, he argued, would have been better served by private enterprise than by public authorities that lacked the resources necessary to make a success of their appointed tasks.

> 'This is London calling' has become a very familiar phrase in most New Zealand homes and perhaps it is this close association that affected the outlook of our political leaders whenever broadcasting was discussed. Successive Governments and spokesmen... continued parroting the slogan 'We are following the B.B.C.' This fetish was... partly responsible for the slow progress made in many fields... an attempt was made to confine our listeners to an imitation framework of the B.B.C.[2]

One former ABC employee argued in similar terms that:

> In following the BBC mould [the ABC had] catered only to its natural audience, the more discriminating minority which found the advertisements and the 'matey' approach of the commercials offensive. This ABC audience expected formality, dilettantism, 'educated southern English' and an appropriate deference of manners. Announcers, dressed in dinner jackets, always called a speaker or an interviewee, 'Sir'. What the early ABC

[2] Ian K. MacKay, *Broadcasting in New Zealand* (Wellington, 1953), 49–50.

> audiences really wanted was to continue indefinitely the 'in' game which catered for a sense of superiority on both sides. It is to its discredit that the ABC chose to go along with this attitude. To make matters worse, what the ABC gave its audience for years was an amateurish shadow of what the BBC was doing so well.[3]

Yet, possibly, there would have been more UK cultural imperialism if broadcasting in Britain had gone private rather than public from the beginning. This is of course a counterfactual that cannot be explored with any degree of confidence, but we should not ignore the signs that private enterprise might have occupied a more substantial position in the empire's broadcasting framework. Marconi played an important role in the preliminary stages of wireless development, and its affiliates remained a significant force in broadcasting in Canada, Australia, Egypt, and elsewhere for many years. Was Marconi the private imperial broadcasting weapon that was never properly deployed? The BBC's public broadcasting monopoly had originally been established in the UK largely with domestic considerations in mind. The imperial consequences were largely unintended. Subsequently, few took time to consider whether a UK public broadcasting monopoly best served imperial interests, but the dominant position of private stations in Australia and Canada was a reminder of the lost opportunities for collaboration. Once the BBC's domestic television monopoly was broken in the mid-1950s, and UK commercial broadcasting companies established, entrepreneurs around the Commonwealth could form a new set of interconnections, paralleling those previously established by the public authorities. Was this too little, too late; an opportunity missed due to the crowding-out of private enterprise? Or did public broadcasting, after all, allow Britain to occupy a protected niche that would not have been available in toe-to-toe commercial competition with American private companies, a chance to remain at least a junior partner in what was becoming a global media order?

Is the idea of cultural imperialism any more useful in considering the BBC's role beyond the dominions? In the 1930s and for most of the 1940s the British broadcasting presence in Africa and Asia was hardly overbearing; indeed, it was weak to non-existent. Even in India, perhaps a little unfairly, Lionel Fielden concluded that despite four years of effort, in 1939 All India Radio remained 'the biggest flop of all time'.[4] BBC involvement in broadcasting in the remaining dependent colonies increased considerably after the Second World War. Radio was undeniably used as a tool of the colonial state, encouraging economic 'development', helping to suppress insurgency, and supporting peaceful but gradual constitutional progress. If this was a form of cultural imperialism (and, again, it was not a flood of British programmes that the managers of colonial broadcasting authorities worried about, but rather an influx of cheap American Westerns and crime dramas) then, crucially, it was one in which the ABC, CBC, New Zealand Broadcasting Service (NZBS), and New Zealand Broadcasting Corporation (NZBC) were also implicated, through their provision of training, advice, and programmes to broadcasters in Africa, Asia, the Caribbean, and the Pacific. When the NZBC and ABC moved towards closer

[3] Ellis Blain, *Life with Aunty: Forty Years with the ABC* (Sydney, 1977), 14.
[4] Lionel Fielden, *The Natural Bent* (London, 1960), 204.

cooperation with broadcasters in Asia, facilitated by the new Asian Broadcasting Union (inaugurated in 1964 with Charles Moses as its first secretary-general), was this an attempt to find a new framework for international collaboration after the collapse of the British world-system, or a means to exert continued influence over less-developed neighbours?

The evaporation during the 1960s of the Britannic rhetoric that had accompanied earlier broadcasting collaboration in part reflected the wider retreat of British identities in Canada and, to a lesser extent, in Australia and New Zealand. More significant though was the transformation of the Commonwealth as an arena for international cooperation. Decolonization meant that the Commonwealth Broadcasting Conferences primarily became a chance for the BBC, CBC, ABC, and NZBS/NZBC to advise fledgling public broadcasters in newly independent countries. In this context, officers could talk about Commonwealth or international cooperation, but there was little room for the old language of Britishness. The older members of the conference were not entirely happy with the new Commonwealth order, and they tended to pursue bilateral cooperative projects rather than multilateral ones that would involve complicated and potentially difficult transactions with African and Asian authorities whose commitment to public-service principles seemed uncertain. Problems encountered with news exchange agreements were representative of a wider gap. The fact that the old 'British' countries found it necessary to form a separate Radio Projects Group, made sure that the word 'Commonwealth' appeared nowhere in its remit, and courted collaboration with American rather than African or Asian public broadcasting authorities, is instructive.

What of the impact of empire on the BBC as a domestic British institution? Clearly, this was not negligible. Of considerable importance was the BBC's attempt to parlay its home broadcasting monopoly into the right to control all external broadcasting from Britain. This claim was not always accepted by policymakers, but it did provide the basis for the formidable array of External Services built up under BBC auspices from the early 1930s onwards. It meant that the same British organization that was broadcasting to home audiences was also serving the empire. While the BBC largely kept home and overseas audiences separate, it was a relatively easy matter to serve them with the same programmes when desired, such as on great royal and other imperial ceremonial occasions. During the Second World War the BBC had a particular incentive to collapse some of the boundaries between home and overseas listeners, to bring the truly imperial nature of the war effort home to audiences in Britain and around the world. Broadcasting in Britain did not have to be organized this way, with the BBC retaining responsibility for domestic and overseas services. The fact that it was meant that men with overseas experience (some of whom had a broadcasting career in the dominions behind them) rose to key domestic positions of responsibility within the BBC: men like Rooney Pelletier and Robert McCall, and to a lesser extent Charles Curran and Hugh Carleton Greene.

Yet it would be wrong to exaggerate the extent to which empire transformed the BBC at home. Was it really the empire that caused the BBC to take on an external broadcasting role? Specifically in 1932, yes, but otherwise surely overseas broadcasting from Britain would have started later in the decade anyway, as the European war of words intensified? Similarly, the Cold War as much as the empire was the guarantee of the BBC's continued overseas presence in the post-war years. Moreover, internal barriers to the spread of 'colonial' influences were erected within the BBC, particularly by successive controllers of the domestic services, who time and again refused to carry material from the dominions on the grounds that it was of a poor standard. Finally, we should not exaggerate the relative importance assigned by the BBC to its imperial role. America and Europe generally loomed much larger on the BBC's world map. In short, the evidence presented in this study has tended to support Peter Marshall's claim that, while some British institutions were 'very willing' to take on an imperial role, they seldom allowed this to lead to any 'fundamental change' in their domestic position.[5] If the External Services had been closed down overnight, would there have been any difference the next day in how the BBC went about discharging its home duties?

Many BBC officers seemed enthusiastic about the promotion of empire at home, despite the gatekeeping role played by controllers of the domestic services. On a personal level, cooperation with broadcasters from Canada, Australia, and New Zealand usually seemed to be natural and easy, facilitated by a myriad common cultural reference points. Many BBC officers who found themselves appointed to an imperial role did all they could to preserve Britain's broader links with the dominions. Yet insofar as this involved making UK listeners and viewers aware of the empire in general, and enthusiastic about the British world in particular, few policy- or programme-makers ever felt that they had much success. BBC officers continued to believe that audiences were ignorant, apathetic, or hostile when it came to the empire. Nothing the BBC did seemed to change this impression. The archival evidence underlines this point again and again, and surely needs to be taken into account in the current historiographical debate about British popular attitudes towards empire. Overseas expansion reshaped British culture and society much less than enthusiasts for empire hoped it would.

Media historians have to some extent put ideas about cultural imperialism aside in the last few decades, preferring instead to write of cultural hybridity, or to engage in more recent debates about transnationalism and globalization. If we want to place public broadcasting in the British world in this latter context, we must recognise that radio and television developed at a time of broader British imperial decline. For Britain, broadcasting seemed to offer a means to shore up an old, disintegrating world-system, rather than to create a new one. It is thus hard to present twentieth-century Commonwealth broadcasting collaboration as an episode in a progressive modern history of globalization. Collaborative Commonwealth connections were formed at a time when, thanks to the triple-shock of the First World War, the great depression, and the Second World War, earlier processes of transnational integration seemed to have been thrown into reverse. Contemporaries sought to use broadcasting as a means

[5] P. J. Marshall, 'Imperial Britain', *Journal of Imperial and Commonwealth History (JICH)*, 23/3 (September 1995), 379–94, quotes at 382.

to hold together an essentially Victorian British world-system that was unravelling before their eyes.

In the inter-war years the alternative seemed to be chaos, a world of violent national antagonisms. In the post-war years it was a different world-system, in which the forces of cohesion in the Western bloc were cemented by the fruits of American commercial culture. To some extent, Britain might continue to occupy a leading role in this new order, and the hope that Britain could preserve its overseas interests by working with the US, rather than become a mere American pawn, helped console policymakers as they adjusted to a reduced world role. 'Under the shadow of cold war, a once British Empire modulated strategically into an Anglo-American field of influence, and thence into a predominantly American commitment.'[6] Yet if the changing balance of British and American power generally involved accommodation and cooperation rather than all-out rivalry, this did not mean that former imperial masters greeted their loss of influence and prestige with enthusiasm. Neither were they happy to see the remnants of British overseas cultural expansion absorbed into a great American neo-colonial order.

Broadcasting might have played a significant role in uniting the British world, but for three decades it also helped divide it from the other main branch of the English-speaking world. The gulf was only really bridged in the 1960s, as far as Britain, Australia, and New Zealand were concerned—Canada had made the leap long before—as America became the dominant supplier of imported television programmes. Without the resources or political backing necessary to create a state-subsidized television transcription service, from the late 1950s the BBC had to engage with the international trade in television programmes on a commercial basis. It subsequently sold many television programmes to the old dominions, benefiting from enduring links with public broadcasting authorities overseas and continuing cultural affinities that still connected the component parts of the former British world.[7] Yet, ultimately, much now came to depend on decisions made by purchasers, on an essentially commercial basis, rather than on any lingering sense that the BBC and other public broadcasters had a role to play in binding together Britons at home and overseas. In this new, commercial world, the BBC also had to accept a subordinate position, a long way behind the big US exporters of television programmes. Both the nature and the extent of the BBC's presence in the British world had fundamentally changed.

This transformation was made complete in the decades that followed by the effects of new communications and recording technologies, deregulation, and the concentration of ownership of media enterprises in diverse territories in the hands of a small number of giant commercial global companies. The status of the BBC was further eroded at home, while overseas it was obliged fully to commercialize its operations. BBC Enterprises became BBC Worldwide, today the biggest exporter

[6] Wm. Roger Louis and Ronald Robinson, 'The Imperialism of Decolonization', *JICH*, 22/3 (September 1994), 462–511, quote at 473.

[7] Tom O'Regan, 'The International Circulation of British Television', in Edward Buscombe (ed.), *British Television: A Reader* (Oxford, 2000).

of British television programmes.[8] The BBC also sold more of its radio programmes overseas on a commercial basis, rather than as a subsidized exercise in cultural diplomacy. The BBC Transcription Service was subject to heavy cuts, and eventually (as BBC Radio International) commercialized as part of BBC Worldwide. In 2010, on the grounds that the world financial crisis necessitated massive retrenchment of state expenditure, the UK Foreign Office announced that financial support for the BBC World Service would be withdrawn. In response, the BBC considered accepting online advertising to make up some of the shortfall, and to allow it to sustain its overseas broadcasting activities.[9] The debate over the proper relationship between external broadcasting, state subsidization, and the free market, ongoing since the 1930s, had now entered a new phase.

[8] Jeanette Steemers, *Selling Television: British Television in the Global Marketplace* (London, 2004), 7–10. Michael Tracey, *The Decline and Fall of Public Service Broadcasting* (Oxford, 1998).

[9] 'Threat of Revolt Reprieves BBC Services', *The Guardian* (London), 27 October 1981. 'BBC World Service Considers Hosting Ads on Some Foreign-language Websites', *The Guardian*, 21 October 2010.

Archival and Manuscript Sources

UK

BBC Written Archives Centre, Caversham Park, Reading

Bodleian Library, Oxford
Sir Hugh Carleton Greene papers

Commonwealth Broadcasting Association archive, London

Manchester Archives and Local Studies, Manchester Central Library
Lord Simon of Wythenshawe papers

National Archives, Kew
CAB32—Cabinet Office: Imperial and Imperial War Conferences: minutes and memoranda
CAB124—Cabinet Office: Offices of the Minister of Reconstruction, Lord President of the Council and Minister for Science: records
CO323—Colonial Office: colonies, general: original correspondence
DO35—Dominions Office and Commonwealth Relations Office: original correspondence
FCO26—Foreign and Commonwealth Office and predecessors: Information, News and Guidance Departments: registered files (I and P series)
INF1—Ministry of Information: files of correspondence

CANADA

Library and Archives Canada, Ottawa
MG26 JI—William Lyon MacKenzie King papers
MG27 III B20—C. D. Howe fonds
MG30 D297—Graham Spry papers
MG30 D67—E. Austin Weir papers
MG30 E186—William Ewart Gladstone Murray fonds
MG30 E250—Ernest L. Bushnell papers
MG30 E298—Andrew Gillespie Cowan fonds
MG30 E326—René Landry papers
MG32 B5—Brooke Claxton fonds
R770—James Burns McGeachy fonds
R10120—Matthew Halton fonds
RG33 14—Royal Commission on Radio Broadcasting fonds
RG41—Canadian Broadcasting Corporation fonds
RG42—Records of the Department of Marine and Fisheries, Radio Branch

Library of the University of British Columbia Special Collections Division, Vancouver
Alan Plaunt fonds
Alan Thomas oral history interview with Charles A. Bowman

University of Toronto Archives and Records Management, Toronto
Vincent Massey papers

AUSTRALIA

Australian Broadcasting Commission (ABC) Document Archives, Ultimo, Sydney

Mitchell Library, State Library of New South Wales, Sydney
MSS 5454—Leslie Rees papers
MSS 5636—Clement Semmler papers
MSS 6275—E. T. Fisk papers

National Archives of Australia
Sydney
C1574 T1 and T2—ABC Head Office: correspondence files re television
C2218/6—ABC Head Office: correspondence files re television programmes
C2327 T1—ABC Head Office: general correspondence files
SP312/1—ABC Head Office: files re war correspondents
SP314/1—ABC Head Office: files re News Department and news contracts prior to the introduction of the independent news service
SP341/1—ABC Head Office: general correspondence and policy files
SP341/2—ABC Head Office: unregistered correspondence files
SP368/1—ABC Head Office: artist files
SP613/1—ABC Head Office: general correspondence including administration, policy, and artist contract files
SP724/1—ABC Head Office: general correspondence including administration, policy, and artist contract files
SP987/1—ABC Head Office: general correspondence
SP1299/2—ABC Head Office: correspondence files re television programmes
SP1423/1—ABC Head Office: correspondence files, radio programmes
SP1489/1—ABC Head Office: unregistered correspondence files of the chairman and commissioners
SP1558/2—ABC Head Office: central files
SP1687/1—ABC Head Office: correspondence re radio programmes
ST1790/1—ABC Head Office: general correspondence
ST3051/2—ABC Head Office: correspondence re television programmes
Melbourne
B2111—ABC Victorian Branch: correspondence files
B2112—ABC Victorian Branch: correspondence files
MP272/2—Shortwave Division, Department of Information: general correspondence files re overseas broadcasts
MP341/1—Postmaster-General's Department: general correspondence
Canberra
A461/10—Department of External Affairs/Prime Minister's Department: correspondence files
SP112/1—Department of Information: general correspondence files

National Film and Sound Archive, Canberra
Keith Barry papers
Talbot Duckmanton papers

National Library of Australia, Canberra
MS 1924—Herbert and Ivy Brookes papers
MS 2823—Sir Keith Murdoch papers
MS 3181—Sir Richard Boyer papers
MS 4738—Arthur Calwell papers
MS 5539—William James Cleary papers
MS 7290—T. W. Bearup papers
MS 7826—Sir James Darling papers
MS 8436—Chester and Edith Wilmot papers

University of Melbourne Archives, Melbourne
Tom Hoey papers

NEW ZEALAND

Alexander Turnbull Library, National Library of New Zealand, Wellington
MS-Group-0860—Trevor Richards papers
MS-Group-1999—Ian Keith Mackay papers
MS-Papers-6488-1—Godfrey Gray papers

Archives New Zealand, Wellington
AADL 563; AADL 564; AAFK 890; and AAFL 563—Broadcasting Corporation of New Zealand and predecessor agencies

Hocken Library, Dunedin
MS 0982—John Thomas Paul papers
MS 0985—William Downie Stewart papers

Radio New Zealand Sound Archive, Christchurch

SOUTH AFRICA

University of South Africa Library, Pretoria
United Party Archives

US

National Archives and Records Administration, College Park, Maryland
RG208—Records of the Office of War Information

Index

Lightning Source UK Ltd.
Milton Keynes UK
UKOW07n1852121214

243056UK00010B/167/P